石油工业概论

OVERVIEW OF THE PETROLEUM INDUSTRY

（英汉对照）

詹　静◎著

中国石化出版社

图书在版编目（CIP）数据

石油工业概论：英汉对照 / 詹静著.
—北京：中国石化出版社，2017.7(2024.7 重印)
普通高等教育“十三五”规划教材
ISBN 978-7-5114-4435-6

Ⅰ①石… Ⅱ.①詹… Ⅲ.①石油工业—高等学校—教材—英、汉 Ⅳ.①TE

中国版本图书馆CIP数据核字（2017）第165914号

中国石化出版社出版发行
地址：北京市东城区安定门外大街 58 号
邮编：100011　电话：(010)57512446
发行部电话：(010)57512575
http://www.sinopec-press.com
E-mail:press@sinopec.com
北京捷迅佳彩印刷有限公司印刷
全国各地新华书店经销
*
787×1092毫米 16开本 17.5 印张 354千字
2017年8月第1版 2024年7月第2次印刷
定价：68.00元

前　言

目前，出版物市场上缺乏图文并茂、生动形象地介绍石油工业的通识性教材和读物，针对这种情况，本书通过参考国内外相关资料编排了大量图片，生动地展示了石油工业主要领域——石油勘探、开发、储运、炼制和石油化工中的一些基本概念、理论和过程，以及石油工业的发展。本书涉及石油行业的方方面面，并创新性地增加了新技术在石油工业中的应用。在吸取同类教材优点的基础上，更加注重章节内容的紧密衔接和循序渐进，有利于读者加强对内容的理解和掌握，形成对整个石油工业的整体认识。

本书采用中英双语的方式介绍石油工业科普知识，生动直观、图文并茂，易于中国学生和外国留学生理解和掌握石油工业概况、理论原理和操作流程，同时为入门石油行业打下基础。此外，书中还展示了大量当今石油工业中应用的新技术、新工艺，以带给读者时下最新的资讯。

本书为石油工业通识类课用书，也可作为相关专业中国学生和外国留学生教材和阅读材料。

本书为笔者十多年教学经验的总结，希望对读者有所裨益。同时，笔者也乐意收到关于本书的任何反馈意见，以此促进水平的提高。

由于石油工业发展日新月异，涉及面也十分广泛，有些数据难以及时得到更新和完全反映，书中不妥之处，敬请各位同行和读者批评指正。

本书获“西安石油大学优秀学术著作出版基金”资助。书中大量图片引自网络，笔者已尽力将其出处标示出来，在此对书中图片原作者表示深深地感谢。

祝阅读愉快！

CONTENTS/目录

1 INTRODUCTION/简介

1.1 A Brief History of the Petroleum Industry ··· 1

1.2 Petroleum, Petroleum Industry, Petroleum Engineering and Petroleum Engineer ······ 6

1.3 The Field Life Cycle ······ 8

1.1 石油工业简史 ······ 1

1.2 石油、石油工业、石油工程、石油工程师 ······ 6

1.3 油气田开发过程 ······ 8

2 PETROLEUM GEOLOGY/石油地质

2.1 Structure of the Earth ······ 12

2.2 Rocks ······ 14

2.3 Origin of Petroleum ······ 22

2.4 Organic Source ······ 23

2.5 Generation of Oil ······ 24

2.6 Migration of Oil ······ 27

2.7 A Petroleum Accumulation ······ 29

2.1 地球的结构 ······ 12

2.2 岩石 ······ 14

2.3 石油的起源 ······ 22

2.4 有机质来源 ······ 23

2.5 油气的生成 ······ 24

2.6 油气的运移 ······ 27

2.7 石油的聚集 ······ 29

3 PETROLEUM EXPLORATION/石油勘探

3.1 Introduction ······ 34

3.2 Petroleum Geophysical Exploration ······ 36

3.3 Petroleum Geochemistry Exploration ··· 43

3.1 简介 ······ 34

3.2 石油地球物理勘探 ······ 36

3.3 石油地球化学勘探 ······ 43

4 WELL DRILLING/钻井

4.1 Drilling History ······ 45

4.2 Well Planning ······ 49

4.3 Site Preparation ······ 51

4.4 Rig Types and Rig Selection ······ 55

4.1 钻井历史 ······ 45

4.2 钻井设计 ······ 49

4.3 井场准备 ······ 51

4.4 钻井类型及选择 ······ 55

4.5 Drilling Systems and Equipment ··· 62
4.6 Routine Drilling Procedures ········· 86
4.7 Non-Vertical Wells Drilling ········· 96

4.5 钻井系统和装备 ····················62
4.6 常规钻井程序 ·······················86
4.7 非直井钻井 ··························96

5 FORMATION EVALUATION/地层评价

5.1 Drilling Operations Log ··········· 111
5.2 Coring ····························· 124
5.3 Drill-Stem Test (DST) ··········· 126
5.4 Initial Potential (IP) Test ········ 127
5.5 Abandonment & Reclamation ··· 127

5.1 钻时录井 ························ 111
5.2 取心 ······························ 124
5.3 中途测试 ························ 126
5.4 IP测试 ··························· 127
5.5 废井&复原 ······················ 127

6 WELL COMPLETION & CEMENTING/完井与固井

6.1 Downhole Strings ················· 130
6.2 Cementing ························· 137
6.3 Well Completion ················· 138

6.1 井下管柱 ························ 130
6.2 注水泥 ··························· 137
6.3 完井方式 ························ 138

7 RESERVOIR PROPERTIES/油气藏性质

7.1 Properties of Reservoir Rock ······ 146
7.2 Reservoir Pressure ················· 157
7.3 Reservoir Fluids ··················· 158
7.4 Oil Reservoirs—— Primary Drives ······································· 185
7.5 Gas Reservoirs ···················· 196
7.6 Oil and Gas Recovery Methods ··· 198

7.1 油藏岩石性质 ···················· 146
7.2 油藏压力 ························· 157
7.3 油藏流体 ························· 158
7.4 油藏天然能量驱动机理 ······································· 185
7.5 气藏 ······························· 196
7.6 油气采收率 ······················ 198

8 FIELD APPRAISAL/油气田评价

8.1 Production Testing ················ 201
8.2 Reservoir Simulation ·············· 203
8.3 Oil and Gas Reserves ············· 203

8.1 生产测试 ························· 201
8.2 油藏数值模拟 ···················· 203
8.3 油气储量 ························· 203

9 PRODUCTION & DEVELOPMENT/开发与生产

9.1 NaturalFlow in a Well ………… 209
9.2 Artificial Lift ………………………… 210
9.3 Stimulation………………………… 219
9.4 Enhanced Oil Recovery………… 224

9.1 自喷井 ……………………………… 209
9.2 人工举升 ………………………… 210
9.3 增产措施 ………………………… 219
9.4 提高采收率 ……………………… 224

10 SEPARATION & TRANSPORTATION & STORAGE/油气分离、储存与运输

10.1 Oil Separation ………………… 239
10.2 Gas Separation………………… 245
10.3 Water Separation ……………… 250
10.4 Transportation & Storage……… 252

10.1 原油分离 ……………………… 239
10.2 天然气分离 …………………… 245
10.3 产出水处理 …………………… 250
10.4 油气储运 ……………………… 252

11 PETROLEUM REFINING AND ITS PRODUCTS/石油炼化及其产品

11.1 Distillation …………………… 263
11.2 Primary Refining and Secondary Refining ……………………… 266
11.3 Refinery Products……………… 268

11.1 蒸馏 …………………………… 263
11.2 初次加工和二次加工 ………… 266
11.3 石油炼化产品 ………………… 268

REFERENCES/参考文献……………………………………………… 271

INTRODUCTION 1
简 介

1.1 A Brief History of the Petroleum Industry

The petroleum industry, contrary to most people's understanding, did not come about overnight, yet the industry did not take hundreds of years to evolve into what it has become today either. Petroleum based products have been present throughout early European history with uses ranging from healing ointments to combustible weapons of war. Eventually a small oil industry began to develop in Eastern Europe around 1854 that focused primarily on peasant-dug shafts to obtain crude oil. This oil was then refined to abstract kerosene, which was then used in cheaply manufactured lamps. By 1859 a thriving kerosene oil business had been established. Total European crude production in 1859 had been estimated at 36,000 barrels, primarily from Galicia and Romania. More than anything, the Eastern European industry lacked the technology for drilling.

This problem was solved through the determination of one man, George Bissell. Bissell had the inspiration and insight to realize that the same method used for salt "boring" could be used to "bore", or drill, for oil deposits buried beneath the earth. Crude oil, or "rock oil" as it was called, was then a by-product

1.1 石油工业简史

石油工业的形成，既非一蹴而就，又非历史百年。欧洲历史早期，石油就被用来作为疗伤的药膏或是战争中的火攻武器。1854年左右，在东欧出现了石油工业的萌芽，人们采用农用的杵来掘井采油。采出来的原油被用来提炼煤油，用作廉价煤油灯的燃料。到了1859年，煤油生意已经逐渐兴盛起来。当年欧洲全年原油产量达到了36000bbl(1bbl=0.159m^3)，主要来自西班牙的加利西亚以及罗马利亚。当时，东欧关键欠缺的是钻井技术。

这个问题被一个名叫乔治•比斯尔的人通过不懈努力解决了。比斯尔意识到可以采用开凿盐井的方式来开凿油井。原油当时被称为“岩石油”，被当作凿盐过程中带来的令人生厌的副产品。比斯尔吸引到一些投资人后，成立了宾西法

of the salt drilling process and considered a nuisance. Bissell recruited investors in his plan and formed the Pennsylvania Rock Oil Company. Their first objective was to find a promising area of land on which to drill, and then to find someone capable to do the drilling. The man they propositioned was one Edwin L. Drake who happened to meet one of the key investors in the Pennsylvania Rock Oil Company while living in the same hotel. This investor was James Townsend, a New Haven banker, who was impressed with Drake's demeanor and even encouraged him to buy stock in the company. Townsend sent Drake to obtain the title to the prospective oil land on a farm in Titusville, Pennsylvania in 1857. The next step was to begin the process of drilling for the oil. Unfortunately for Drake the local "salt borers" could not be relied upon to stay loyal to the job, and most disappeared or found some other way to get out of the arrangement. It was not until the spring of 1859 that Drake found good fortune in acquiring a driller, a man by the name of William "Uncle Billy" Smith, along with his two sons. Smith was a local blacksmith who made tools for the salt drillers and knew a good amount about what needed to be constructed for the oil drilling operation. The team was fastly draining the monetary resources of the original investors, and most of them stopped their financing of the operation. Townsend was the only investor who still believed in the project, even to the point of paying the venture's bills out of his own pocket. In the same week that Townsend finally lost faith and sent a dispatch to Drake to close up the operation, nothing short of a miracle occurred. On August 27, 1859, the drill dropped into a crevice and then slid another six inches; Drake struck oil (Figure 1.1).

Unlike the oil gushers, geysers, or "oil fountains" most people commonly associate with striking oil, Drake's discovery had to be hand pumped out of the

利亚石油公司。他们的第一个目标是找到一处有钻探前景的地区，然后找到合适的钻井人员来完成这一工作。这时，一名叫埃德温•L•德雷克的人出现了。他与宾西法利亚石油公司的一位主要投资人在宾馆一处偶遇后，被该投资人推进来主持钻井工作。推荐他的这名投资人名叫詹姆士•汤森，是一位纽黑文的银行家。他被德雷克的言谈举止所打动，甚至劝说他购买公司的股票。1857年，汤森派德雷克在宾西法利亚的蒂图斯维尔征得了一块土地，接下来就开始打井。可惜不尽如人意的是，当地的钻盐工人对工作不够忠诚，很多人中途就打了退堂鼓，找别的活儿去了。直到1859年的春天，公司才幸运地找到了满意的钻井人员——“比利大叔”威廉•史密斯及他的两个儿子。史密斯是当地的一名铁匠，专门打制钻盐工具，对于石油需要哪些工具颇有一些认识。工作开展起来以后，投资人的钱被项目一天天地迅速吸进去，最后大部分人都停止了对项目的投资。汤森成了唯一相信该项目一定会成功的投资人，他甚至不惜自掏腰包对项目进行投资。最终，连汤森也失去了信心，发函让德雷克停止作业。就在后面那一周，奇迹出现了。1859年8月27日，钻杆掉到一个裂缝中，又下滑了6ft（1ft=0.3048m），钻到了石油（图1.1）。

Figure 1.1 Historic Drake Well—1859
(Source: http://explorepahistory.com/)

ground. This event began the mad rush for oil that would engulf the nation in a new era of industrial achievement. The only comparable event as to what took place would be the California Gold Rush. Wells sprung up around the Titusville area like wildfire, but the oil still had to be hand pumped out of the ground, limiting the amount of oil that could be pumped to about 50 bbl/d (barrels per day). All of this changed in 1861 when the world's first flowing well, or geyser, was struck, flowing at about 3000 bbl/d. In 1860, production wavered around 450,000 barrels, then jumped to 3 million barrels in 1862. Unfortunately, production outpaced demand and prices dropped from $10 per barrel to less than 10 cents per barrel within the year, ruining many producers. On the other

德雷克打的这口井没有自喷能力，原油需要用手工泵抽到地面。该井成功掀起了一场堪比当年加利福利亚“淘金热”的原油“淘金热”。这一热潮带来了一个新的工业时代。油井如雨后春笋般在蒂图斯维尔地区出现，不过这些油井由于不具备自喷能力，产量仅徘徊在50bbl/d左右。1861年，事情出现了转折，人们打到了世界第一口自喷油井，产量达到了3000bbl/d。1860年全年的产量仅徘徊在45×10^4bbl，1862年这一数字就飞跃到了300×10^4bbl/d。不幸的是，由于

hand, this swing of prices led to Pennsylvania Oil's dominance in the marketplace by undercutting coal-oils and other illuminants. Demand quickly caught up with supply due to the widespread crossover to Pennsylvania Oil from coal and animal fat products（Figure 1.2）.

供大于求，油价直线下降，1862年当年的油价从最初的10美元/桶一路跌到了10美分/桶以下，许多生产商深受其害。另一方面，这种大降价也从煤成油及其他照明用油那里抢到了相当一部分市场，对宾西法尼亚石油公司的石油需求旋即增长起来（图1.2）。

Figure 1.2　Drake Well—2012
(Source:http://www.answers.com/topic/drake-well)

The easily acquired fortune and the relatively open market allowed for just about anyone to set up their own production facility. All that was needed was a little startup capital and someone who knew how to drill. These aspects also provided for the frequent fluctuations in the retail price of oil products. Overproduction was a result of a lack of restrictions, for it was not widely understood what exactly went on in these mysterious chambers and veins under the earth. Petroleum geology was a yet unheard of science; for it was still believed that oil discovery was based on luck combined with a natural talent for "sniffing out" oil. It was not until later in the Oil Age that geologists were frequently consulted for information on rock formation, compositions, etc. in relation to the location of oil deposits. This was especially apparent in the case of Patillo Higgins, a one-armed mechanic and lumber merchant from Beaumont, Texas. Higgins believed that oil would be found under a large hill named Spindletop. On the numerous outings

开发的市场以及巨大的利润使得稍有资金及钻井技术的个体商人纷纷加入钻油行业。由于缺乏秩序，导致过度生产，使得石油零售价格变得飘忽不定。当时的人们还根本不了解原油生产过程中地下发生的情况，石油地质还是一门闻所未闻的学科。人们认为找到原油凭的是运气和"嗅"出石油的天赋。直到后来，地质学家们才受邀加入进来，深入研究了油气储藏与地层构造及组成的关系。一个典型的例子是来自德克萨斯州博蒙特的机

Higgins took to the hill, he saw that many of the springs had gas bubbling up into them, and upon poking a stick into the ground near one particular spring he lit the gas that escaped. Every geologist that stopped to survey the area proclaimed that Higgins had no idea what he was talking about and refuted his claim of the presence of oil. Higgins could find no one to support or help him now that most members of respected society had publicly disavowed him. He would not give up, however, and proceeded to advertise for someone else to drill. He received a response from a Captain Anthony F. Lucas, an experienced prospector of salt in sulfur found in geological structures known as salt domes. The pair then obtained help from the firm of James Guffey and John Galey, who had developed the first major oil field in the mid-continent, in Kansas. This experience paid off with Spindletop, which was a very large salt dome. On January 10,1901, almost two years after their first attempt, Lucas and Higgins struck what was be the largest geyser yet. Lucas 1 on Spindletop erupted at 75,000 bbl/d. This sparked the southwestern oil boom, starting with other salt domes across the Gulf Coast of Texas and Louisiana, then spreading into Oklahoma with the vast discoveries at Glenn Pool in 1905, and finally North Texas.

The surplus of oil that was created was only partially due to the increased discoveries of new oil producing locations. “Flush production” was the true father of overproduction. Flush production consisted of racing of racing to produce the most oil out of a well due to the close competition of other producer pumping out of the same location. The lack of geological knowledge combined with the ignorance of exactly how a well worked helped to support this system. Producers had to constantly pump oil out of a well; otherwise their neighboring competitor might produce more oil out of the same deposit. This constant race led to instability in pricing and wasted

械师和木材商人帕瑅罗•希金斯，他深信在一座称为“纺锤顶”的山下埋藏着石油。在对该山无数次的实地考察中，他注意到山泉中常有气泡冒出。有一次，他拿一根木棍捅进一个山泉旁的土地，有天然气溢出。然而，每一名到该地考察过的地质专家都认为希金斯在痴人说梦，没有人相信那儿真的有石油。由于被权威集团所否定，希金斯找不到一个支持他的人。然而，在坚持不懈的努力下，通过广告的方式，他最终找到了一个来帮助他打井的人，这个人叫安东尼•F•卢卡斯。此人在盐丘硫黄钻探领域很有经验。两人此后得到了詹姆士•葛菲和约翰•加利合办的公司的赞助。该公司此前已经在堪萨斯州开发了美国中部第一个大油田。该公司的经验在钻探纺锤顶油区的过程中发挥了重要作用。1901年1月10日，在付出了两年努力后，卢卡斯和希金斯打到了当时为止最大的一口自喷井——卢卡斯1井。该井每天的产量达到75000bbl。这一事件掀起了美国西南部的石油潮，先是从德克萨斯州和路易斯安娜州墨西哥湾沿岸地区的其他盐丘开始，一路扩展到了俄克拉荷马州，并于1905年在那里发现了巨型的Glenn Pool油田。最后，原油潮扩展到了德克萨斯州北部。

原油产量的增加并不是完

oil due to excess production.

Transportation of oil was chaotic at best during the beginning stages of the industry, for teamsters would blockade transportation routes to the railroads, charging outrageous rates for through passage. This led to the construction of a wooden pipeline system, which transported oil more efficiently and cheaply to the railroads. By 1866, pipelines were hooked up to most of the wells in the Oil Regions, feeding into a larger pipeline gathering system that connected with the railroads.

The advent of electricity threw the oil industry into turmoil yet again in the late 19th century; people now had a much better alternative to petroleum-based illumination at a competitive price. Electricity soon dominated the illumination market within a decade of its inception because of its ease of use and lack of user supervision. Fortunately for the industry, the automobile powered by an internal combustion engine gained credibility in Europe after a Paris-Bordeaux-Paris race in 1895, in which the remarkable speed of 15 miles per hour was achieved. Gasoline, the former by-product of refining kerosene, now had a profitable use. The use of oil products as fuels began at about the same time as the inception of the automobile. Factories, ships, and trains were beginning to transfer over from coal fuel to oil due to its more efficient nature.

全来自新油田的发现，其相当一部分来自破坏性争夺开采。由于大家都在同一地区开采，在缺乏相关地质知识和油井生产知识的情况下，每个人都竭力开采，多多益善，以防对手占了便宜。这种无序竞争和过度生产导致了油价不稳及产品浪费。

在石油工业的起步时期，运输也是极其混乱的。在往铁路运输的过程中，遭遇沿路设卡，被索取高额过路费的情况时有发生。针对这种情况，一套木质管线被建造起来，用来高效、廉价地将石油输往铁路线。截至1866年，美国产油区已经建立了一套管网系统，并有干线连接至铁路车站。

19世纪后半叶，电力行业的发展，特别是电灯的普及，使人们纷纷离开照明用油，导致石油生意一度低迷。幸好在1895年巴黎至波尔多往返的汽车赛上内燃机汽车以15mile/h（1mile/h= 0.44704m/s）英里的杰出表现使得石油业找到了新的市场。汽油，这种早期炼油业的副产品，摇身一变成了主打产品之一。此后，工厂、轮船及火车都相继从煤炭转向了石油这种高效的能源燃料。

1.2 Petroleum,Petroleum Industry, Petroleum Engineering and Petroleum Engineer

Petroleum is a formal or technical word for crude oil and gas. It was derived from the Latin words for rock (petra) and for oil (oleum). So the basic meaning of the

1.2 石油、石油工业、石油工程、石油工程师

石油是原油和天然气的正式名称。它由拉丁语中的岩石和油两个词的组合而来。所

word is oil found in the rock. The advanced definition is Petroleum is a substance, usually liquid or gas, consisting of organic molecules composed of hydrogen and carbon atoms. Thus the general name *hydrocarbon* is often used.

Petroleum= Rock Oil

Petroleum Industry has three main components. The *upstream sector* includes more than 1000 exploration and production companies as well as hundreds of associated businesses such as seismic and drilling contractors, service rig operators, engineering firms and various scientific, technical, service and supply companies. The *midstream sector* includes oil and gas pipeline systems that connect producing and consuming areas. Other facilities extract Sulphur and natural gas liquids, store oil and gas products and transport products by truck, rail or tanker. The *downstream sector* consists of refineries, gas distribution utilities, oil product wholesales, service stations and petrochemical companies (Figure 1.3).

Petroleum Engineering is involved in the exploration and production activities of petroleum at the *upstream* end of the energy sector. The diverse topics covered by petroleum engineering are closely related to the earth sciences. Petroleum engineering topics include geology,

以，石油这个词的基本意思就是在岩石中找到的油。更进一步对石油这个词的定义是石油是一种常态液体或气体的物质，由包含碳和氢原子的有机分子组成。所以，也经常用石油的另一个名字碳氢化合物。

石油=岩石中的油

石油工业可分为三大组成。上游包括一千多家勘探生产公司以及成百家附属公司，如地震队、钻井队、修井队、工程公司和不同行业、不同技术、不同服务及供给公司。中游包括石油、天然气管道系统以连接生产和消费领域。也包括其他一些设施，如除硫、除天然气液烃、储存石油和天然气产品，以及卡车、铁路、油罐车运输设备。下游包括炼油厂、油气分销商、服务站和石化公司（图1.3）。

石油工程是一种包含石油勘探和开发活动处于能源部门上游的工程。石油工程包含多学科内

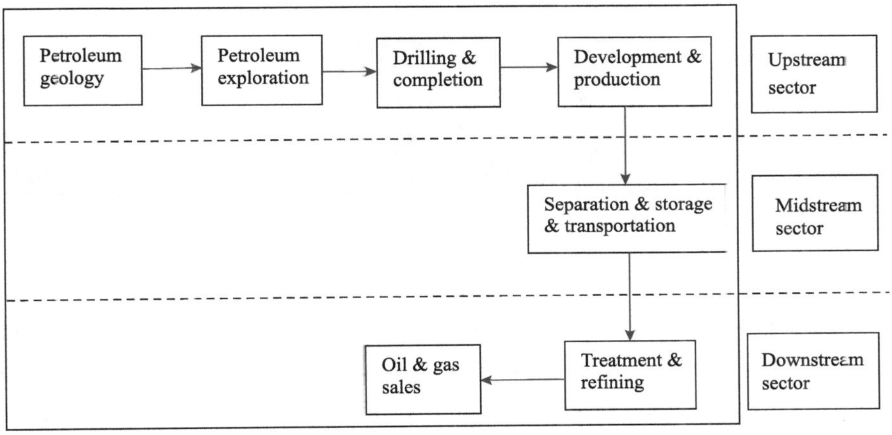

Figure 1.3　A Petroleum Industry Chain

petroleum geochemistry, geophysics, oil drilling, well logging, well completion, oil and gas production, reservoir development and pipelining.

容并与地球科学紧密联系，它包括地质学、地球化学、地球物理学、钻井、测井、完井、油气生产、油气藏开发和输送。

Petroleum Engineer isn't just one job. It's many different specialties, each with its own unique challenges and rewards. You can be a *drilling engineer*, working with geologists and contractors in designing and supervising drilling operations, many of which are multi-million-dollar ventures. You can work as a *production engineer*, develop processes and equipment to optimize oil and gas production. Or you can become a *reservoir engineer* and help determine ideal recovery processes, estimate the number of wells that can be economically drilled, and simulate future performance using sophisticated computer models. Teamwork in petroleum industry plays an important role in exploration and exploitation. Well-organized teamwork will result in high oil recovery and high profit/investment ratio(Figure 1.4).

石油工程师并不是一种单一的工作。它有很多专业，每一项都有自己独特的挑战和回报。可以是钻井工程师，与地质工程师和承包商合作设计和监督钻井工作，这是数百万美元的赌注。也可以是生产工程师，开发设计最优化油气生产的流程和设备。还可以是油藏工程师帮助确定理想的采收率过程，估计经济化钻井的油井数，用先进的计算机模拟油藏未来的趋势。团队工作在石油行业中很重要，特别是在勘探和开采过程中。组织好的团队工作直接导致高采收率和高回报比（图1.4）。

Figure 1.4　A Petroleum Engineer
(Source: http://educatingengineers.com)

1.3　The Field Life Cycle

This section provides an overview of the activities carried out at various stages of field development. Each activity is driven by a business need related to that particular phase (Figure 1.5).

1.3　油气田开发过程

本小节概述不同阶段油气田开发过程及每一开发过程阶段所需的商业运作（图1.5）。

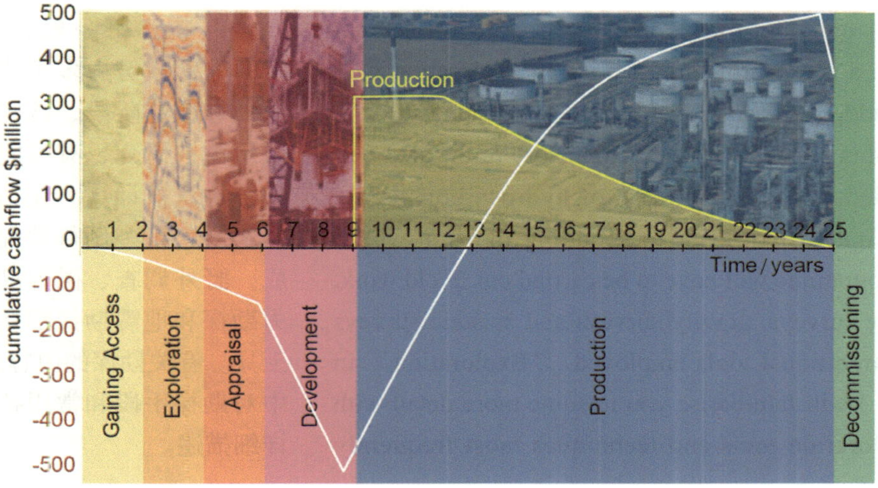

Figure 1.5 The Field Life Cycle and Typical Cumulative Cash Flow
(Source: Hydrocarbon Exploration and Production)

1.3.1 Exploration Phase

Despite the development of new exploration techniques has improved geologists' understanding and increased the efficiency, exploration remains a high-risk activity.

Traditionally, investments in exploration are made many years before there is any opportunity of producing the oil (Figure 1.6).

1.3.1 勘探阶段

尽管新的勘探技巧的发展提高了地质人员对地下的认知和效率，但是勘探仍然是高风险的活动。

传统上，在勘探投资多年后才可能产油（图1.6）。

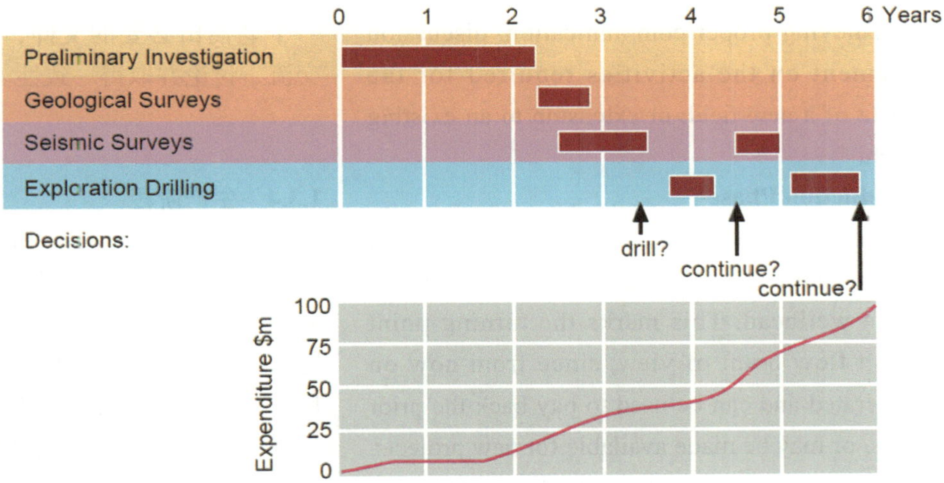

Figure 1.6 Phasing and Expenditure of a Typical Exploration Programme
(Source: Hydrocarbon Exploration and Production)

It is common for a company to work for several years on a prospective area before an exploration well is 'spudded' – an industry term for starting to drill. During this period the geological history of the area will be studied and the likelihood of hydrocarbons being present quantified. Prior to spudding the first well a work programme will have to be carried out. Field work, magnetic surveys, gravity surveys and seismic surveys are the traditional tools employed. 'Exploration' in Chapter 3 will familiarise you in some more detail with the exploration tools and techniques most frequently employed.

一个公司一般在一个有潜力的区域努力多年才开始钻探井。在这阶段，该区域的地质历史被研究，有可能的碳氢化合物量被计算。在钻第一口井之前，许多例行的程序被完成。野外调查、磁力勘探、重力勘探和地震勘探都是传统的工具。有关勘探的内容在第3章中对勘探手段和常用技巧将做详细描述。

1.3.2 Appraisal Phase

Once an exploration well has encountered hydrocarbons, considerable effort will still be required to accurately assess the potential of the find. The amount of data acquired so far does not yet provide a precise picture of the size, shape and producibility of the accumulation.

1.3.2 评价阶段

一旦勘探井钻遇碳氢化合物，将做大量的工作用以评价可能找到的油气区域。截至目前，数据还远不能提供地下油气聚集区准确的大小、形状和生产能力。

1.3.3 Development Planning

Based on the results of the feasibility study, and assuming that at least one option is economically viable, a field development plan (FDP) can now be formulated and subsequently executed. The plan is a key document used for achieving proper communication, discussion and agreement on the activities required for the development of a new field, or extension to an existing development.

1.3.3 谋划开发方案

基于可行性研究，假定至少有一种经济可行性，油田开发方案将被制定和执行。开发方案是开发一个新油田或完善一个老油田必要的文件，方便交流、讨论并取得一致性。

1.3.4 Production Phase

The production phase commences with the first commercial quantities of hydrocarbons (first oil) flowing through the wellhead. This marks the turning point from a cash flow point of view, since from now on cash is generated and can be used to pay back the prior investments, or may be made available for new projects. Minimising the time between the start of an exploration campaign and 'first oil' is one of the most important

1.3.4 生产阶段

生产阶段从第一口商业产量油井出油开始。从现金流看，它标志着产生效益开始偿还投资，或用作可行性新项目。最小化从探井开始到第一口出油井时间是风险投资的第一要务。

goals in any new venture.

Development planning and production are usually based on the expected production profile which depends strongly on the mechanism providing the driving force in the reservoir. The production profile will determine the facilities required and the number and phasing of wells to be drilled. The production profile shown in Figure 1.7 is characterised by three phases:

开发方案和生产很大程度上基于油藏驱动机理，生产指标决定地面设施和钻井数。从图1.7中可看出生产阶段可分为三个阶段：

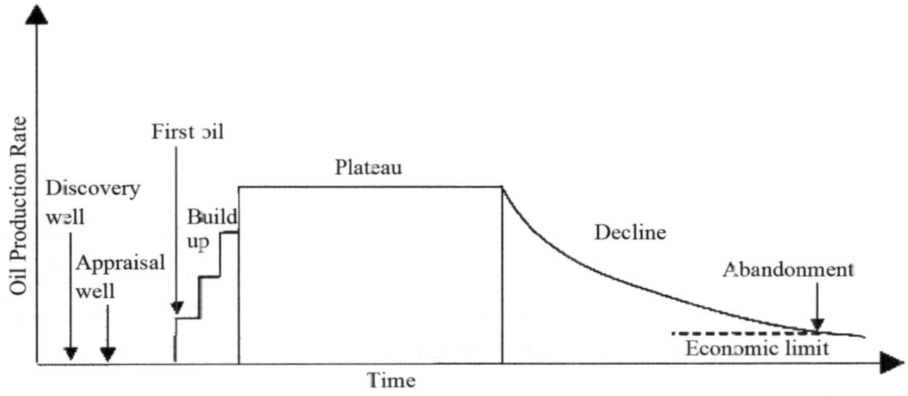

Figure 1.7 The Production Profile
(Source: www.theoildrum.com)

◆ Build-up period. During this period newly drilled producers are progressively brought on stream.

◆ Plateau period. Initially new wells may still be brought on stream but the older wells start to decline. Production facilities are running at full capacity, and a constant production rate is maintained. This period is typically 2~5 years for an oil field, but longer for a gas field.

◆ Decline period. During this final (and usually longest) period, all producers will exhibit declining production.

◆ 生产上升期。该阶段新井促进产量。

◆ 平台期。新井增加产量，但是老井产量衰减。生产设备已全交付，产量恒定，平台期对于油田一般为2～5年，对于气体更长。

◆ 下降期。这是最后阶段，通常也是最长的阶段，生产井开始出现产量下降。

1.3.5 Decommissioning

The economic lifetime of a project normally terminates once its net cash flow turns permanently negative, at which moment the field is decommissioned.

1.3.5 停产

当油田的经济化寿命结束，资金流永久为负，进入油田停产阶段。

2 PETROLEUM GEOLOGY
石油地质

2.1 Structure of the Earth

Gravity continued to shape the planet as it formed. Heavier materials, such as iron and nickel, concentrated in the earth's center, forming the core of the present-day earth (Figure 2.1). Form analysis of earthquake tremors passing through the earth, seismology has determined that this core is mostly in a liquid state, but with a solid center.

2.1 地球的结构

重力形成地球内部的形态。越重的物质如铁、镍集中到地心（图2.1）。从穿越地球的地震波分析，地心绝大多数为以固体介质的液态。

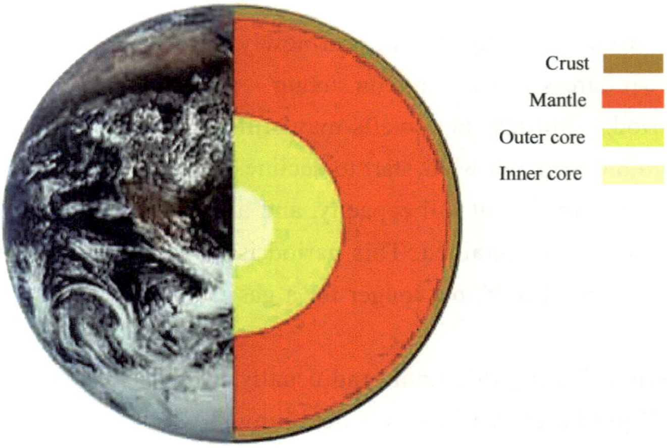

Figure 2.1 Diagram of the Earth's Interior
(Source: http://www.rocksandminerals4u.com/earths_interior.html)

Surrounding the core is the mantle, composed largely lighter silicates. Heat and pressure keep it partially melted so that it is not quite a solid and not quite a liquid. It therefore behaves as a plastic,

包裹着地心的是地幔，主要为轻质硅酸盐。热和压力使地幔部分融化，所以说地幔既不完全是固体也不完全是液体。地幔

extruding under pressure.

在压力下表现出可塑性。

Finally, the solid crust occupies the surface of the sphere. Its materials are the lightest of all. The crust essentially floats on the denser mantle and the mantle floats on the denser core. The force of gravity therefore gravity-segregates the earth, with the densest material in the center, and the density decreasing from the center outward(Figure 2.2).

最外面的是固态的地壳。地壳的组成物质最轻，基本上是漂浮在地幔上的，而地幔又漂浮在更重的地心上。因此，依靠重力将地球分层，最重的为地心，越往外密度越低（图2.2）。

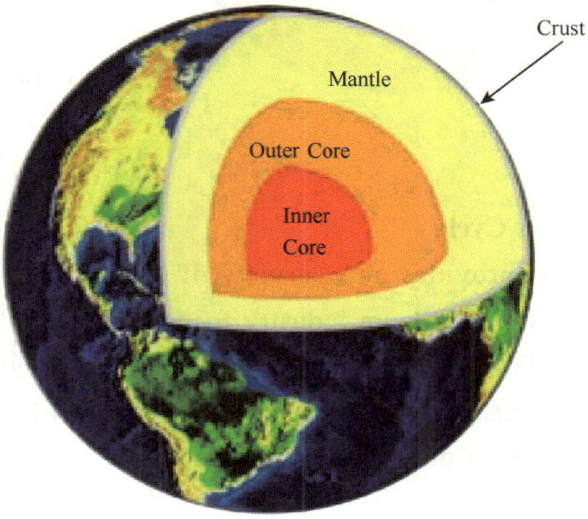

Figure 2.2 The Earth's Interior
(Source: http://easyscienceforkids.com/all-about-earths-crust/)

To complete the inventory of the earth's parts, there is the hydrosphere (water in the oceans, lakes, rivers, etc.) and the atmosphere. These are also held in place by gravity.

完整的地球圈还有水层和大气层，这也靠重力分层。

The earth has a superficial resemblance to a tomato. Each has semi-liquid material beneath a solid skin. Their proportions are similar, with the earth being 8000 miles in diameter with a crust less than 30 miles thick.

实际上地球与西红柿很像，在固态表皮下有半液体的物质。它们的比例也很像，地球直径8000miles(1mile=1609.344m)，而地壳不到30miles。

The diameter of the core is about 4300miles (6900 km), and iron is probably the chief ingredient. The mantle is nearly 1800 miles (2900 km) thick and makes up about 84% of the volume of the Earth. Since the volume of the core is about 16%, the crust actually makes up a very small part of the Earth as a whole.

地心直径大概为4300miles，而铁为其中的主要成分。地幔大概1800miles厚，约占地球体积的84%。地心占地球体积的16%，所以地壳实际只占地球很小的一部分。

From the Figure 2.3, can you see where we drill a well?

由图2.3可知在何处钻井。

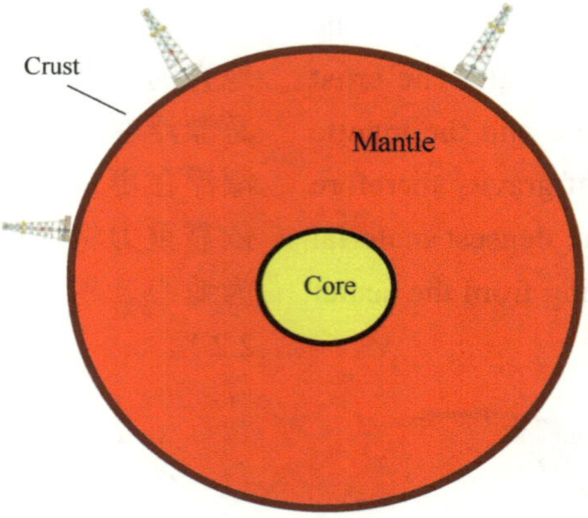

Figure 2.3 Drilling the Crust

2.2 Rocks

2.2.1 The Rocks Cycle

This section provides an overview of the Earth Rock. For the petroleum industry focus in more detail on sedimentary rock(Figure 2.4).

The earth's crust is composed of rocks, and rocks are composed of different minerals pushed together and combined(Figure 2.5). Igneous(fire-formed) rock is formed when liquid magma coming up the volcano and then cools into a solid(Figure 2.6, Figure 2.7). This rocks create new mountains(Figure 2.8). Once these mountains cool, the external force drives, the climate changes that cause weathering and erosion, begin to breaking down and tear down the mountains. Winds, water and ice break down the rocks of the mountains and then move away,

- *The Earth*
  - *Crust*
    - *Sedimentary Rock*
      - *Clastic Rock*
      - *Carbonate Rock*
      - *Evaporite Rock*
    - *Igneous Rock*
    - *Metamorphic Rock*
  - *Mantle*
  - *Core*

Figure 2.4 The Earth Rock

2.2 岩石

2.2.1 岩石圈

这一节对地球岩石做了概述介绍。对于石油行业而言，更多地关注沉积岩（图2.4）。

地壳由岩石构成，岩石由不同矿物组成并包裹在一起（图2.5）。火成岩由火山爆发喷发并冷却成固体（图2.6、图2.7）。火成岩构成了新的山体（图2.8）。一旦火成岩山冷却后，受外部力量驱动，气候改变导致风化和剥蚀，山上岩石破碎山体毁坏。风、

eventually moving it towards and into the oceans. These particles are deposited layers sediment that overtime turn into sedimentary rock. Finally sedimentary rocks get pushed down into the deep levels and under high pressure and high temperature they are changed into metamorphic rock(Figure 2.9, Figure 2.10). Eventually metamorphic rocks are pushed even deeper until they are melted and turned into magma. This is the rock cycle(Figure 2.5,Figure 2.11,Figure 2.12). In these rock cycle, there are three classification of rocks: igneous rock, metamorphic rock and sedimentary rock.

水、冰川将山体破损下的岩石搬运走，最终运到海洋中。这些碎粒沉积下来，又慢慢地变成沉积岩。最后沉积岩埋藏越来越深，在高压和高温作用下变成变质岩（图2.9、图2.10）。最终，埋藏的变质岩又融化转变成岩浆。这就是岩石圈循环（图2.5、图2.11、图2.12）。在这岩石圈中有三类岩石：火成岩、变质岩和沉积岩。

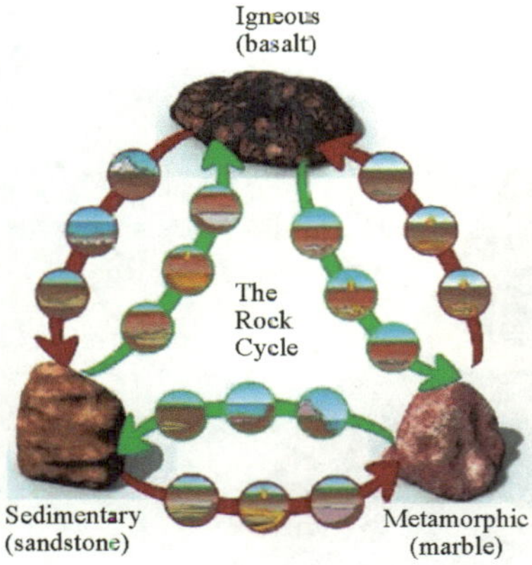

(a)

(b)

Figure 2.5 The Earth Rock Cycle (Source: Overview of the Oil Industry, Richard Lau)

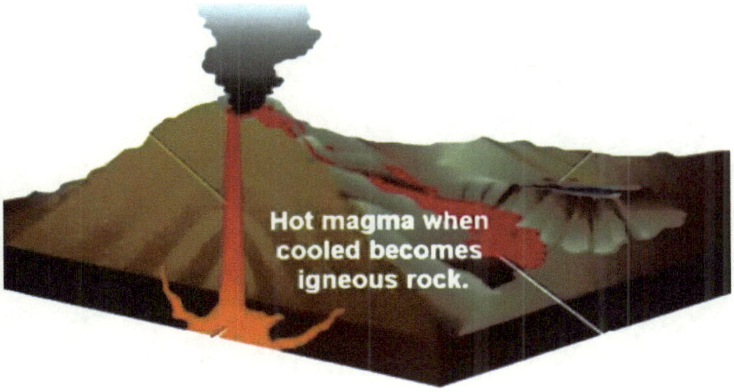

Figure 2.6 Magma & Igneous Rocks
(Source: Overview of the Oil Industry, Richard Lau)

Figure 2.7 Igneous Rocks
(Source:http://www.geography4kids.com/files/earth_rocktypes.html)

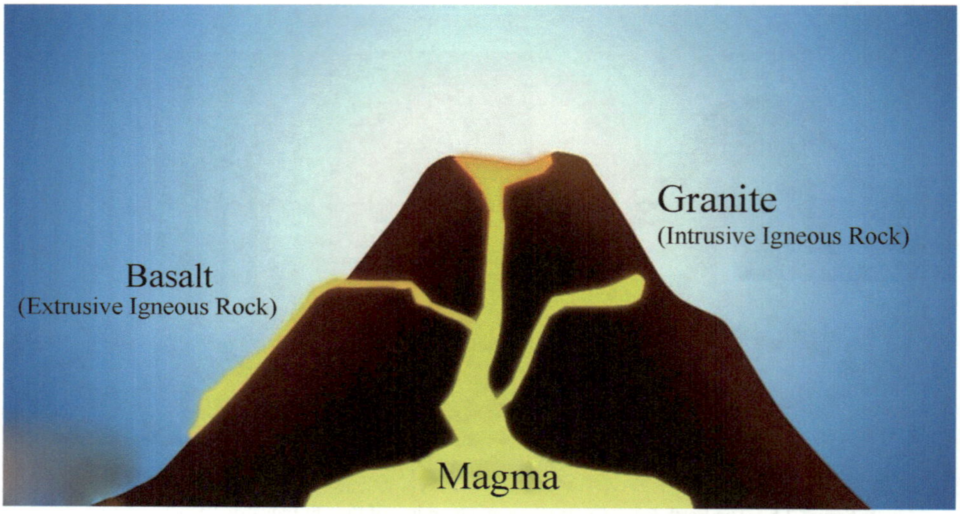

Figure 2.8 Igneous Rocks
(Source: www.makemegenius.com)

Figure 2.9 Metamorphic Rock
(Source:http://geology.com/rocks/amphib dite.shtul)

Figure 2.10 Metamorphic Rocks
(Source: http://libraty thinkquet.org/os aug/00461/meta. htmuakemegenius.com)

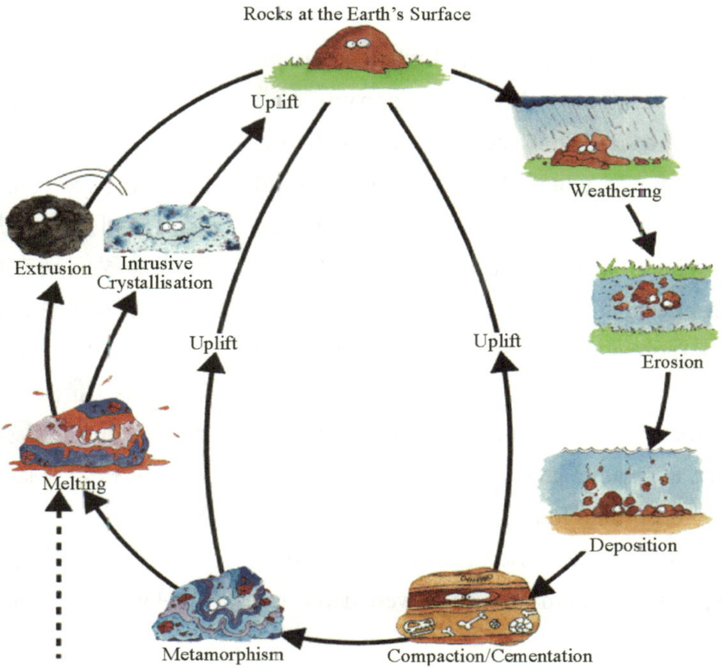

Figure 2.11 Rocks at the Earth's Surface
(Source: http://www.oum.ox.ac.uk/thezone/rocks/cycle/)

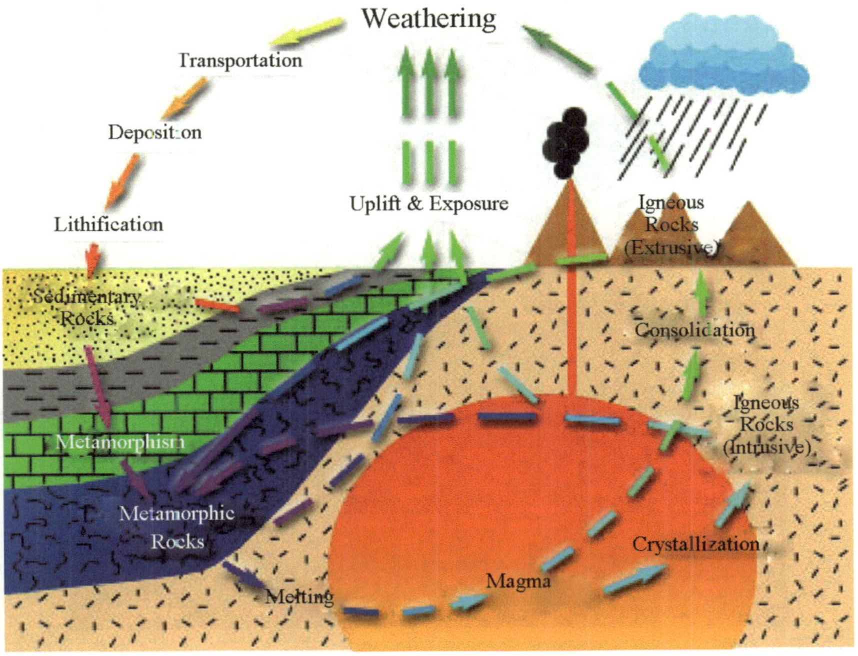

Figure 2.12 The Rock Cycle
(Source: http://www.craven.k12.nc.us/AWE/linkpages/4thprojectpgs)

2.2.2 Sedimentary Rocks

Sedimentary rock types are created when sediment compresses. A river flows through a canyon and picks up a bunch of silt. That sediment and silt runs downstream and deposits where the river ends. It could be in a flood plain or a valley, but we're using a coastline as an example. When that material gets to a flood plain or a valley or a beach, it sits there(Figure 2.13). Over millions of years, more and more sediment builds up and compacts. That compacted sediment eventually becomes a type of rock. Examples of sedimentary rock include sandstone, amber, anthracite, and limestone.

Petroleum is most often found in a sedimentary basin. A sedimentary basin is a depressed area of the Earth's crust where tiny plants and animals lived or were deposited

2.2.2 沉积岩

沉积岩由沉积压缩形成，一条流过峡谷的小河带走一堆淤泥，这些沉积物和淤泥随河流向下流动并沉积在河流末端。这可能是平原和山谷，也可能是我们现在的海岸线。当流到海岸等时，沉积下来（图2.13）。百万年来，越来越多的沉积物堆积并被压实。这些压实的沉积物最终变成了沉积岩。沉积岩有砂岩、琥珀、煤和石灰岩。

绝大多数的石油在沉积盆地中被发现。沉积盆地是地壳

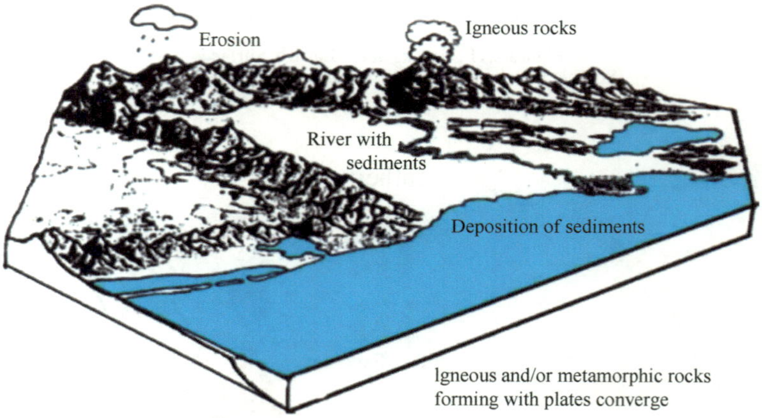

(a)

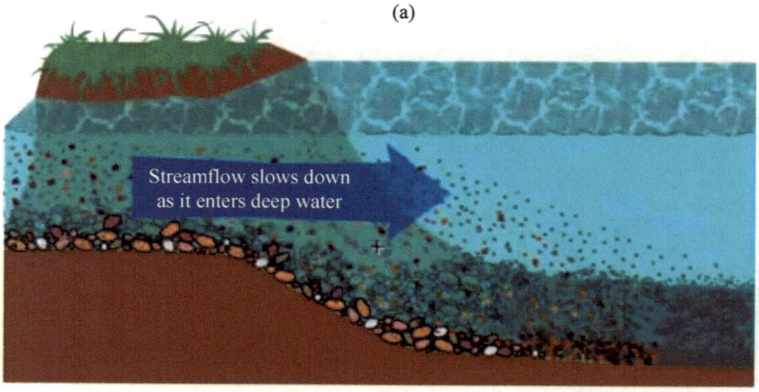

(b)

Figure 2.13 Sediment
(Source: Overview of the Oil Industry, Richard Lau)

with mud and silt from stream and rivers. These sediments eventually harden to form sedimentary rock.

Sedimentary rocks are grouped into three major classes: *clastic, carbonate* and *evaporitic*.

Clastic rock are those that are formed by the accumulation and cementation of sedimentary particles derived from weathered fragments of preexisting rocks. Weathering processes, such as freezing and thawing, rain, wind, and other similar events, break down the parent rock into small particles that can then be transported by wind and rain runoff. Streams carry the mud, sand, and gravel from the source area down to its final resting place, being that a stream channel, floodplain, lake, or ultimately the sea. There it accumulates, is buried and compacted by later-arriving sediments, and cemented to form sedimentary rocks. The mud compacts to shale or mudstone, the sands are cemented by silica or calcite to form sandstones, and the gravels become conglomerates. Sandstones, because of the inherent porosity between their grains, often become excellent reservoirs for oil or natural gas(Figure 2.14, Figure 2.15).

中凹陷区域，盆地中微生植物和动物生存，死后被来自河流、小溪的淤泥、黏土掩埋并沉积下来。这些沉积物最终变硬形成沉积岩。

沉积岩分为三大类：碎屑岩、碳酸盐岩和蒸发岩。

碎屑岩由其他原岩风化破碎的微粒汇集压缩而成。风化过程如霜冻、融雪、下雨、刮风等类似情形将原岩破碎成能被风或雨搬运走的微小颗粒。溪流将黏土、砂、砾从源头搬运到最终沉降地，可能是河道、平原、湖泊或终极的海洋。然后汇集、埋藏，又被后来的沉积物挤压胶结形成沉积岩。这些黏土被压缩成页岩或泥岩，这些砂粒被硅石或钙质胶结形成砂岩，这些砾石变成砾岩。砂岩由于内部颗粒间的孔隙性通常变成油或天然气最佳的储集场所（图2.14、图2.15）。

Figure 2.14　Sandstone Sedimentary Rock
(Source: http://www.minimegeology.com/home/mgeo/page_34)

Figure 2.15　Sandstone-close up
(Source:http://www.earthscienceeducation.com/)

Carbonate rocks are *limestones*(Figure 2.16) and dolomites(Figure 2.17). They usually form in warm seawater at shallow depths, ankle deep to about 6 m, where various plants and animals thrive. The hard, usually calcareous parts of the organisms pile up on the seafloor over time, forming beds of lime particles. Algae, simple plants, are one of the greatest contributors of lime particles, but any shelled animal may contribute whole or fragmented shell to the pile. Reefs, banks of lime mud, and lime sand bars are commonly found preserved in rocks.

碳酸盐岩为灰岩（图2.16）和白云岩（图2.17）。它们一般在浅海温暖的海水中形成，从脚踝到大概6m深处有大量的植物和动物繁衍。有机质中的硬质部分通常为钙质在海底堆积形成的灰质颗粒，而贝壳类动物将颗粒连接成堆。暗礁、灰岩泥质堆积、灰岩砂砾时常在岩石中找到。

Figure 2.16 Calcite Limestone
(Source:http://www.dnr.mo.gov/geology/geosrv/imac/limestone.htm & http://www.daviddarling.info/encyclopedia/L/limestone.html)

Figure 2.17 Dolomite
(Source:http://geology.com/minerals/dolomite.shtml)

Carbonate sediments are subject to many processes in becoming a rock. If lime mud is exposed to the air, it simply dries out and almost overnight becomes a natural *concrete*, and lime sands become cemented by the evaporation of lime-rich seawater to form *beach rock*. Rainwater, however, begins to destroy the rock and forms porosity in the process. If not exposed to the air, lime sediments continue to accumulate, perhaps to great thicknesses. Under these conditions, they compact under the weight of new sediments and eventually are cemented to form limestone. These rocks are rarely porous unless other processes become involved.

Limestone is composed of calcium carbonate (calcite or aragonite), thus the general rock term *carbonate* is used. Magnesium, a common element in seawater, can replace some of the calcium within the crystal structure of calcite. This often happens by various processes that are not fully understood; the resulting rock is known as *dolomite* (calcium-magnesium carbonate). The process of changing limestone to dolomite produces somewhat smaller crystals, so the resulting rock has tiny pores between the new crystals. This kind of porosity, much like that in sandstone, often contains oil and natural gas.

Evaporite rocks are formed by direct precipitation mineral by evaporation of seawater. Resulting rocks are ordinary salt (halite-sodium chloride), gypsum (calcium sulfate with some water), and various forms of potash salts. When gypsum is buried to considerable depths, the water is expelled from the crystals and *anhydrite* (meaning simply without water), a harder crystalline rock, results. Evaporites are not porous, although they are readily dissolved by water and are not source rocks for petroleum. However, they may be formed in highly stagnant water where black mud, rich in organic matter, may also be deposited, and so are commonly associated with good source rocks. Because they are impermeable,

碳酸盐沉积物经过很多过程最终变成碳酸盐岩。假如石灰泥暴露在空气中，它会脱水，几乎一夜之间就变成天然水泥，石灰砂由富含石灰的海水蒸发胶结并形成海岸岩石。雨水开始破坏岩石，并在破坏的过程中形成孔隙。假如没有暴露在空气中，石灰沉积物继续聚集，可能变得很厚。在这些情形下，受新的沉积物的重量压实，最终胶结形成石灰岩。在压实过程中很少有孔隙形成，除非有其他的作用。

石灰岩由钙质碳酸盐组成，因此统称为碳酸盐岩。镁，一种在海水中很常见的元素，能取代方解石结晶中的钙质。这一过程中的许多具体细节并没有完全被理解，但结果是方解石变成了白云石。在方解石变成白云石的过程中产生不少较小的晶体，因此新旧岩石中形成不少微孔隙。这种孔隙类似砂岩中的孔隙，可以存储石油和天然气。

蒸发岩由矿物直接蒸发掉海水形成。原岩一般为盐类、石膏。当石膏埋藏一定深度，海水被从结晶体挤出形成硬石膏，一种更坚硬的岩石。蒸发岩没有孔隙，尽管它们很容易被水融化，但它不是形成石油的母岩。它们在富含有机质的滞水中形成，让人误认为是母岩。蒸发岩不渗透，经常形成油藏封隔层。

evaporates often form seals on other reservoir rocks.

2.3 Origin of Petroleum

There is still some disagreement on the origins of petroleum and the mechanisms by which it is concentrated into commercially size accumulation. Probably the most controversial issue is whether petroleum has an inorganic(abiogenic) or an organic origin.That is, whether it is derived from chemical reactions between minerals or from tissue created by living organisms.

The inorganic thesis has been particularly advocated by geologists from Eastern European countries, but has other adherents as well.

Methane can be generated in the laboratory by applying heat and pressure to naturally occurring minerals, so it's likely that some petroleum has an inorganic origin.

$$\text{Heavy metal carbides} + H_2O \xrightarrow[\textit{High Pressure}]{\textit{High Teperature}} \text{Metallic oxide} + \text{Hydrocarbons}$$

According to the abiogenic hypothesis, petroleum was formed from deep carbon deposits, perhaps dating to the formation of the Earth. Then petroleum may originate from carbon-bearing fluids that migrate upward from the mantle. The presence of huge amounts of methane on Saturn's moon Titan and in the atmospheres of Jupiter, Saturn, Uranus and Neptune is cited as evidence of the formation of hydrocarbons without biology.

The hypothesis was first proposed by Georg Agricola in the 16th century and various abiogenic hypotheses were proposed in the 19th century, most notably by Prussian geographer Alexander Von Humboldt, the Russian chemist Dmitri Mendeleev and the French chemist Marcellin Berthelot. Abiogenic hypotheses were revived in the last half of the 20th century by Soviet scientists who had little influence outside the Soviet Union because most of their research

2.3 石油的起源

关于石油的起源和聚集成藏的机理还存在分歧。其中，最大的分歧点在于石油的来源是有机的还是无机的，即石油来源于矿物成分的化学反应还是有机物组织。

无机成因理论的主要支持者是东欧国家的地质学家。

甲烷在实验室里能通过加热和加压方式在矿物内生成，因此部分人认为是无机成因。

$$\text{重金属碳化物} + \text{水} \xrightarrow[\text{高压}]{\text{高温}} \text{金属氧化物} + \text{石油蒸汽}$$

依据无机成因假设，石油在碳圈深处形成，可能发生在地球形成之初。之后，随地球的冷却而被吸附并凝结在地壳上部，在沿裂隙向地表运移的过程中聚集起来，形成油气藏。在土星、木星、天王星、海王星及土星的卫星泰坦上都有发现甲烷气体，而无生物。

无机成因假设首先在16世纪由德国自然科学家格奥尔格•阿格里柯拉提出，而后在19世纪又有不同的无机成因假设提出，其中最著名的是普鲁士的地理学家亚历山大•冯•洪堡和俄国化学家门捷列夫和法国化学家贝特洛。无机成因假设在20世纪后半叶被前苏联科学家所接受，但因为他们的研究都是用俄语出版的，所以在前苏联以外的国家和地区影响很

was published in Russian. The hypothesis was re-defined and made popular in the West by Thomas Gold who published all his research in English.

小。直到汤姆士•戈登将此无机成因理论被重新定义并用英语介绍给西方才被熟知。

Although the abiogenic hypothesis was accepted by many geologists in the former Soviet Union, it fell out of favor at the end of the 20th century because it never made any useful prediction for the discovery of oil deposits. The abiogenic origin of petroleum has also recently been reviewed in detail by Glasby, who raises a number of objections, including that there is no direct evidence to date of abiogenic petroleum (liquid crude oil and long-chain hydrocarbon compounds). Geologists now consider the abiogenic formation of petroleum scientifically unsupported, and they agree that petroleum is formed from organic material. However, some argue that the abiogenic theory cannot be dismissed yet because the mainstream theory still has to be established conclusively.

尽管无机成因假设被前苏联的地质学家所接受，但在20世纪末该假设就不被多数人接受了。因为该理论在预测发现油气藏沉积过程中从没有提供有用的意见。无机成因假设最近被盖兹比详细提起，他举出大量反证。不过现在的地质学家还是认为无机成因假设无科学支撑，他们认同石油来源于有机物。然而，有人认为无机成因说不能被驳掉，因为主流的有机成因说也还需要确凿证据。

The organic thesis, generally speaking petroleum was originated from the dead body of tiny plants and animals buried deeply underground with the condition of high temperature and pressure for millions of years.

有机成因说认为石油由千百万年前微小动植物死去的尸体埋藏在高温高压的地下演变而来。

The latter opinion is little doubt in the minds of most petroleum geologists that most petroleum is formed from *organic* matter, tiny particles of plant and animal debris that accumulated with mud deposited in stagnant bodies of water. These ideas receive firmer support from geochemical studies of sedimentary organic matter.

有机成因说在绝大多数的石油地质工作者中没有争论，他们认为石油来自有机生物，微小动植物的残骸和泥沙一起沉积在停滞的水体中聚集演变而来。有机成因说也得到研究沉积有机物的地球化学家的支持。

2.4 Organic Source

2.4 有机质来源

Generally speaking petroleum was originated from the dead body of tiny particles of plant and animal debris that accumulated with mud deposited in stagnant bodies of water and buried deeply underground with the condition of high temperature and pressure for millions of years(Figure 2.18).

一般来说，微小动植物死后的遗骸在地下百万年的高温高压作用下形成了石油（图2.18）。

形成石油的原始有机物是开放大海里的浮游生物，也有

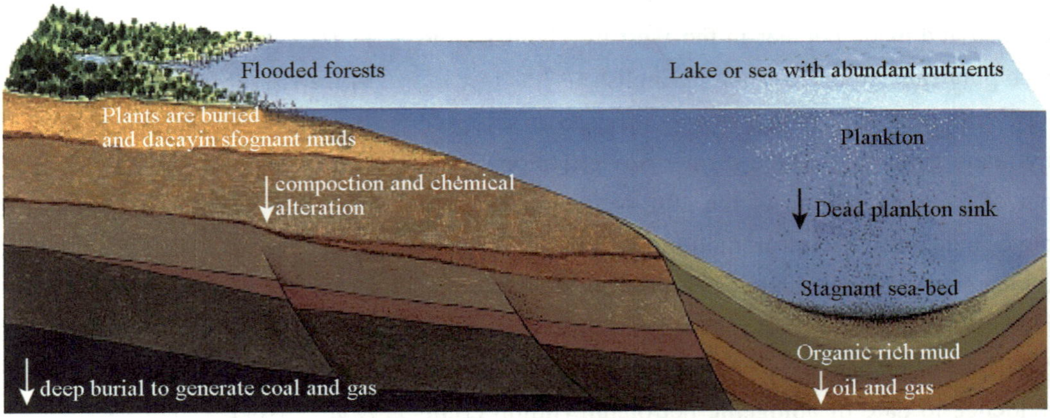

Figure 2.18 Origins of Oil and Gas
(Source: oilandgasuk.co.uk)

The original organic material was mostly microscopic marine plants and animals that lived in open seas, but much of it may have been derived from land plants and carried by streams to the site of deposition. Despite stories you might have heard, dinosaurs did not contribute to the original soup. Organic material is easily destroyed by exposure to air. Deposition of the material must be rapidly buried in waters containing little or no oxygen due to water stagnation. Where water is oxygen deficient and chemical reactions take oxygen atoms away from molecules, it is said to be a reducing condition, a requirement for the preservation of organic matter. Also, low-oxygen environments greatly decrease the number of scavengers that might otherwise consume and destroy the organic morsels. Usually the proper conditions occur in quiet marine basins, lagoons, and rapidly buried delta deposits, but similar conditions prevail in some lakes.

部分来自大陆的植物被溪流带到沉积地。不管你所听过的关于石油来源的任何故事，恐龙并不是石油的来源。暴露在空气中的有机物很容易被摧毁。沉积下来的有机物必须迅速被掩埋在缺氧的滞水中。滞水中氧气缺乏，形成还原环境有利于保存有机质。同时，缺氧环境极大地减少了食腐动物的数量，从而减少消耗有机质。有利的环境通常是安静的海底盆地、潟湖、迅速掩埋的三角洲和湖泊。

2.5 Generation of Oil

Generation depends on three main factors:

◆ the presence of organic matter rich enough to yield hydrocarbons;

◆ adequate temperature;

2.5 油气的生成

油气生成需要三个主要因素：

◆ 丰富的有机质用于产生碳氢化合物；

◆ and sufficient time to bring the source rock to maturity.

Pressure and the presence of bacteria and catalysts also affect generation. Generation is a critical phase in the development of a petroleum accumulation.

As these organic materials become more deeply buried, heat and pressure transform them into solid, liquid or gaseous hydrocarbons known as fossil fuels——coal, crude oil or natural gas. Coal is generally formed form the remains of terrestrial (land-based) plants. Oil is typically derived from marine (water-based) plants and animals, mainly algae(Figure 2.19), which have been gently cooked for at least one million years at a temperature between 50℃ and 150℃. Natural gas can be formed from almost any marine or terrestrial organic materials, under a wide variety of temperatures and pressures.

The conversion of sedimentary organic matter into petroleum is termed maturation. The resulting

◆ 合适的温度；

◆ 足够的时间使生油岩成熟。

压力、细菌、催化剂作用等也影响油气生成。油气生成是油气成藏过程中最主要的过程。

当有机质被深深地掩埋，压力和温度将它们转化为固态、液态或气态的碳氢化合物即化石能源——煤、原油和天然气。煤通常由陆生植物的残骸形成。原油来源于水生动植物，通常为海藻（图2.19）。这些微生物在温度为50～150℃的环境下被“烹煮”至少一百万年后形成原油。天然气可以来源于陆生或是水生的有机物，温度和压力范围变化很广。

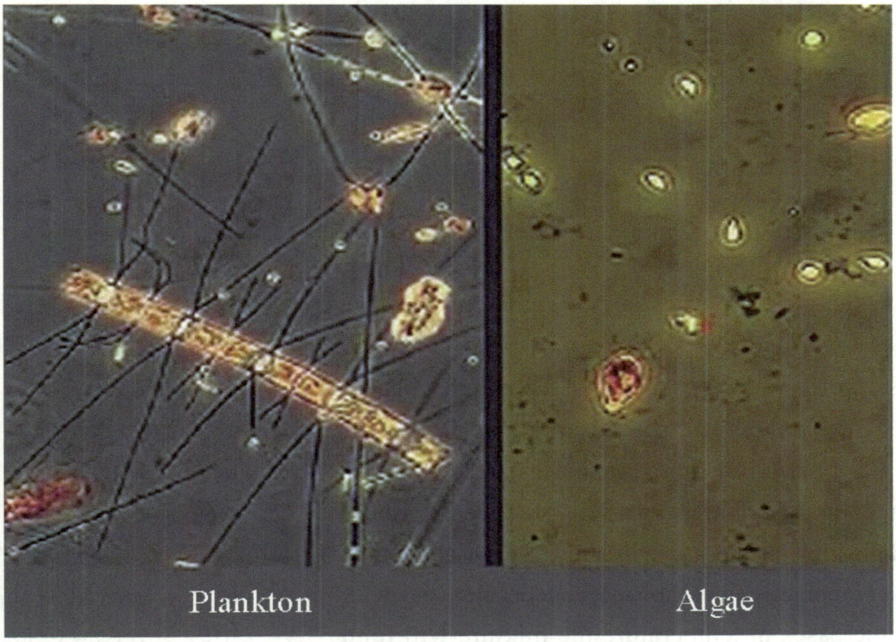

Figure 2.19 Plankton & Algae
(Source: Overview of the Oil Industry, Richard Lau)

products are largely controlled by the composition of the original matter. In the maturation process, which starts with the conversion of mainly kerogen into petroleum; but in very small amounts below a temperature of 50℃(kerogen: organic rich material which will produce hydrocarbon on heating). When kerogens are present in high concentrations in shale, and have not been heated to a sufficient temperature to release their hydrocarbons, they may form oil shale deposits(Figure 2.20).

沉积的有机物转化为石油有一个专有名词叫成熟。有机质的成分控制了转化后的产品。在成熟过程中，转化主要来源干酪根（有机质富含物在加热下可以产生碳氢化合物）转化为石油，极小部分转化也可以发生在50℃以下。干酪根在页岩中有很高的浓度，当温度不够高转化为碳氢化合物时，干酪根可以形成油页岩沉积（图2.20）。

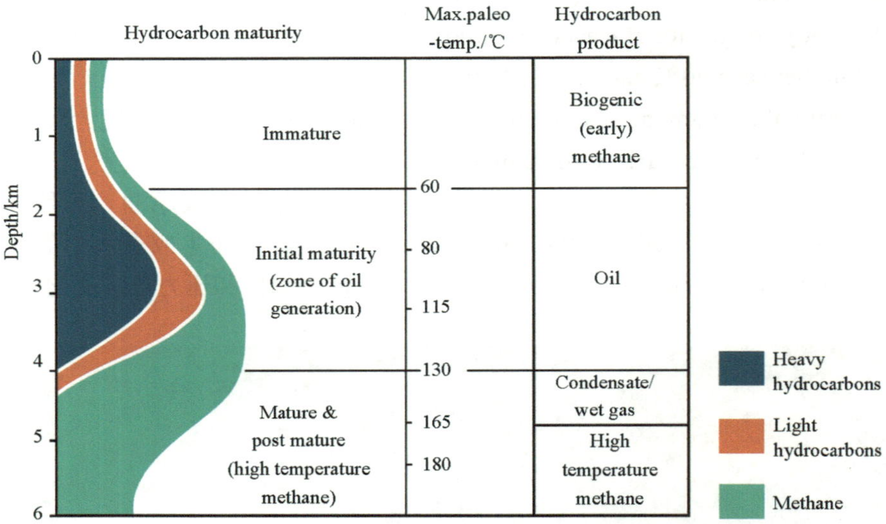

Figure 2.20 Hydrocarbon Maturation
(Source: Hydrocarbon Exploration and Production)

The temperature rises as the sediment package subsides within the basinal framework. The peak conversion of kerogen occurs at a temperature of about 100℃. If the temperature is raised above 130℃ for even a short period of time, crude oil itself will begin to ‘crack’ and gas will start to be produced. Initially the composition of the gas will show a high content of C_4~C_{10} components (wet gas and condensate), but with further increases in temperature the mixture will tend towards the light hydrocarbons (C_1~C_3, dry gas).

在沉积盆地内温度随埋藏深度增加而增加。干酪根转化的峰值温度是100℃。而温度超过130℃即使是很短的时间，原油也会自身裂化，天然气由此产生。刚开始转化的气还有高浓度的C_4～C_{10}成分即湿气和凝析气成分。随着温度升高，天然气中轻组分C_1～C_3成分即干气成分开始增加。

The most important factor for maturation and hydrocarbon type is therefore heat. The increase of temperature with depth is dependent on the geothermal gradient which varies from basin to basin. An average value is about 3℃ per 100m of depth.

因此，石油成熟度和碳氢化合物类型的决定因素是温度。各个沉积盆地温度随深度增加的地温梯度是不同的。平均值为每增加100m温度增加3℃。

2.6 Migration of Oil

2.6 油气的运移

The movement of hydrocarbons from their source into reservoir rocks.

◆ The movement of newly generated hydrocarbons out of their source rock is primary migration, also called expulsion.

◆ The further movement of the hydrocarbons into reservoir rock in a hydrocarbon trap or other area of accumulation is secondary migration.

◆ Migration typically occurs from a structurally low area to a higher area in the subsurface because of the relative buoyancy of hydrocarbons in comparison to the surrounding rock.

◆ Migration can be local or can occur along distances of hundreds of kilometers in large sedimentary basins, and is critical to the formation of a viable petroleum accumulation.

The maturation of source rocks is followed by the migration of the produced hydrocarbons from the deeper, hotter parts of the basin into suitable structures. Hydrocarbons are lighter than water and will therefore tend to move upwards through permeable strata(Figure 2.21).

Two stages have been recognised in the migration process. During primary migration the very process of kerogen transformation causes micro-fracturing of the impermeable and low porosity source rock which allows hydrocarbons to move into more permeable strata. In the second stage of migration the generated fluids move more freely along bedding planes and faults into a suitable reservoir structure. Migration can occur over considerable lateral distances of several tens of kilometers.

油气的运移指油气从生油岩运动到储集岩。

◆ 新生成的油气运移出生油岩的运动叫初次运移，也叫排出运移。

◆ 油气进一步运移到储集岩圈闭或其他储集空间叫二次运移。

◆ 因为油气相对于周边岩石的浮力作用，运移常发生在低构造点向高构造点运动

◆ 运移可以发生在当地，也可以发生在成百上千公里以外，运移距离对发现油气成藏有重要影响。

随着生源岩的成熟，生成的碳氢化合物会由又深又热的沉积底部向上运移到合适的地方存储下来。碳氢化合物比水轻，因此会沿可渗透层向上运移（图2.21）。

运移过程分两步，初始运移和二次运移。初始运移是干酪根在不渗透层和低孔隙度生油岩中转化为碳氢化合物，运移到可渗透的储集层。二次运移是油气自由进入岩石层理、断层，进入理想的储集空间。运移能发生在几十千米的距离外。

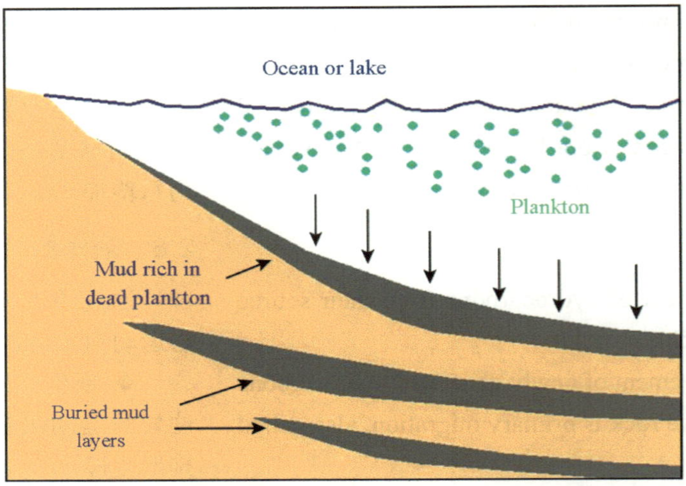

Figure 2.21 Oil is a Fossil Fuel that is Formed from Plankton that is Deposited in the Sediments of Oceans and Lakes, and then Buried Deeply in the Earth. Photo Credit: J.Bratton
(Source: http://oceanexplorer.noaa.gov/explorations/03mex/logs/sept25/sept25.html)

Due to the force of gravity and the pressure created by the overlying rock layers, oil and natural gas seldom stay in the source rock in which they are formed. Instead, they move through the underground layers of sedimentary rocks until they either escape at the surface or are trapped by a barrier of less permeable rock.

So these reservoirs are often long distance away from the original source. A seep occurs when hydrocarbons migrate to the Earth's surface(Figure 2.22). Over time, huge amounts of these hydrocarbons

因为重力和上覆岩层压力的作用，原油和天然气很少待着生油岩的原地。它们一直在地下岩层中运移着，直到逃逸到地表或是被不渗透的盖层遮挡圈闭下来。

因此，油藏经常在远离生油岩的地方。一个原油露头发生在碳氢化合物运移到地表（图2.22）。大量的碳氢化合物

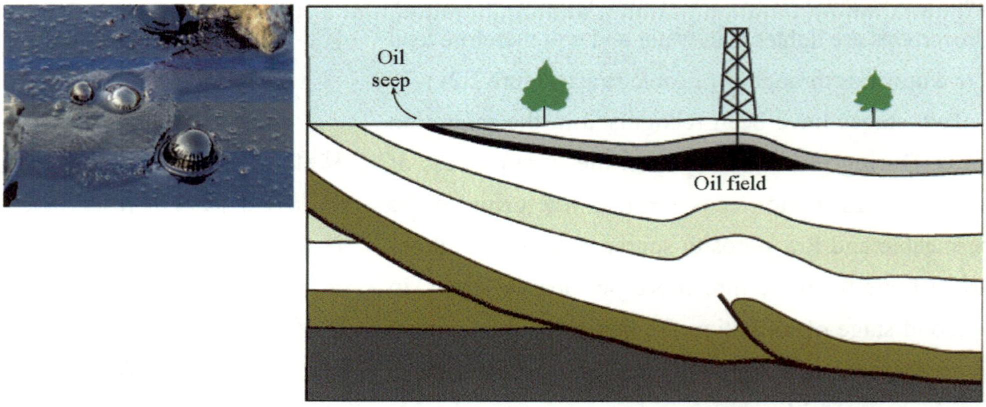

Figure2.22 Oil Seep
(Source: http://www.geo.wvu.edu/~jtoro/petroleum/Review1.htm)

have escaped into the atmosphere. Flowing water can also wash away hydrocarbons. Sometimes only the lighter, more volatile compounds are removed, leaving behind reservoirs of heavier types of crude oil(Figure 2.23).

逃逸到大气层中，流水也可带着部分碳氢化合物，特别是轻质的可挥发组分流走，留下重组分原油在露头中（图2.23）。

2.7 A Petroleum Accumulation

2.7 石油的聚集

A petroleum accumulation consists of a mature source rock, migration pathway, reservoir rock, trap and seal(Figure 2.24). Appropriate relative timing of formation

石油能够聚集成藏需要成熟的生油岩、运移通道、储集岩、圈闭、盖层（图2.24），

Figure2.23 Oil Seep of HeiYouShan in Xin Jiang Province, China
(Source: http://my.dili360.com/home/space.php?uid=1498389&do=blog&id=69079)

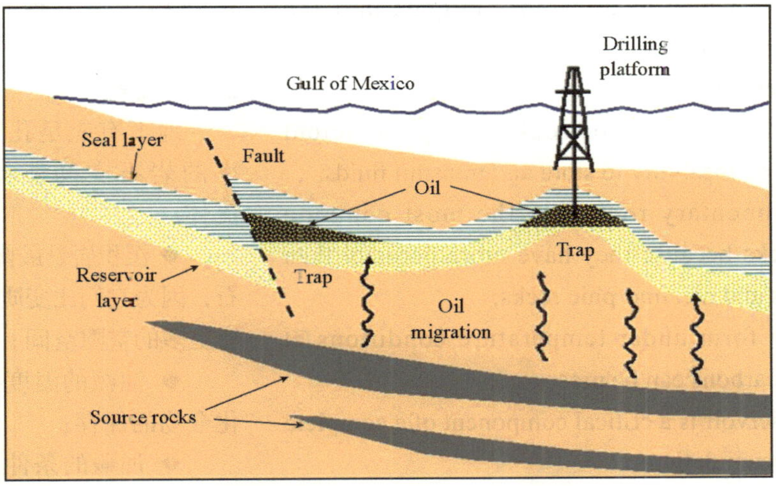

Figure 2.24 Cross-section of an Oil Deposit
Known as a "Play" in the Industry, Including Its Source Rock, Reservoir, Trap, and Seal. The Best Place to Drill to Recover this Oil is at Its Thickest and Shallowest Point
(Source: http://oceanexplorer.noaa.gov/explorations/03mex/logs/sept25/sept25.html)

of these elements and the processes of generation, migration and accumulation are necessary for hydrocarbons to accumulate and be preserved(Figure 2.25).

以及生成、运移、聚集的合适时机和保存条件（图2.25）。

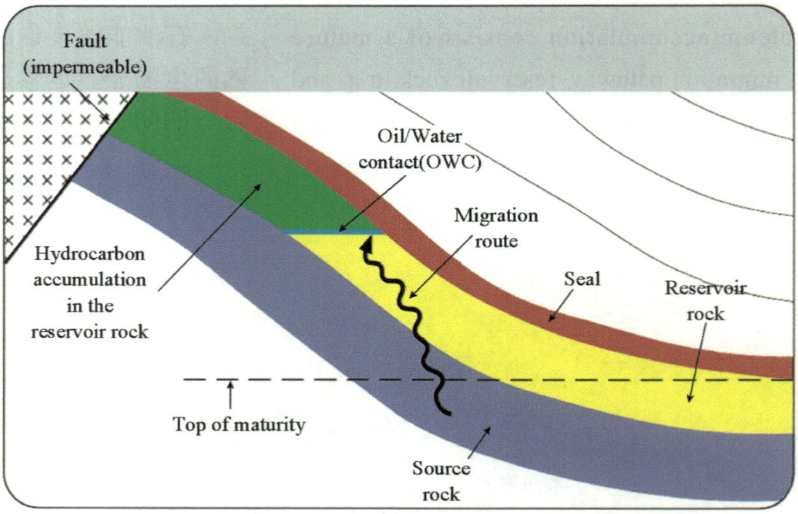

Figure 2.25 Generation, Migration and Trapping of Hydrocarbons
(Source: Hydrocarbon Exploration and Production)

2.7.1 Accumulation

The phase in the development of a petroleum system during which hydrocarbons migrate into and remain trapped in a reservoir.

2.7.1 成藏

成藏是一个用以描述碳氢化合物运移并保存在油藏的名词。

2.7.2 Reservoir

A subsurface body of rock having sufficient porosity and permeability to store and transmit fluids.

◆ Sedimentary rocks are the most common reservoir rocks because they have more porosity than most igneous and metamorphic rocks;

◆ they form under temperature conditions at which hydrocarbons can be preserved.

◆ A reservoir is a critical component of a complete petroleum accumulation.

2.7.2 油藏

地下有充足孔隙度和渗透率的岩石空间能储存和输送流体。

◆ 沉积岩是最普通的油藏岩石，因为它们比变质岩和火成岩有更多的空隙空间；

◆ 油藏的温度要适宜碳氢化合物的保存；

◆ 油藏的条件是石油能够聚集成藏的关键因素。

2.7.3 Seal (cap rock)

An impermeable rock that acts as a barrier to further migration of hydrocarbon liquids.

Rocks that forms a barrier or cap above and around reservoir rock forming a trap such that fluids cannot migrate beyond the reservoir. Commonly:shale, mudstone, anhydrite or salt.

A seal is a critical component of a complete petroleum accumulation.

2.7.4 Trap

A configuration of rocks suitable for containing hydrocarbons and sealed by a relatively impermeable formation through which hydrocarbons will not migrate.

Traps are described as(Figure 2.26):

2.7.3 盖层

盖层为一不渗透层，能阻挡油气进一步运移。

能形成遮挡或成为盖层阻挡流体进一步运移出圈闭的岩石通常是：页岩、泥岩、硬石膏岩和盐岩。

盖层是油气聚集成藏中至关重要的一环。

2.7.4 圈闭

圈闭是能储存碳氢化合物并能阻挡其进一步运移的地质构造，它包含相对不渗透层。

圈闭有两种（图2.26）：

- *Trap*
  - *Structural Trap*
    - *Anticline*
    - *Fault*
    - *Salt Dome*
  - *Stratigraphic Trap*
    - *Pinchout*
    - *Unconformity*
    - *Reef*

Figure 2.26 Trap in Petroleum Geology

◆ Structural traps (Figure 2.27)

Hydrocarbon traps that form in geologic structures such as folds and faults;

◆ Stratigraphic traps(Figure 2.28)

Hydrocarbon traps that result from changes in rock type or pinch-outs, unconformities, or other sedimentary features such as reefs or buildups.

A trap is an essential component of a petroleum accumulation.

◆ 构造圈闭（图2.27）

由地质构造如断层、褶皱形成的圈闭；

◆ 非构造圈闭（图2.28）

由岩石类型转化形成的圈闭，如尖灭、不整合等。

圈闭是油藏成藏的必要因素。

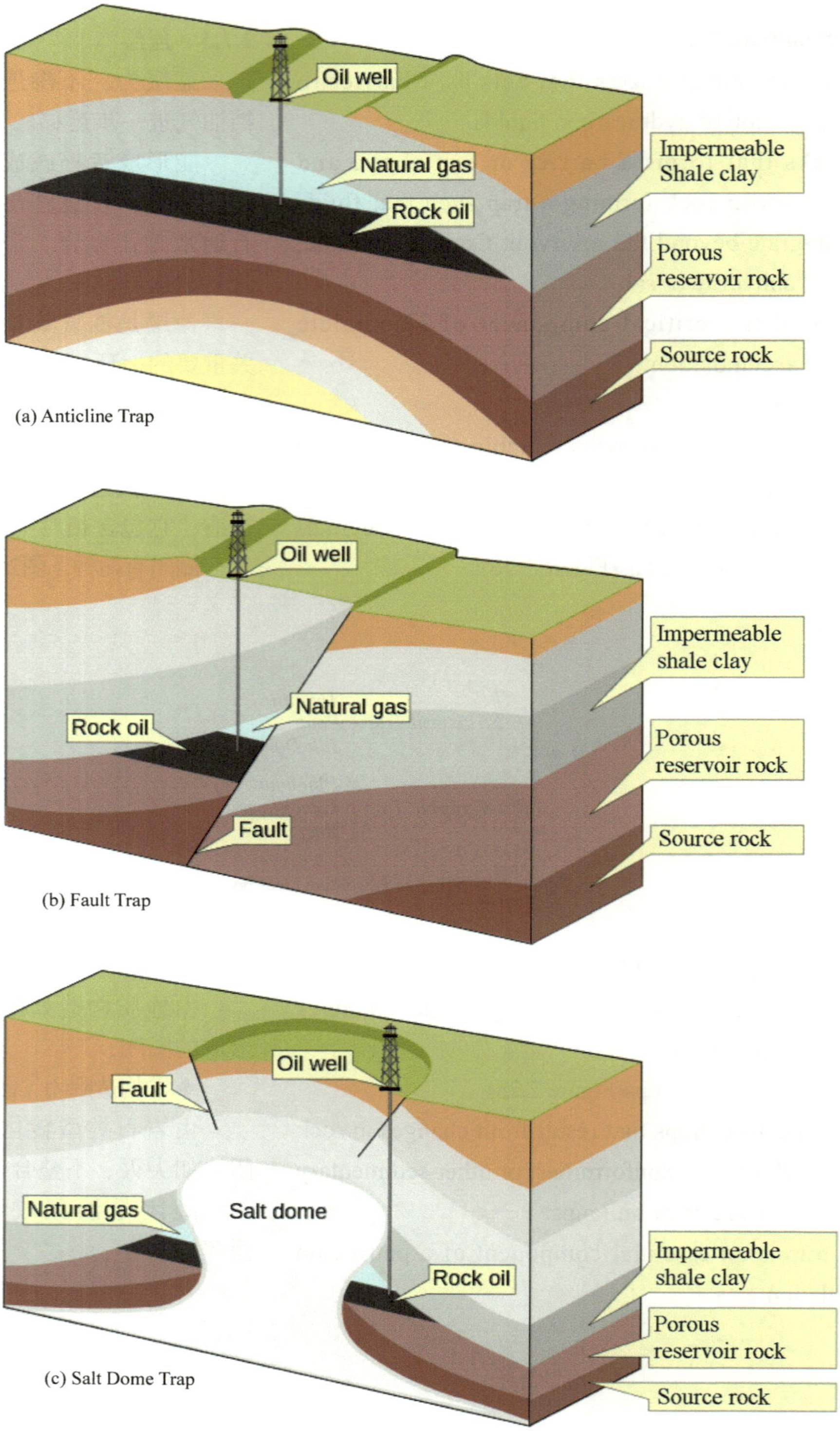

Figure 2.27 Structural Traps
(Source: https://en.wikipedia.org/wiki/Structural_trap)

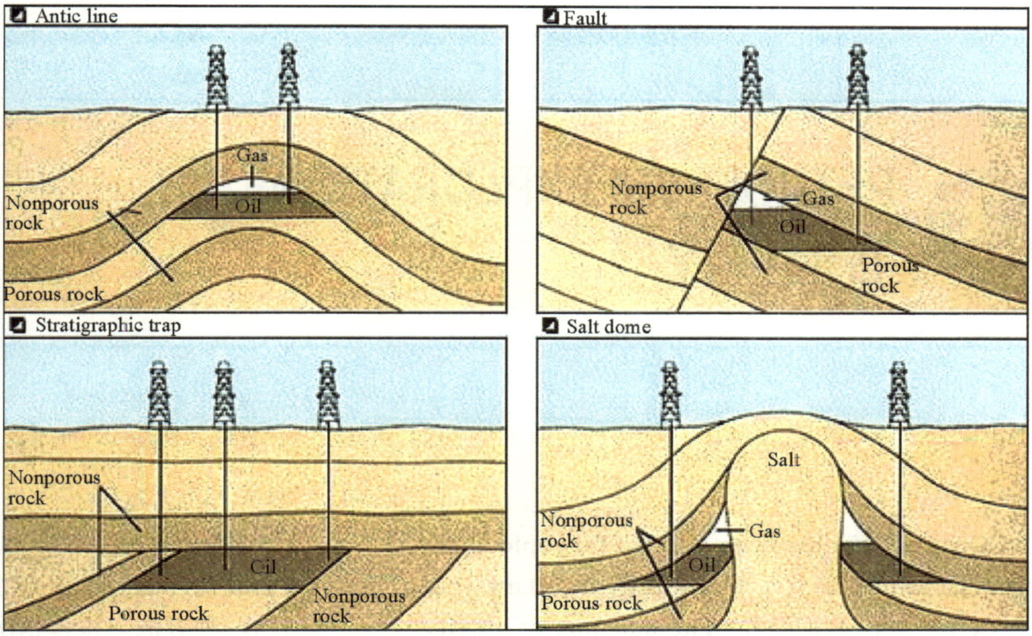

Figure 2.28 Four Types of Structural and Stratigraphic Traps
(Source: http://www.geomore.com/oil-and-gas-traps/)

3 PETROLEUM EXPLORATION 石油勘探

3. 1 Introduction

Once an area has been selected for exploration, the usual sequence of technical activities starts with the definition of a basin. The mapping of gravity anomalies and magnetic anomalies will be the first two methods applied. In many cases this data will be available in the public domain or can be bought as a 'non exclusive' survey.

Next, a coarse two-dimensional (2D) seismic grid, covering a wide area, will be acquired in order to define leads, areas which show for instance a structure which potentially contains an accumulation (seismic methods will be discussed in more detail in the next section). Recently electro-magnetic techniques have also been deployed at this stage to assist in the delineation of basins and the identification of potential hydrocarbon accumulations(Figure 3.1).

A particular exploration concept, often the idea of an individual or a team will emerge next. Since at this point very few hard facts are available to judge the merit of these ideas they are often referred to as 'play'. More detailed investigations will be integrated to define a 'prospect', a subsurface structure with a reasonable probability of containing all the elements of a petroleum accumulation, namely source

3.1 简介

一旦选定一块区域进行勘探，重力勘探和磁力勘探是首先选用的两种方法。很多情况下这两种数据可以通过公共数据库查得或是购买获得。

接下来是一个覆盖大区域的粗网格二维地震勘探，用以找到潜在的油气区。电磁技术也在此时期用以辅助确定盆地轮廓和潜在的油气聚集地（图3.1）。

一个由个人或团队提出的勘探设想此时将形成。在缺乏数据支持的前提下，潜在的油气区块只是一个设想。随着调查数据的增多，这个设想变得越来越翔实，包括确定地下生油岩种类、成熟度、运移路径、储集岩和盖层等。

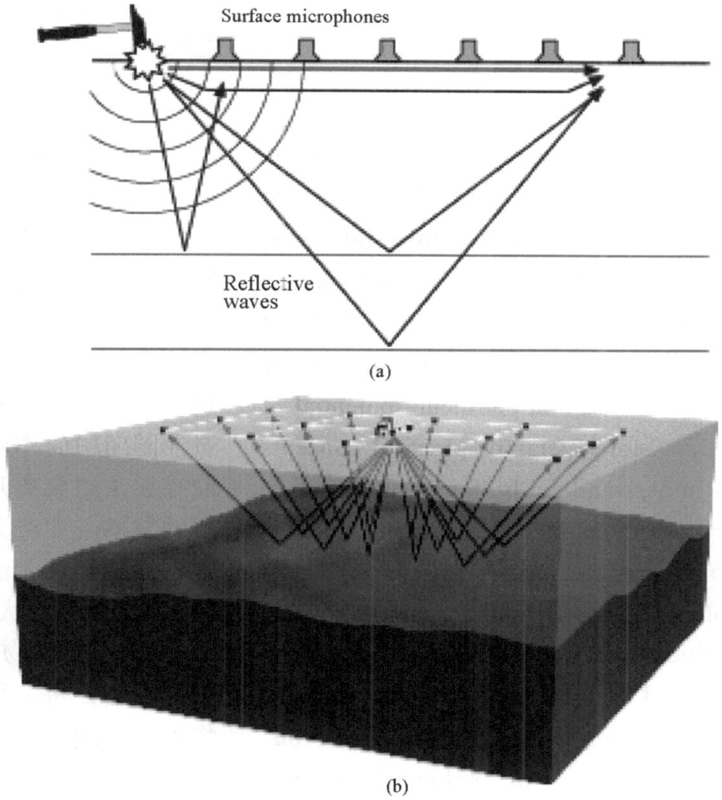

Figure 3.1 Seismic Reflection
(Source: http://www.natgas.info/html/gasexploration.html)

rock, maturation, migration, reservoir rock and trap.

Eventually, only the drilling of an exploration well will prove the validity of the concept. A 'wildcat' well is drilled in a region with no prior well control. Wells either result in discoveries of oil and gas, or they find the objective zone to be water-bearing in which case they are termed 'dry'.

最终，只有钻井才能明确勘探设想的油气区块是否真有油气。"野猫井"专指该区块此前没有钻过井。如果没有钻到油气，只钻到含水层的井叫"干井"。

Exploration activities are potentially damaging to the environment. The cutting down of trees in preparation for an onshore seismic survey may result in severe soil erosion in years to come. Offshore, fragile ecological systems such as reefs can be permanently damaged by spills of crude or mud chemicals. Responsible companies will therefore carry out an *environmental impact assessment* (EIA) prior to activity

勘探活动可能对环境构成威胁。比如陆上地震勘探过程中会砍伐树木导致接下来几年的水土流失。海上原油或泥浆的泄漏会对脆弱的生态系统造成永久的破坏。相关公司在进行勘探前首先要进行环境影响

planning and draw up contingency plans should an accident occur.

评估并起草应急方案用以应对危机。

3.2 Petroleum Geophysical Exploration

Geophysics, the application of physics to the study of the earth' s subsurface, respond to variations in physical properties of its rocks, fluids and voids. In its broadest application, geophysics makes a major contribution to understanding the earth' s crust, especially through the application of plate tectonic theory, the genesis and petroleum potential of sedimentary basins. More specially, physical concepts are required to understand folds, faults and diapirs, and hence their roles in petroleum entrapment. Modern petroleum exploration is unthinkable without the aid of Petroleum Geophysical survey in finding potential traps. Nor could any finds be evaluated effectively without Petroleum Geophysical wireline well logs to measure the lithology, porosity, and petroleum content of a reservoir.

There are several means of exploring the Earth by Petroleum Geophysical methods. Each of these techniques exploits fundamental physical aspects of Earth materials such as electrical, magnetic, acoustical, or gravitational properties. While these techniques do not allow detailed examination of the rocks beneath us, they often enable geologists and geophysicists to infer the most likely properties of large volumes of rocks.

3.2 石油地球物理勘探

地球物理勘探指应用物理手段研究地下情况，包括地下岩石、流体和孔隙。其最广泛的应用在于研究地壳，特别是应用地壳板块理论研究沉积岩中的石油起源。而且应用物理理论理解褶皱、断层和底辟在石油圈闭中的作用。现在石油勘探无法想象离开地球物理探测的结果。那将无法找到潜在的石油圈闭，无法确定地下岩性、孔隙和油藏大小。

石油地球物理勘探有很多种方法，每一种方法都应用物理确定地下的某一性质，包括电性、磁性、声性和重力性等。这些物理技巧虽不能让我们直接检测地下岩石，却可以帮助地质学家和地球物理学家更好地理解地下岩石的特性。

3.2.1 Gravity Surveys

The Earth' s gravitational attraction varies slightly from one place to another on the Earth' s surface. Some of this variation occurs because the Earth is not a perfect sphere, and some is related to differences in elevation on the Earth' s surface. While these variations in gravity are predictable and can be calculated for each spot on the Earth' s surface, other variations in gravity, such as those caused by unknown geologic features are not predictable(Figure 3.2).

3.2.1 重力勘探

地球重力在地表上会表现出不同地点具有不同重力加速度值。这不同值是因为地球并非是完美的球形，这与地表海拔有关。地表每一点的重力变化在已知地质特征下是可以计算和预测的，而在未知地质特征下将无法预测（图3.2）。

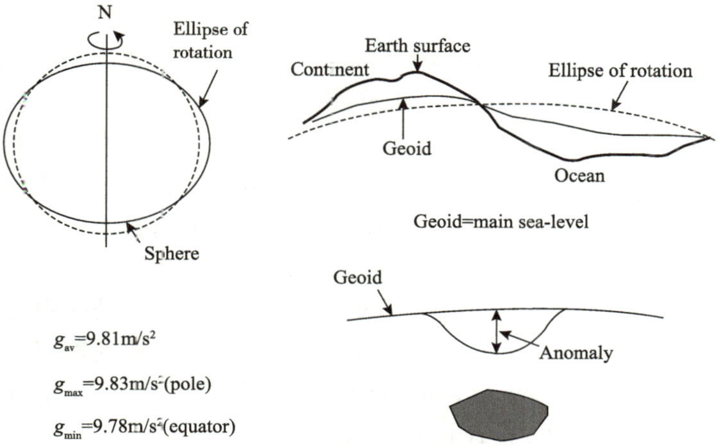

Figure 3.2 The Earth's Gravitational Variation
(Source: Hydrocarbon Exploration and Production)

Gravity measurements are made with an instrument known as a gravity meter, and maps can be produced that show differences in pull of gravity across the area. These variations are useful in locating geologic faults and ancient volcanoes, for example. They can also indicate the presence of geologic basins that are filled with unusually large thickness of sedimentary rocks(Figure 3.3).

通过重力测量计进行重力测量，从而绘制该区域的重力差异图。重力差异图在确定断层和古代火山时很有用。同时，也用作确定地质盆地和沉积岩厚度（图3.3）。

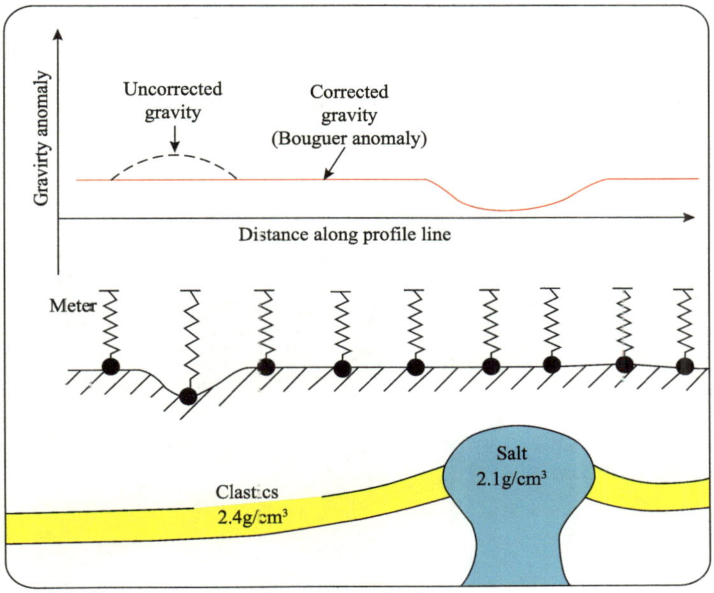

Figure 3.3 Principle of Gravity Surveying
(Source: Hydrocarbon Exploration and Production)

3.2.2 Magnetic Surveys

The Earth' s magnetism varies from place to place, much as the gravity varies(Figure 3.4). The magnetic method detects changes in the earth's magnetic field caused by variations in the magnetic properties of rocks. In particular, basement and igneous rocks are relatively highly magnetic. If they are located close to the surface they give rise to anomalies with a short wavelength and high amplitude. The method is airborne (plane or satellite) which permits rapid surveying and mapping with good areal coverage. Like the gravity technique this survey is often employed at the beginning of an exploration venture(Figure 3.5).

3.2.2 磁力勘探

地球磁力也和地球重力一样每一点有不同的值（图3.4）。地磁勘探监测地球磁场由地下岩石磁性导致的变化。特别是基岩和火成岩有较高的磁性。它们会引起磁性异常。从航空地磁勘探中可以大范围地调查。地磁勘探和重力勘探用于勘探早期（图3.5）。

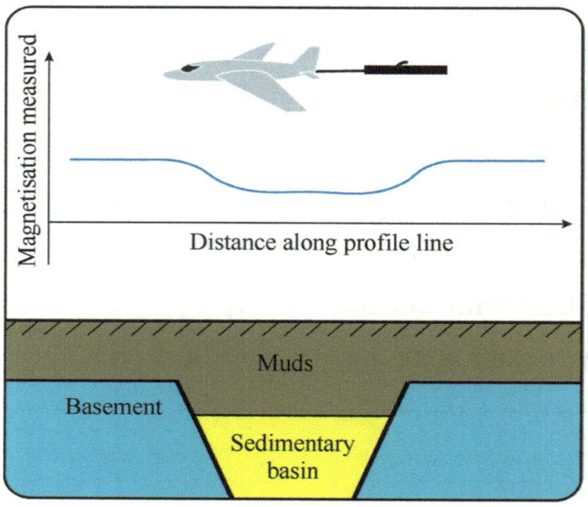

Figure 3.4 Principle of Magnetic Surveying(Source: Hydrocarbon Exploration and Production)

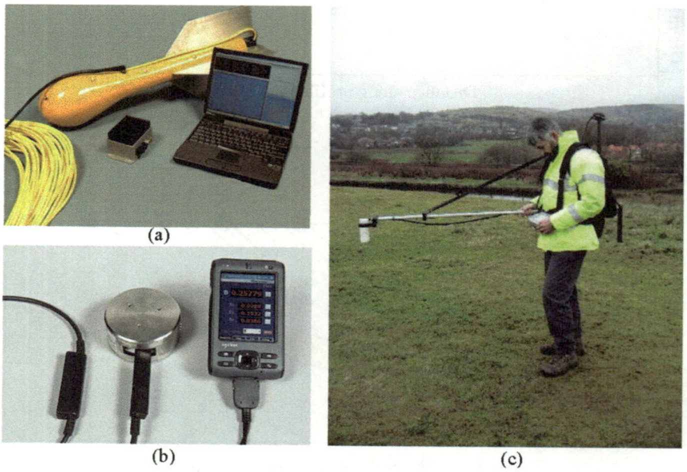

(a) (b) (c)

Figure 3.5 Magnetic Survey Instruments (Source: passc.net)

3.2.3 Seismic Surveys

From being a predominantly exploration focused tool, seismic surveying has progressed to become one of the most cost effective methods for:

◆ exploration for delineating structural and stratigraphic traps;

◆ field appraisal and development for estimating reserves and drawing up FDPs;

◆ production for reservoir surveillance such as observing the movement of reservoir fluids in response to production.

Seismic acquisition techniques vary depending on the environment (onshore or offshore) and the purpose of the survey. In an exploration area a seismic survey may consist of a loose grid of 2D lines. In contrast, in an area undergoing appraisal, a 3D seismic survey will be shot. In some mature fields a permanent 3D acquisition network might be installed on the seabed for regular (6~12 months) reservoir surveillance, called ocean bottom stations (OBS) or ocean bottom cables (OBC).

Seismic waves are essentially sound waves that underground at 3 to 6 km per second, depending upon the type of rock through which they pass. Seismic surveys involve generating sound waves which propagate through the earth's rocks down to reservoir targets. The waves are reflected to the surface, where they are registered in receivers, recorded and stored for processing. The resulting data make up an acoustic image of the subsurface which is interpreted by geophysicists and geologists(Figure 3.6).

A source of seismic waves emits into the ground, commonly by explosion, truck-mounted vibrators, mass drop, or projectile impact.Waves are radiated spherically away from the source. One ray originating at the source will pass through subsurface layer and part of them reflects back at every formation interface, then the echo return to the receiver at the surface. In the case of a

3.2.3 地震勘探

在勘探过程中最重要的是地震勘探，地震勘探能：

◆ 描述地下构造圈闭或地层圈闭；

◆ 用于油田评价、储层描述和开发方案；

◆ 用于油藏监测，如观测流体随生产进行的移动。

地震勘探手段随环境（陆上或海上）、调查目的的不同而不同。在勘探早期地震勘探可以是粗网格的二维模式，在生产评估阶段可以用三维地震调查。在其些成熟区块，三维地震网格安装的海底每6～12个月进行一次常规地震监测，叫做海底站监测或是海底线缆监测。

地震波其实是以3～6km/s的速度在地下传播的声波，其具体传播速度由途经的岩石决定。地震勘探其实就是地面产生的声波向地下传播，再返回到地面由接收器接收、记录、储存的过程。接收记录下的数据反映了地下岩石的声学性能，由地球物理学家和地质学家进行解释（图3.6）。

地震波震源通常是爆炸、车载震动器、落重物、抛射物影响。从震源出发地震波球形传播。从震源开始向下传播的地震波穿过地下岩层，部分遇界面后反射回来，由地面回声接收器所接收。以向下传播到

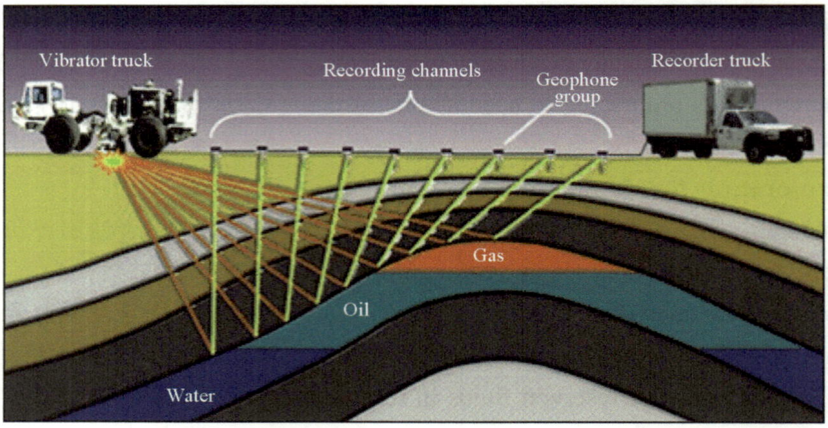

Figure 3.6 Seismic Reflection

single flat-lying layer and a flat to topographic surface, the path of least time will be form a reflecting point midway between the source and the receiver with the angle of incidence on the reflecting layer equal to the angle of reflection from the reflecting layer(Figure 3.7).

某一平层又反射回地面平层为例，从震源途经反射点到接收器的传入角应该等同于反射层中的反射角（图3.7）。

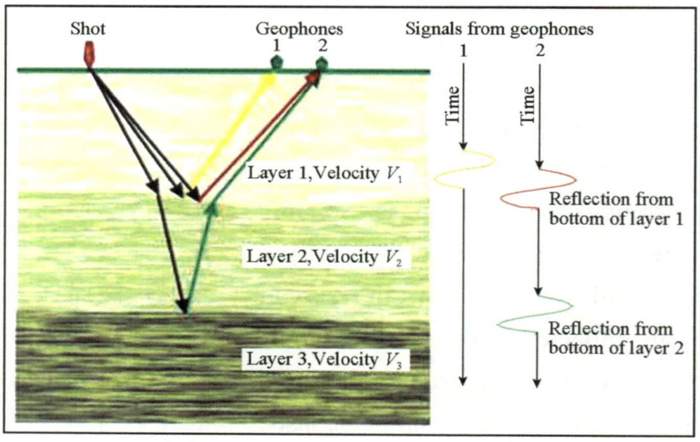

Figure 3.7 Seismic Reflection
(Source: cflhd.gov)

In a seismic survey, the Petroleum Geophysical contractor' s crew lays out a line or several lines of sensitive receivers, called geophones or jugs, on the ground. Then explosions or mechanical vibrations are created on the surface. The geophones record the energy reflected back as seismic waves from rock layers at various depths(Figure 3.8).

在地震勘探中，石油地球物理方面的承包商在地面上纵横布线，这种带接收器的电线叫地震检波器。然后在地面产生爆炸或机械震动的震源。检波器将记录地震波传回的不同深度下岩层的信息（图3.8）。

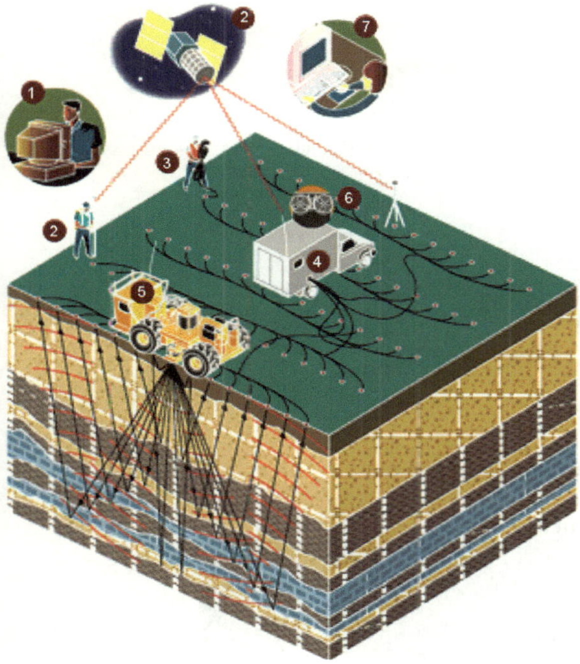

1. Surveyors use satellite based global positioning system (GPS) to find precise locations for vibration source and geophones.
2. Seismic crew lays out cables and geophones.
3. Cables are attached to recording system.
4. Vibrator trucks generate a controlled vibration force of up to 32 tones at each source point.
5. Underlying geologic structures reflect some of the vibrations back to the surface where geophones convert them into electrical impulses that are recorded on magnetic tape.
6. Magnetic tapes are sent for processing and interpretation of the recorded data.
7. A graphic 3-D representation.

Figure 3.8 Vibroseois 3D Seismic Method
(Source: aadnc-aandc.gc.ca)

Like the rest of the petroleum industry, explorationists have adopted ever-higher environmental standards for seismic surveys. For example, improved methods for clearing and using narrower cutlines have reduced the impacts on soil, water, plants and wildlife in forested areas. In other sensitive areas such as mountainous terrain, seismic crews use helicopters or even packhorse to limit surface disturbance. In offshore exploration, air guns using compressed air have replaced dynamite as a better, safer energy source which also minimizes the impact on marine life(Figure 3.9). A marine vessel records the reflected energy from a towed array of hydrophones(Figure 3.10). In shallow waters, the hydrophones may be laid out on the seabed. To reduce environmental impact, many contractors today use the mechanical vibroseis method to send energy waves from a heavy, vibrating vehicle into the earth(Figure 3.11).

像石油行业的其他作业一样，勘探者采用更高的环境标准要求地震勘探。例如，作业中减少对土壤、水源、植被和野生动物的影响。在一些敏感山区，采用直升飞机或马驮方式减少地面干扰（图3.9）。在海上勘探时，用空气压缩枪取代炸药，从而减少对海洋生物的影响。在海上记录反射波的装置叫海上检波器（图3.10）。在浅海，海上检波器直接布置在海底。用机械震源震动地下产生震动波，从而减少对环境的影响（图3.11）。

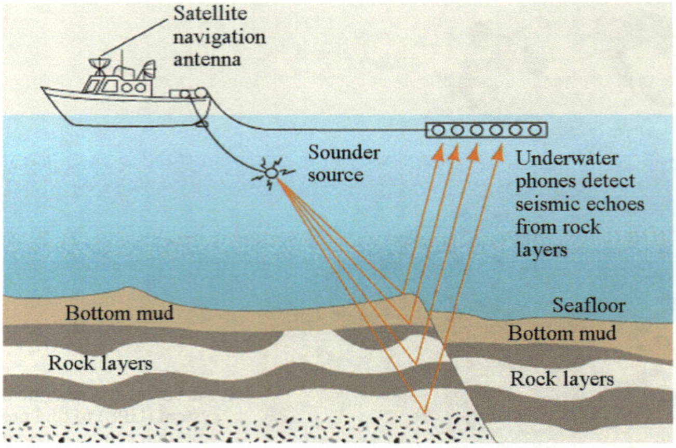

Figure 3.9 Seismic Reflection
(Source: earth.northwestern.edu)

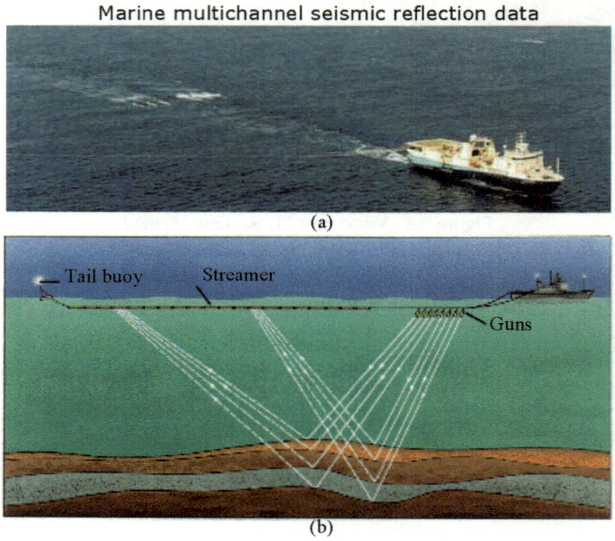

Figure 3.10 Marine Reflection Seismology
(Source: woodshole.er.usgs.gov)

- Petroleum Exploration
  - Geophysical Exploration
    - Gravity Surveys
    - Magnetic Surveys
    - Seismic Surveys
      - 2D
      - 3D
    - Others
  - Geochemistry Exploration
  - Others

Figure 3.11 Petroleum Exploration Methods

3. 3 Petroleum Geochemistry Exploration

Petroleum Geochemistry, the application of chemistry to the study of rock, has many uses in petroleum exploration. In other words, Petroleum Geochemistry is the application of chemical principles to the study of the origin, migration, accumulation, and alteration of petroleum.

Petroleum geochemical techniques include: ①identify source rocks and determine the amount, type, and maturation level of the organic matter; ② evaluate the potential timing of petroleum migration from the source rock; ③assess the potential migration pathways; and ④correlate petroleum compounds found in reservoirs, leaks, and surface seeps to find new pools of petroleum(Figure 3.12~Figure 3.14).

3.3 石油地球化学勘探

石油地球化学指应用化学知识研究岩石并指导石油勘探。换句话说，石油地球化学即应用化学原理研究石油的起源、运移、聚集和改变。

石油地球化学的工作包括：①确定烃源岩、确定有机质的数量、类型和成熟水平；②估算从烃源岩运移的时间；③估摸运移路径；④结合石油成分寻找油藏、露头、泄漏点及新的石油聚集带(图3.12～图3.14)。

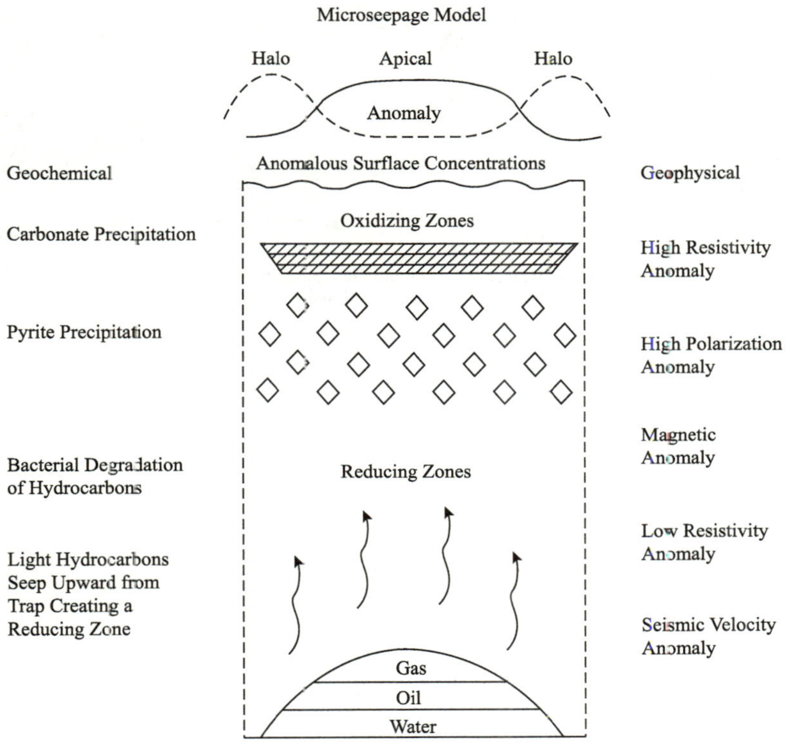

Figure 3.12 Generalized Model of Hydrocarbon Microseepage and Hydrocarbon-induced Effects on Soils and Sediments. (Source: http://tle.geoscienceworld.org/content/19/3/258.figures-only)

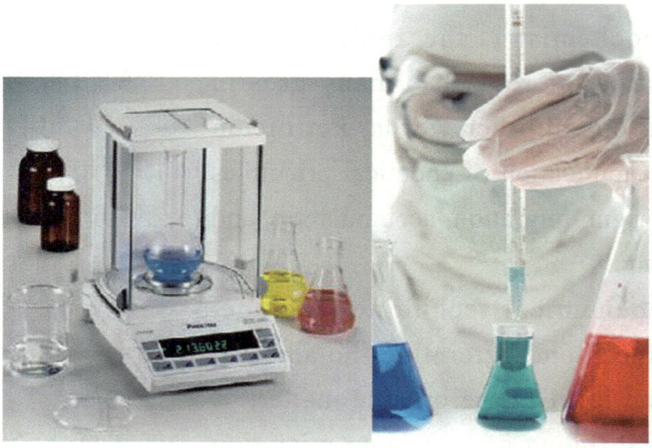

Figure 3.13 Petroleum Geochemistry Analytical Testing
(Source: http://www.actlabsint.com/)

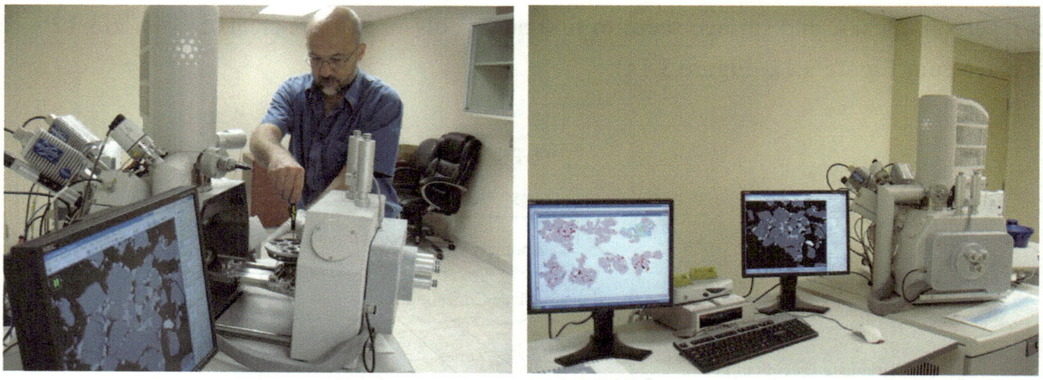

Figure 3.14 Petroleum Geophysical & Petroleum Geochemistry Exploration Analytical Testing
(Source: http://www.actlabsint.com/)

WELL DRILLING 4
钻 井

4.1 Drilling History

The early need for water and salt prompted the Chinese to develop a percussion-type drilling apparatus to replace the practice of digging wells by hand. Around 2000 years ago, Chinese in Sichuan province originated deep drilling. The primary motive for deep drilling in China was the search for salt. Even as recently as 1965, 16.5% of China's salt supplies came from brine pumped out of deep boreholes, making this source of supply second only to sea salt (Figure 4.1).

4.1 钻井历史

早期为满足水和盐的需要，促使中国人发展冲击钻具取代人工凿井。大约在2000年前，中国四川省就开始了钻深井。最初的钻深井的目的是为了寻找盐。直到1965年，16.5%的中国盐来自深井的卤水，成为仅次于海盐的第二大盐供给（图4.1）。

Figure 4.1 Ancient Sketch Originally from "the Annals of Salt Law of SiChuan Province" Shows the Structure of the Ancient Derrick
(Source: http://www.pvisoftware.com/blog/ancient-drilling-technologies)

The ancient percussive cable drilling system was called "churn". The derrick had height of about 33 ft and all parts of the rig were made from wood (mainly bamboo). A large wooden drum had 16 ft in diameter and was used to perform round trips (Figure 4.2). Rocking movement of the balancing beam created the percussive impulses on the bit, which sometimes weighted as much as 300 lbs. By alternately lifting this tool and letting it fall, the Chinese achieved a well depth of 2000 ft. Wonder the rate of penetration (ROP)? 2 feet per day! The Orientals were willing to work as long as 3 years to complete a well.

古代的冲击钻井叫打桩。井架高度约33ft高，钻机全由木质（主要是用竹子）做成。一个直径为16ft的木制大鼓用于绞线（图4.2）。摇动平衡杆造成钻头的冲击运动，钻头质量可达300lb（1lb=0.45359237kg）。提起放下这木制的钻机，中国人可钻达2000ft的地下。钻速多少？每天2ft！这些东方人愿意花3年去钻一口井。

Figure 4.2 A wood Well Tower in Traditional Salt Drilling Wells in Zigong & All Kinds of Drilling Wells and Dredging Tools in Ancient Salt Wells in China
(Source: http://www.chinamuseums.com/zigong_Salt.htm)

The deep drilling for brine yielded natural gas (primarily methane) from time to time. The boreholes producing methane were known to the Chinese as "fire wells". So the drilling for natural gas followed and was developed at the same time.

在钻深井汲卤的过程中，经常发现有天然气（主要成分是甲烷）时不时地冒出。这产生甲烷的井中国人称为"火井"。几乎同时，中国人也开始钻这种天然气井。

The bamboo tubes were used as pipelines, carrying both brine and natural gas for many miles, sometimes passing under roads and sometimes going overhead on trestles. Among other uses, natural gas was used to heat evaporation pans of brine to make salt: perfect example

竹子的竹节打通后用作输送管，运送卤水和天然气到几英里以外。有时候输送管在地下，有时候输送管架高在头顶上。在掘井汲卤的过程中又自

of killing 2 birds with 1 stone.

产天然气熬卤取盐，这真是一石二鸟的绝佳例子。

Percussion, or cable tool is the method of alternately lifting and dropping a heavy iron bit to beat the rock into pieces and bail out the particles (Figure 4.3).

钝钻或冲击钻井是靠反复举起和落下沉重的铁钻头去冲击岩石，破坏岩石成碎块，再将碎屑打捞上地面的一种钻井方法（图4.3）。

Until early last century, oil and gas wells were drilled with cable-tool rigs. The hole was made by spudding, which is making a hole by reciprocating a heavy, chisel-shaped bit suspended on a hemp or steel line.The major problem with cable-tools is that there was nothing to prevent oil and gas from blowing out at the surface whenever a reservoir was penetrated. This is wasteful and dangerous, and today would have major environmental repercussions. The rotary drilling system in use today avoids this problem by keeping the hole full of heavy mud that holds back the reservoir fluids (Figure 4.4).

直到20世纪以前，油气井都主要靠这种冲击钻井。用钢绳悬挂的楔形重钻头反复冲击地层钻井。冲击钻井的主要弊病是钻到油气藏后无法防止油气的井喷。这是既浪费又危险的事故，对今天的环境有很大的影响。旋转钻井是现今应用的主要钻井方法，用高比重的泥浆压住油藏流体来避免井喷（图4.4）。

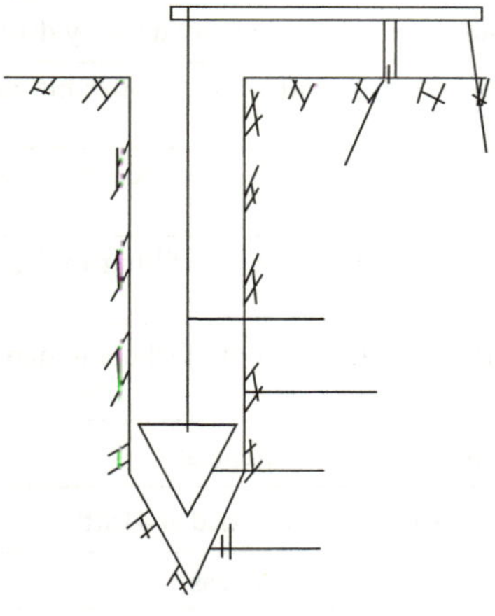

Figure 4.3 Percussion
(Source: The Petroleum Industry by Charles F. Conaway)

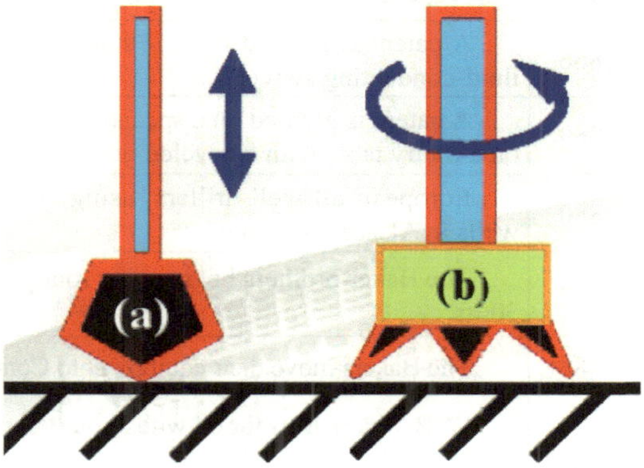

Figure 4.4 (a) Percussion Drilling & (b) Rotary Drilling

Although improvements were made to this concept over a long time period, other individuals eventually conceived of methods for boring a hole into the earth and flushing out the cuttings, as opposed to be beating the rock into pieces and bailing out the particles.

尽管钻井方法随时提高，但是人工钻井还是采用破坏地层岩石，捞取破碎的岩石碎片，再破碎再打捞的方式进行。

The emergence of a commercial use for petroleum in the mid-1800s accelerated the development of equipment and techniques used to drill for oil (Table4.1).

18世纪中叶后对石油的商业需求加速了钻井工具和钻井技术的发展（表4.1）。

Table4.1 Rotary Drilling History

Ancient Times	Wells dug by hand, this practice continued in Burma until the 1900s
250 B. C.	Chinese use percussion drilling to drill wells for salt water, using derrick, tubing, bits, and cemented bamboo casing
1808	Ruffner brothers of West Virginia use a "spring pole" percussion-drilling apparatus to "kick down" a well. The driller attaches a cable and bit to a flexible sapling secured as a lever over a fulcrum. After using their own weight to bend the pole and drop the bit into the hole, they allow the spring pole to lift the bit back up
1829	Steam is used to operate improved cable tool equipment that utilizes derricks, engines, and fishing tools to retrieve lost bits
1844	An English named Robert Beart invents a drilling machine that includes a hydraulic swivel, hollow drilling rods, and circulating fluid
1845	A French engineer named Fauvelle drills a water near Perpignan, France, using a set of hollow boring rods to allow pumped water to flush excavated material from the hole
1848	August Beer, an Austrian professor, suggests the possibility of drilling by a rotary method
1859	Using cable-tool percussion drilling equipment, Col. E. L. Drake completes the first commercial oil well in America, at a depth of 69ft
1860	A French civil engineer named Leschoc uses a power-driven, diamond-studded rotary drill
1866	A patent is granted for a "stone drill" that includes a hollow drill stem, a roller bit, and a fluid-conducting swivel
1869	A patent is granted on a special type of offshore drilling rig. Another inventor receives a patent on a rotary table with a beveled gear drive
1880	European oil well drillers, using versions of Fauvelle's water-flushed drilling tools, drill wells in Alsace and Baku
1882	The Baker brothers begin using rotary equipment to drill for water in South Dakota to depth of 500ft
1888	The Bakers move their equipment to Corsicana, Texas, where it becomes popular
1893	W. B. Sharp drills for oil with a rotary rig near Beaumont; his well is abandoned at 418ft
1897	P. Higgins unsuccessfully drills for oil near Beaumont and then hires Anthony Lucas
1900 ~ 1901	Anthony Lucas and the Hemil brothers drill the discovery well at the Spindletop Field near Beaumont, Texas, using rotary drilling equipment from Corsicana. The well flows 100000BOPD from 1040ft and rotary drilling is on its way
1909	Howard Hughes invents the rotary rock bit, with a rolling cutter.
1918	The world's deepest well, drilled with a cable-tool percussion-drilling rig, is 7386ft
1920s	Combination rigs are developed that use percussion cable tools to drill down to 4000ft and rotary drilling equipment below that depth
1930	The world's deepest well, drilled with a rotary rig, is 10000ft

Edwin Drake drilled the first oil well in Titusville Pennsylvania in 1859, creating the modern petroleum industry technologies that were available at the time. Although the percussion, or cable tool, method of alternately lifting and dropping a heavy iron bit remained popular in the American northeast for a long while, the rotary drilling rig, introduced in the American mid-continent, gradually become the most widely used method worldwide. Since its acceptance in the early 1900s, the turning bit has drilled close to ten billion feet of hole, enough to penetrate the Earth from pole to pole over two hundred times. In the search for hydrocarbons, drilling technology has been constantly called upon to adapt to new and harsher surface and subsurface environments.

1859年，德雷克在宾夕法尼亚特斯维尔开钻了现代石油工业的第一口井。尽管提起落下的冲击钻法在美国东北部盛行了一段时间，从美国中部开始的旋转钻井最终取代了冲击钻井成为全世界主要的钻井方法。旋转钻井自19世纪早期引入，旋转的钻头已经钻过成百上千亿英尺的地层，足足将地球钻穿好两百多次了。在钻取石油的过程中，钻井技术一直是与时俱进的，采用最新的技术适应地面和地下的环境。

4. 2 Well Planning

The drilling of a well involves a major investment, ranging from a few million US$ for an onshore well to 100 million US$ plus for a deepwater exploration well. Well engineering is aimed at maximising the value of this investment by employing the most appropriate technology and business processes, to drill a 'fit for purpose' well, at the minimum cost, without compromising safety or environmental standards. Successful drilling engineering requires the integration of many disciplines and skills.

Successful drilling projects will require extensive planning. Usually, wells are drilled with one, or a combination, of the following objectives:

◆ to gather information;

◆ to produce hydrocarbons;

◆ to inject gas or water to maintain reservoir pressure or sweep out oil;

◆ to dispose of water, drill cuttings or CO_2 (sequestration).

To optimise the design of a well it is desirable

4.2 钻井设计

钻井占油田投资的大头，从几百万美元的陆上钻井到上亿美元的海上深水探井不等。钻井工程着眼于用最适宜的技术和商业运营模式，去钻成本最低又无损安全和环境的井，从而实现最大化投资效益。成功的钻井需要整合多学科知识和技巧。

成功的钻井作业需要做钻井计划。通常钻井不单是向地下钻一个洞，它有多重目的：

◆ 获取地下信息；

◆ 生产碳氢化合物；

◆ 通过注气或注水以保持油藏压力或方便驱油驱气；

◆ 排泄水或岩屑，封存二氧化碳。

最优化钻井前最好对地下

to have as accurate a picture as possible of the subsurface. Therefore, a number of disciplines will have to provide information prior to the design of the well trajectory and before a drilling rig and specific equipment can be selected.

情况有清楚的认识，因此在钻井和设备选择前，设计井轨迹有一系列的行业操作。

The subsurface team will define optimum locations for the planned wells to penetrate the reservoir and in consultation with the well engineer agree on the desired trajectory through the objective sequence. In discussions with production and well engineers maximum hole inclination and required wellbore diameter will be determined. Wellhead locations, well design and trajectory are aimed at minimising the combined costs of well construction and seabed/surface facilities, whilst maximising production (Figure 4.5).

在钻井前定井点需要咨询其他相关工程师以确保设计出满足多种条件的井眼轨迹。生产和钻井工程师讨论决定最大的井斜角和合适的井径大小。以最小成本和最大产出为目标设计最优的井口位置、井眼轨迹（图4.5）。

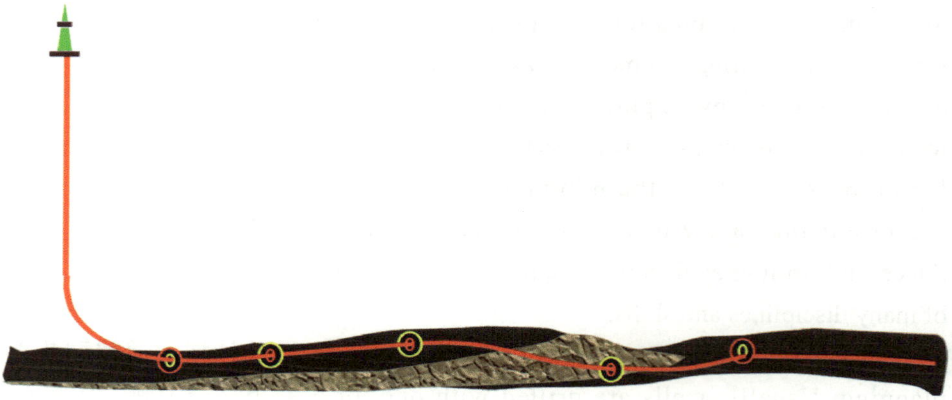

Figure 4.5 Well Trajectory

The accuracy of the parameters used in the well planning process will depend on the knowledge of the field or the region. Particularly during exploration drilling and the early stages of field development considerable uncertainty in subsurface data will prevail. It is important that

钻井设计的准确性取决于对该区域和油田地下的了解程度。特别是在勘探钻井和早期油田开发时对地下情况还不明晰，阐明这些不清楚的情况是钻井设计必

the uncertainties are clearly spelled out and preferably quantified. Potential risks and problems expected or already encountered in offset wells(earlier wells drilled in the area) should be incorporated into the design of the planned well. This is often achieved by using a decision tree approach in the well planning phase. The optimum well design balances risk, uncertainty and cost with overall project value.

须的。潜在的危机和困难在探井时就被预测到并遇见，这些信息都要整合到钻井设计中。决策树常用于此阶段设计中。最优化的钻井设计应该是平衡风险和不确定性的全局投资考量。

The basis for the well design is captured in a comprehensive document. This is then 'translated' into a drilling programme. In summary, the well engineer will be able to design and cost the well in detail using the information obtained from the petroleum engineers, geoscientists and production engineers. In particular, he will plan the setting depth and ratings for the various casing strings, cementing programme, mud weights and mud types required during drilling, and select an appropriate rig and related hardware, for example drill bits.

钻井设计的基础是将已知文件信息翻译成钻井操作程序。也就是说，钻井工程师要利用从石油地质、生产和地球科学家获得的已知信息详细设计和核算钻井成本。特别是，他要设计钻井深度、套管尺寸、泥浆程序、钻井液类型和相对密度，并筛选硬件，比如选择合适的钻头、钻机。

4. 3 Site Preparation

Once the objectives of the well are clear, further decisions have to be made. One decision will be where to site the drilling location relative to the subsurface target and which type of rig to use (Figure 4.6).

4.3 井场准备

一旦钻井目标确定，接下来就要做更多的决定。其中一项是确定钻井的位置和地下的钻遇目标以及钻机的使用（图4.6）。

Figure 4.6 Onshore Sites (Source: http://www.c-a-m.ru/)

4.3.1 Onshore Sites

A site survey will be carried out, from which a number of geotechnical parameters can be established, for example carrying capacity of the soil at the planned location, possible access routes, surface restrictions like built-up areas, lakes, nature reserves, the general to pography and possible water supplies. The survey will allow the adequate preparation of the future location. For instance, onshore in a swamp area the soil needs to be covered with support mats.

The size of the rig site will depend on operational requirements and possible constraints imposed by the particular location. It will be determined by:

◆ the type of derrick or mast (which will depend on the required loads); it must bepossible to rig this up on site;

◆ the layout of the drilling equipment;

◆ the size of the waste pit;

◆ the amount of storage space required for consumables and equipment;

◆ the number of wells to be drilled;

◆ whether the site will be permanent (in case of development drilling).

A land rig can weigh over 200 tons and is transported in smaller loads to beassembled on site (Figure 4.7).

Prior to moving the rig and all auxiliary equipment, the site will have to be cleared of vegetation and leveled. To protect against possible spills of hydrocarbons or chemicals, the surface area of a location should be coated with plastic lining and a closed draining system installed. Site management should ensure that any pollutant is trapped and properly disposed of.

If drilling and service personnel require accommodation at the well site, a camp will need to be constructed. For safety reasons, the camp will be located at a distance from the drilling rig and consist of various

4.3.1 陆上井场

一套相关的勘查将实施，包括确定一系列的地质参数和目标点的土壤承载能力、途经的路线、地表的限制，如已建区域、湖泊、保护区、地形和潜在的水源区等。勘查将为未来相关定井点措施做准备。例如，陆上的沼泽区需要用支撑垫垫起来。

钻井的尺寸将由可选设备以及潜在的限制条件决定，包括：

◆ 井架的类型，是现场搭建还是车载；

◆ 如何布置钻井设备；

◆ 污水池的大小；

◆ 耗材和设备的存储空间；

◆ 需钻的井数；

◆ 井场是租借还是永久所有等。

一个陆上钻机重达200t，能被拆卸成小部件运输到井场再组装起来（图4.7）。

在移动钻井和其他设备之前，井场需铲除植被并填平。为防止碳氢化合物或化学品的泄漏，井场将用塑料衣铺垫并安装封闭的排水系统。井场管理者应保证无污染物外泄并妥善处理。

如果钻井或服务人员要住在井场，则相关的工作房将修建。为安全起见，工作房一般离钻机有一定的距离，并且为可拆卸的活动房。污水、公

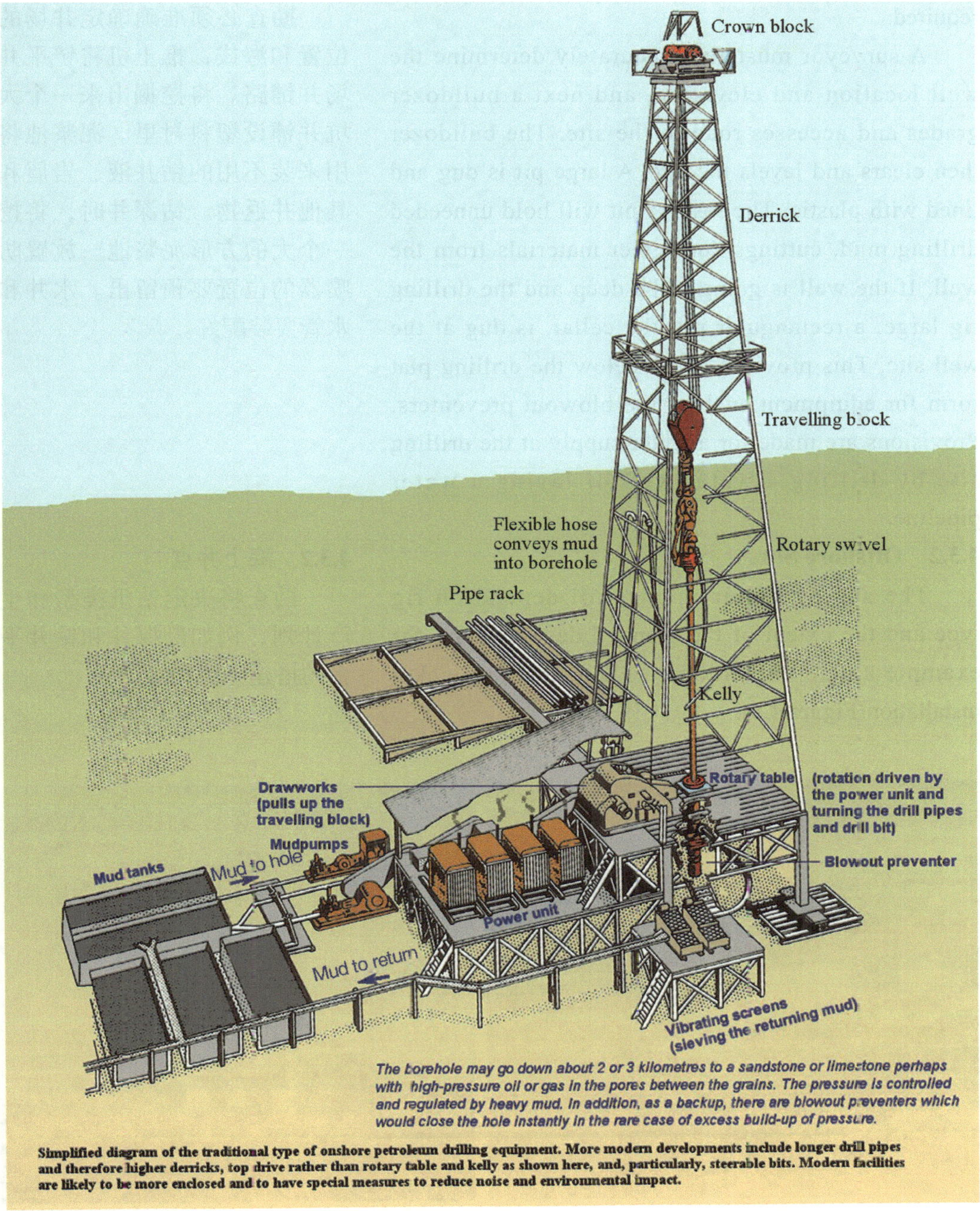

Figure 4.7 Onshore Petroleum Drilling Equipment Diagram
(Source: http://www.nomenclaturo.com/engineering-mining)

types of portacabins. For the camp, waste pits, access roads, parking space and drinking water supplies will be required.

路、停车场和饮用水等相关措施配套好。

A surveyor must first accurately determine the well location and elevation, and next a bulldozer grades and accesses road to the site. The bulldozer then clears and levels the site. A large pit is dug and lined with plastic. The reserve pit will hold unneeded drilling mud, cuttings, and other materials from the well. If the well is going to be deep and the drilling rig large, a rectangular pit, the cellar, is dug at the well site. This provides space below the drilling plat form for equipment such as the blowout preventers. Provisions are made for a water supply at the drilling site by drilling a water-well or laying a water pipeline.

勘查必须准确确定井场的位置和海拔，推土机将铲平井场并铺路。将挖掘出来一个大坑并铺设塑料衬里。泥浆池将用来装不用的钻井液、岩屑和其他井返物。钻深井时，将挖一个大的方形泥浆池。放置防喷器的位置要预留出。水井和水管要装配好。

4.3.2 Offshore Sites

4.3.2 海上井点

The survey requirements will depend on rig type and the extent of the planned development, for example single exploration well or drilling jacket installation(Figure 4.8).

勘查将决定钻机类型和生产计划，例如勘探井和钻井平台的搭建（图4.8）。

(a) Onshore (b) Offshore

Figure 4.8 Onshore and Offshore Sites
(Source: http://in.sodexo.com/inen/solutions/osss/remotesites/remotesites.asp/)

4.4 Rig Types and Rig Selection

A drilling rig is a machine which creates holes in the ground. Drilling rigs can be massive structures housing equipment used to drill water wells, oil wells, or natural gas extraction wells, or they can be small enough to be moved manually by one person and are called augers. They sample sub-surface mineral deposits, test rock, soil and groundwater physical properties, and also can be used to install sub-surface fabrications, such as underground utilities, instrumentation, tunnels or wells. Drilling rigs can be mobile equipment mounted on trucks, tracks or trailers, or more permanent land or marine-based structures (such as oil platforms, commonly called'offshore oil rigs'even if they don't contain a drilling rig). The term "rig" therefore generally refers to the complex of equipment that is used to penetrate the surface of the Earth's crust(Figure 4.9, Figure 4.10).

4.4 钻井类型及选择

钻井是一台用以向地下钻孔的装备。钻机可以是庞大的设备用来钻水井、油井和天然气井，也可以是被叫做地钻的单人手持小装备。他们可以进行地下矿物取样、岩层、土壤和地下水物理性质测试，也可以用于安装地下基建设备，如地下管线和井管等。钻井装备可由卡车、火车、拖车运输，也可以像陆上或海上平台一样永久搭建（例如石油钻井平台，通常被称为海上“石油钻井平台”，即使它们不包含钻机）。因此，钻机指任何刺穿地表的复杂设备（图4.9、图4.10）。

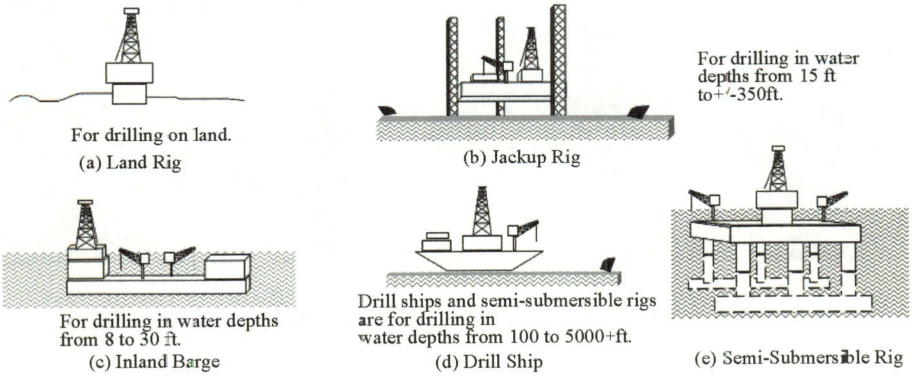

Figure 4.9 Oil Platforms and Rigs
(Source: http://www.ppdm.org/wiki/index.php/Well_Operations_Reference_guide)

- Rigs
  - Onland
    - Transported over Local Roads
    - Transported with Helicopters
  - Offshore
    - Barge-Mounted
    - Jackup
    - TenderSupport
    - Semi-Submersible
    - DrillShip
    - Permanent Platform

Figure 4.10 Drilling Rigs

People in the oil industry group break into eight basic rig types:

land, swamp barges, jacket, jackup, platform, submersible, semisubmersible and drill ship.

Land rigs drill on dry land (Figure 4.11). They are the most common rigs. Crew members can move land rigs on trucks, tractors trailers, barges, helicopters, heavy rolling gear, skids and in rare cases on specialized air pressurized equipment. Small light duty land rigs are pretty simple to move, but heavy duty land rigs can be difficult to move (Figure 4.12).

石油行业业内人士将钻机分为八大种类：

①陆地钻机；②驳船式钻机；③固定式导管架钻机；④海上自升式钻机；⑤海上平台式钻机；⑥海上沉浮式钻机；⑦海上半潜式钻机；⑧钻井船式钻机。

陆地钻机用于干燥陆地上的钻井，是一种最普通的钻机（图4.11）。井队人员通过卡

Figure 4.11 A land Rig
(Source: http://rapaddrilling.com/drilling/)

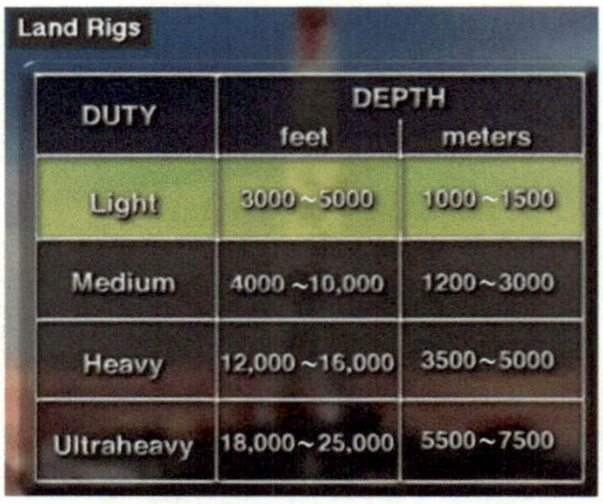

Land Rigs

DUTY	DEPTH feet	DEPTH meters
Light	3000~5000	1000~1500
Medium	4000~10,000	1200~3000
Heavy	12,000~16,000	3500~5000
Ultraheavy	18,000~25,000	5500~7500

Figure 4.12 Land Rig Types
(Source: https://www.youtube.com/watch?v=GVBXwyDPbG8)

Swamp barges operate in very shallow water (less than 20 ft). They can be towed onto location and are then ballasted so that they ‘sit on bottom’. The drilling unit is mounted onto the barge. This type of unit is used in the swamp areas of, for example Nigeria, Venezuela and US Gulf Coast.

Offshore Rig Types (Figure 4.13 ~ Figure 4.15):

Drilling jackets are small steel platform structures which are used in areas of shallow and calm water (Figure 4.16). A number of wells may be drilled from

车、拖拉机、拖车、驳船、直升机、重型盘车装置、爬犁，有时还用特殊的气压装置来运输钻机。小型陆地钻机相对简单且易于移动，而大型陆地钻机则很难移动（图4.12）。

驳船式钻机一般用于浅水区（少于20ft）。它们被拖动到指定位置然后压舱坐在水底。钻井设备安装在驳船上。此类

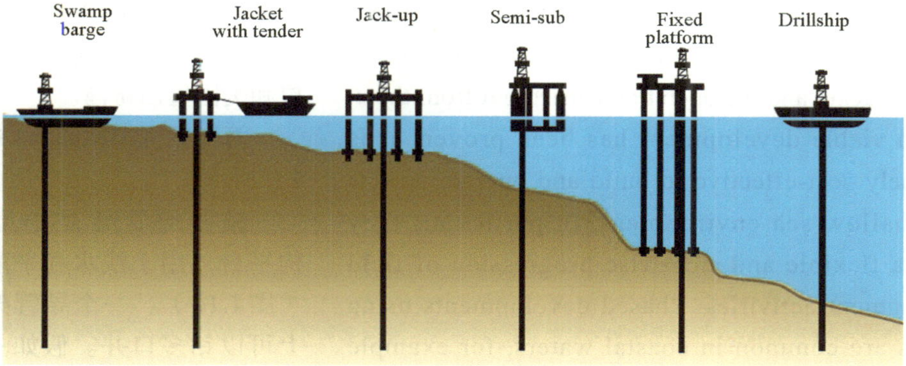

Figure 4.13 Offshore Rig Types
(Source: Hydrocarbon Exploration and Production)

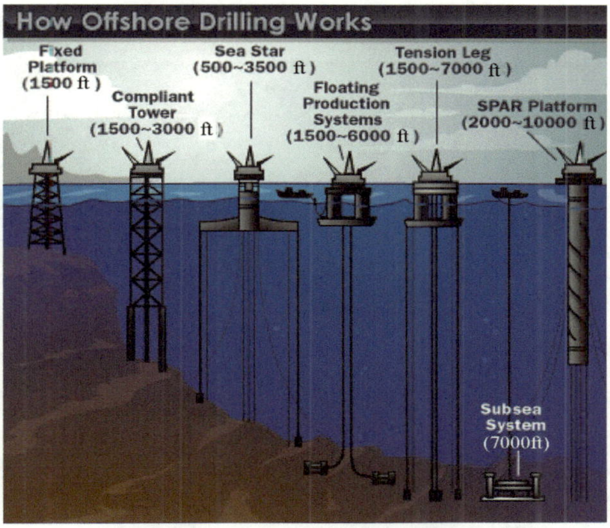

Figure 4.14 Offshore Drilling Platforms
(Source: http://how stuff works.com)

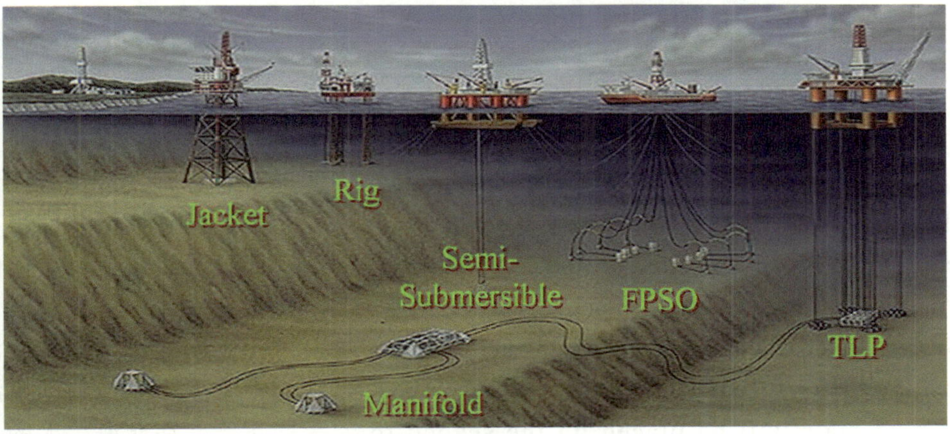

Figure 4.15 Offshore Rig Structures
(Source: http://www.civildesignhelp.info/offshore.html)

one jacket. If a jacket is too small to accommodate a drilling operation, a jack-up rig is usually cantilevered over the jacket and the operation carried out from there. Once a viable development has been proven, it is extremely cost-effective to build and operate jackets in a shallow sea environment. In particular, they allow a flexible and step-wise progression of field development activities. Phased developments using jackets are common in coastal waters, for example South China Sea and the shelf GoM. Wells drilled from large production platforms in the North Sea are drilled in a similar fashion.

驳船式钻机一般用于湿地浅水湾区域，如尼日利亚、委内瑞拉和美国墨西哥湾。

海洋钻机类型如图4.13～图4.15所示。

钻井导管架是小型钢结构平台，用于浅水和平静海域（图4.16）。一个导管架平台上可以钻多口井。假如一个导管架平台太小不能完成钻井工作，常用一台自升式钻机悬臂在导管架上完成钻井。一旦证

Figure 4.16　Jacket with Tender
(Source: http://reftrade.com/wp-content/)

Jack-up rigs are either towed to the drilling location (or alongside a jacket) or are equipped with a propulsion system (Figure 4.17). The three or four legs of the rig are lowered onto the seabed. After some penetration the rig will lift itself to a determined operating height above the sea level. If soft sediment is suspected at seabed, large mud mats will be placed on the seabed to allow a better distribution of weight. All drilling and

明地下有可开发区域，在浅海区域搭建导管架钻井回报率极高。而且导管架还可灵活的、阶梯式开发。因此，阶段性开发过程中用导管架钻井在沿海区域很常见，比如中国南海和墨西哥湾。在北海，大的生产平台钻井也用导管架钻井。

Figure 4.17 Jack-up Rig (Courtesy of Reading & Bates, UK)
(Source: Hydrocarbon Exploration and Production)

supporting equipment are integrated into the overall structure. Jack-up rigs are operational in water depths up to about 450 ft and as shallow as 15 ft. Globally, they are the most common offshore rig type, used for a wide range of environments and all types of wells.

Semi-submersibles are used for exploration and appraisal in water depths too great for a jack-up (Figure 4.18). A semi-submersible rig is a movable offshore vessel consisting of a large deck area built on columns of steel. Attached to these heavy-duty columns are at least two barge-shaped hulls called pontoons. Before operation commences on a specified location, these pontoons are partially filled with water and submersed in approximately 50 ft of water to give stability. A large-diameter steel pipe (riser) is connected to the seabed and serves as a conduit for the drill string. The blowout preventer (BOP) is also located at the seabed (subsea stack).

自升式钻井平台可以被拖到钻井区域（挨着导管架平台）也可以安装推进系统（图4.17）。3条或4条钻井腿下降深入海底。钻进一段时间后，可以根据海面高度提升钻井平台位置。当海底有大量软沉积时，一个大的泥浆垫将安放在海底以更好地平衡钻井平台的重量。所有钻井和支撑装备将整合成一个整体机构。自升式钻井平台用于15～450ft的水域。从全球来看，自升式钻井平台是最普遍的海洋钻机形式，广泛地用于各种海洋环境和各类钻井类型中。

半潜式钻井平台用于勘探或评价钻井水深超过自升式钻井平台的情况（图4.18）。半潜式钻井平台是可移动的海上船只，包

Figure 4.18 Semi-submersible Rig (Courtesy of Stena Drilling)
(Source: Hydrocarbon Exploration and Production)

A combination of several anchors and dynamic positioning (DP) equipment assists in maintaining position. Relocation of the semi-submersible vessel is made possible by the utilisation of tugboats and/or propulsion machinery.

Heavy-duty semi-submersibles, for example Deepwater Horizon (rated 15,000 psi), can handle high reservoir pressures and operate in the most severe metocean conditions in water depths down to 3000m.

Drill ships are used for deep and very deep water work. They can be less stable in rough seas than semi-submersibles (Figure 4.19). However, modern high-specification drill ships such as Discoverer Enterprise can remain stable, and on target during 100 knot winds using powerful thrusters controlled by a DP system. The thrusters counter the forces of currents, wind and waves to keep the vessel exactly on target, averaging less than 2m off her mark, without an anchor.

括钢柱的大型钢铁甲板区域。维系这些重型钢柱的是两只以上驳船外形的浮船。在特定位置钻井时，注满水的浮船浸没大概50ft以保证稳定性。一个大直径的钢管连接到海底充当钻柱的导管。防喷器（BOP）也坐落在海底。

几个锚的组合和动态定位装置辅助定位。半潜式船只的移动可通过拖船或动力推进系统。

重型半潜式平台，比如深水地平线号（150000psi级）能钻高压油藏，能在极端恶劣的海洋环境下钻达水深3000m。

钻井船用于深海和超深海钻井工作。钻井船在恶劣海况下没有半潜式钻井平台稳定（图4.19）。但现代高级别的

Figure 4.19 Drill Ship 'Transocean Enterprise'
(Source: Hydrocarbon Exploration and Production)

Heavy-duty drill ships are capable of operating in water depths up to 3000m.

In some cases, oil and gas fields are developed from a number of platforms. Some platforms will accommodate production and processing facilities as well as living quarters. Alternatively, these functions may be performed on separate platforms, typically in shallow and calm water. On all offshore structures, however, the installation of additional weight or space is costly. Drilling is only carried out during short periods of time if compared to the overall field life span and it is desirable to have a rig installed only when needed. This is the concept of tender-assisted drilling operations.

In *tender-assisted drilling*, a derrick is assembled from a number of segments transported to the platform by a barge (Figure 4.20). All the supporting functions such as storage, mud tanks and living quarters are located on the tender, which is a specially built spacious

钻井船如探索企业号能在100节风速下，用强有力的动力推进器实现定位。推进器能抵消风力、海浪、潮流的作用，不用锚就能保持钻井船的稳定，平均不超过2m的误差。

重型钻井船能钻达水下3000m。

在某些情况下，油气田通过钻井平台进行开发。一些钻井平台上有生产和处理设备，还有生活区域。一些钻井平台只有单一的设备，如在浅水区域和海浪平静区域的平台。在海上结构上额外安装设备提供空间是费钱的。钻井作业相对于整个油田生命过程只占很短的时间，钻机只在需要时安

Figure 4.20 Tender-assisted Drilling
(Source: Hydrocarbon Exploration and Production)

barge anchored alongside. It is thus possible to service a wholefield or even several fields using only one or two tender-assisted derrick sets. Inrough weather, barge type tenders quickly become inoperable and unsafe since theplatform is fixed whereas the barge moves up and down with the waves. In thesecases and in the hostile environment of the North Sea, a modified semi-submersiblevessel may serve as a tender. Currently, purpose-built semi-submersible tenders arebeing introduced for some future field developments.

装，因此就有辅助船钻井作业的概念。

在辅助船钻井作业中，井架由驳船分段运输到平台上（图4.20）。其他钻井作业如泥浆罐、生活区等安装在辅助船上，辅助船由大空间的驳船建造，用锚停在旁边。一艘或两艘辅助船井架可服务整个油田或几个油田。在恶劣天气下，驳船类型的辅助船因随波浪起伏迅速变的不可操作和不安全。特别是在北海恶劣的环境下，中型半潜式船可以作为辅助船。目前，引进了专门建造的半潜式辅助船为未来油田的开发服务。

4.5 Drilling Systems and Equipment

4.5.1 Rotary Systems

Basically with the rotating system, the drill bit

4.5 钻井系统和装备

4.5.1 旋转系统

旋转系统指的是钻头旋转

turns, cutting the rocks in making the hole. As the bit rotates, the drilling fluid is forced down the drill pipe to jet nozzles and drill bit cuts the rocks. These jet fluid clean the bottom of the hole, up the debris away from the bit and then up to the surface (Figure 4.21).

切削岩石进而钻进成井。当钻头旋转时，钻井液从钻杆向下流从喷嘴喷出并切削岩石，这喷出的钻井液清洗井底并携带岩屑返出地表（图4.21）。

Rotary rigs drill the vast majority of wells today, including all medium and deep wells. The rotary rig consists of five main systems (Figure 4.22). These include a rotating system, a hoisting system, a circulating system, a power(an engine) system, and a pressure control system.

目前，旋转钻机钻了当今绝大多数的井，包括中深井和深井。旋转钻机由五部分组成（图4.22）：①旋转系统；②提升系统；③循环系统；④动力系统；⑤压力控制系统。

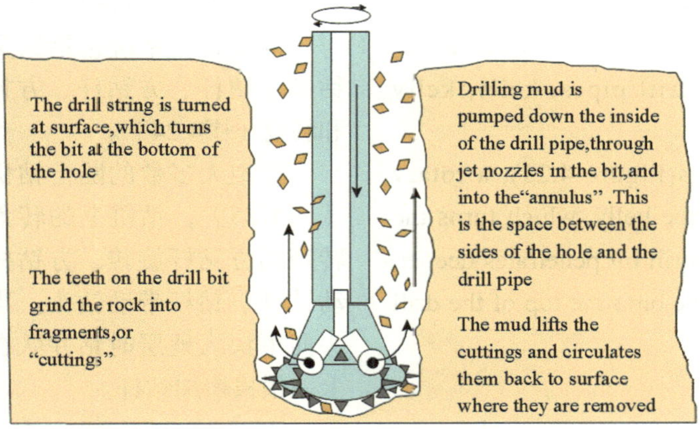

Figure 4.21 Rotating Drilling
(Source: http://www.ppdm.org/wiki/index.php/Well_Operations_Reference_guide)

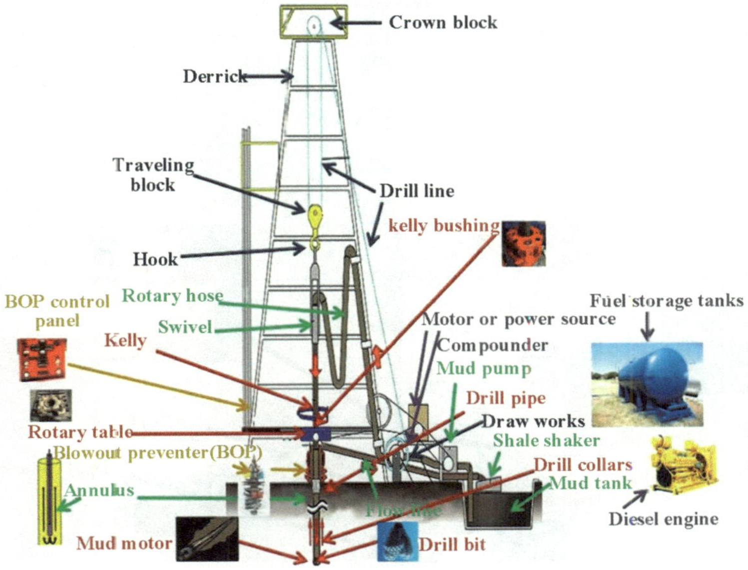

Figure4.22 Five Main Systems on a Rig
(Source: Overview of the Oil Industry, Richard Lau)

There are three ways to turn the drill bit or three types of rotary systems: ①Conventional Rotary System; ② Top Drive Rotary System; and ③ Down Hole Mud Motor Rotary System (Figure 4.23).

有三种方式带动钻头旋转：①常规旋转系统；② 顶驱系统；③底驱系统（图4.23）。

$$Drilling\begin{cases} Percussion\ Drilling\ (Cable\quad Tool) \\ Rotary\ Drilling\begin{cases} Coventional\ Rotary\ System \\ Top\ Drive\ Rotary\quad System \\ Downhole\ Mud\ Motor\ System \end{cases} \end{cases}$$

Figure 4.23 Drilling Methods

4.5.1.1 Conventional Rotary System

Conventional rotary system (Figure 4.24)includes:

drill bit, drill collars, drill pipes, kelly, kelly bushing and rotary table.

On most land-based rigs(Figure 4.25), a rotary table on the rig floor rotates the kelly, which turns the drill pipe and drill bit. As the drill bit penetrates deeper, the crew threads additional pipe onto the top of the drill string.

4.5.1.1 常规旋转钻井

常规旋转系统包括钻头、钻铤、钻杆、方钻杆、方补心和转盘（图4.24）。

在绝大多数的陆上钻机中（图4.25），钻机上的转盘旋转带动方钻杆旋转，方钻杆带动钻杆、钻杆带动钻头。当钻头向下深入地层时钻井队在地上一根根增加钻杆。

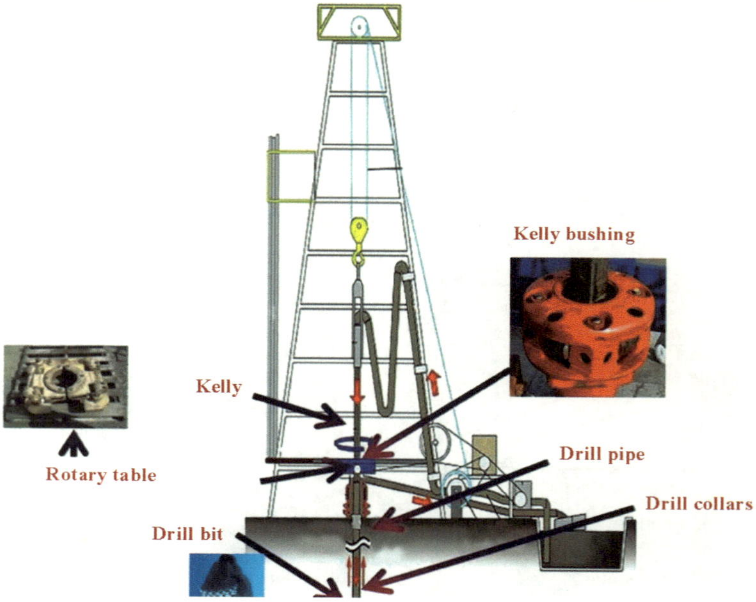

Figure 4.24 Conventional Rotary System
(Source: Overview of the Oil Industry, Richard Lau)

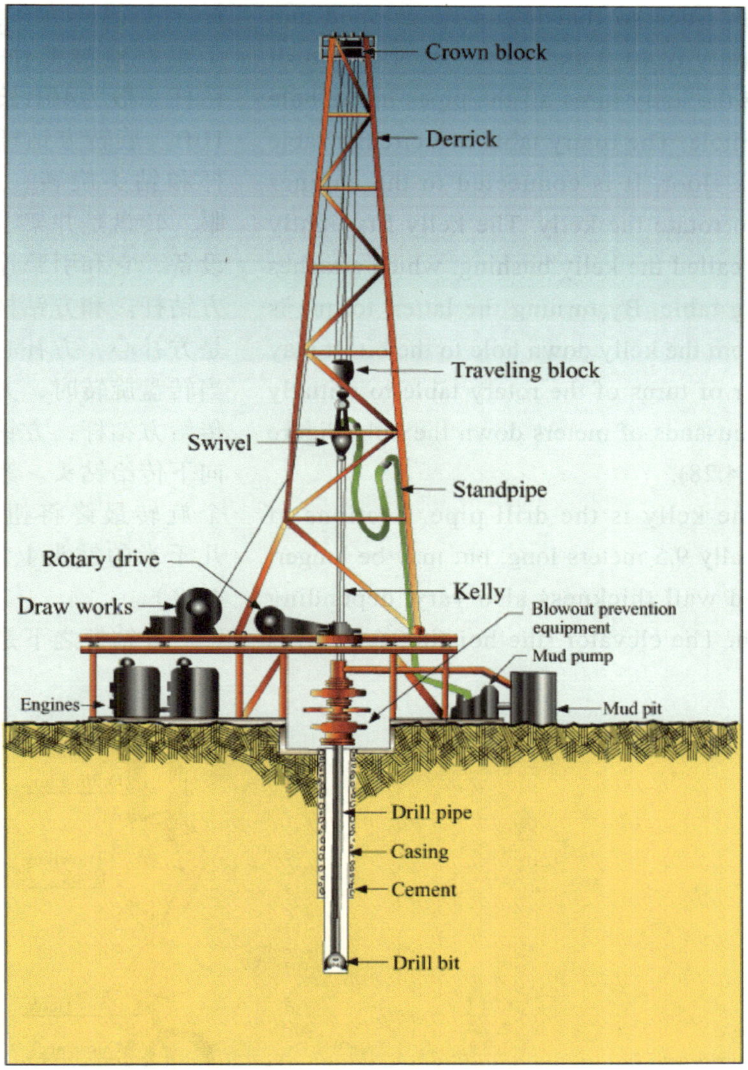

Figure 4.25 A land Rig
(Source: https://www.pinterest.com/)

◆ **Drill string**

Between the bit and the surface, where the torque is generated, we find the drill string. The drill string is the rotating pipes and its attachments. This includes the swivel, kelly, drill pipe, drill collars, and bit.

Suspended on the hook directly below the swivel. The swivel allows the pipe below it to rotate while the weight of the pipe is suspended from the derrick. Below the swivel is a four or six-sided pipe called the kelly.

◆ **钻柱**

钻柱位于钻头和地表之间，用于传导扭矩。钻柱由钻杆和辅助件组成，包括水龙头、方钻杆、钻杆、钻铤和钻头。

大钩下悬挂着水龙头。水龙头及其下旋转的钻柱重量都悬挂在井架上。水龙头下是四边或六边的空心管柱，叫方钻

The kelly has sides so that it can be gripped and turned. The kelly is 40 ft and 10 ft longer than drill pipe. Turning the kelly turns all the pipes in the hole and drills the hole. The rotary table is a circular table in the derrick floor. It is connected to the engines and is used to rotate the kelly. The kelly fits tightly into a device called the kelly bushing, which attaches to the rotating table. By turning the latter, torque is transmitted from the kelly down hole to the bit. It may take a number of turns of the rotary table to initially turnthe bit thousands of meters down the hole(Figure 4.26 ~ Figure 4.28).

Below the kelly is the drill pipe. Sections of pipe are typically 9.5 meters long, but may be longer; diameters and wall thickness also vary, depending on well depth. The elevator (the hoisting system) is

杆。方钻杆为多边形，是为了更好地被抓住并旋转。方钻杆一般为40ft长，比钻杆长10ft。旋转方钻杆从而带动钻杆和钻头旋转，进而钻成井眼。转盘是井架上的圆形旋转设备，它和引擎连接用于旋转方钻杆。和方钻杆密切配合的是方补心，方补心连着转盘。当转盘旋转时，方补心将扭矩传给方钻杆，方钻杆再将扭矩向下传给钻头。经过转盘的多个旋转最终将扭矩传到井下几千米的钻头上（图4.26~图4.28）。

方钻杆之下是钻杆。典型

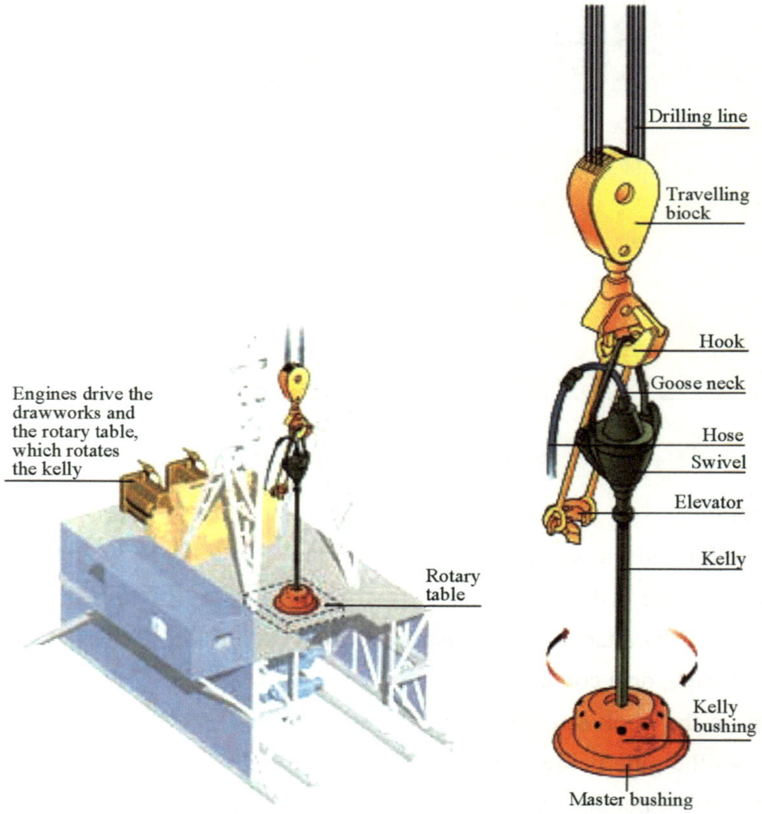

Figure 4.26 The Conventional Rotating Drilling System
(Source: http://www.drillingfunds.com/rotating.html)

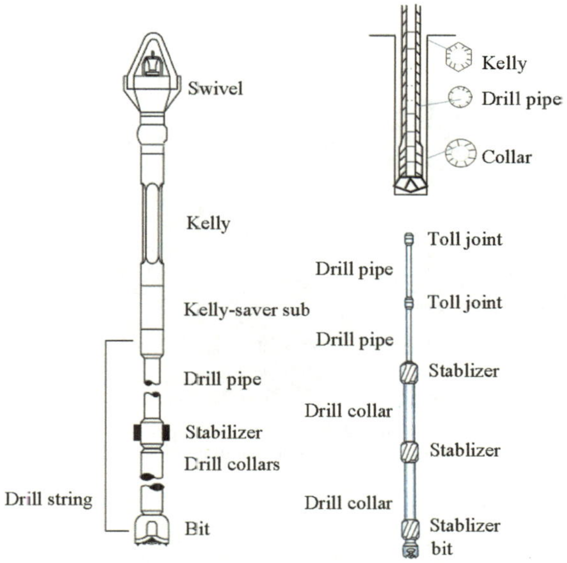

Figure 4.27 The Drill String (Schematic)
(Source: Hydrocarbon Exploration and Production)

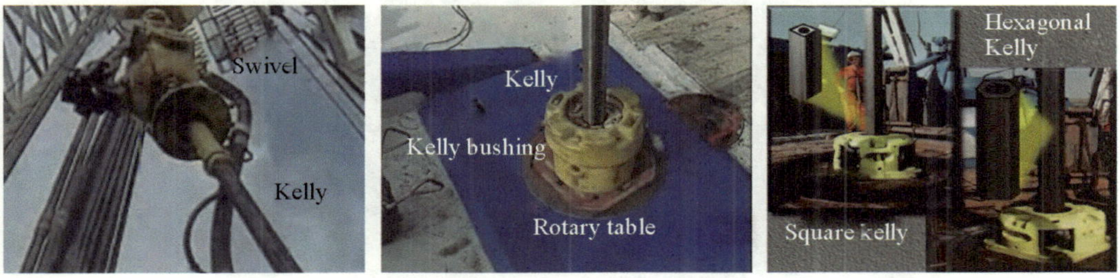

Figure 4.28 Kelly (Source: Kelly and Top Drive)

a device used when lifting the drill string out of the wellbore(Figure 4.29).

Below the drill pipe are larger-diameter, thick-walled, high-density and heavy lengths of pipes called drill collars(Figure 4.30).

Drill collars (DCs) weigh more than drill pipe and are designed to put weight on the bottom of the drill pipes (Figure 4.31). This helps control drilling and prevents the pipe from kinking and breaking. That is say, they keep the drill string in tension (avoiding buckling) and provide weight onto the bit.Two to twenty drill collars are often used. Stabilizers are added to the drill string at intervals to hold, increase or decrease the hole angle.

的钻杆有9.5m长，或许更长，直径和壁厚随井深不同而不同。起吊机用于提升钻柱出井筒（图4.29）。

钻杆之下是大直径、厚壁、高强度钢、重尺寸的钻铤（图4.30）。

钻铤比钻杆重是为了向下施加钻压在钻杆上，从而帮助钻进并避免钻杆折断和扭结（图4.31）。也就是说，钻铤提升钻杆的抗折强度并将钻压施加在钻头上。井场上一般用

Figure 4.29　5-inch drill pipes
(Source: http://www.rigmanufacturing.com)

Figure 4.30　Drill Collar
(Source: http://www.everypart.com/)

Figure 4.31　Drill Collars
(Source: http://www.westerndrilling.ca/products/drill-colar)

◆ Drill bit

The bit screws into the bottom of the drill collars(Figure 4.32 ~ Figure 4.34). The most common bit is the tricone bit, which has three rotating cones(Figure 4.35). The cones have teeth that are designed to chip and flake away the rock as the bit is rotated. Some tricone bits have hard tungsten carbide buttons instead of teeth. Another type of bit is the diamond or button bit, which has diamonds embedded in the bottom and sides. Different bits are used for different hardness formations. The button bit is the best for drilling limestones.

2～20根钻铤。稳定器加在钻柱之间，用于稳定、增加或减少井斜角。

◆ 钻头

钻头旋紧在钻铤下方（图4.32～图4.34）。最常用的钻头是三牙轮钻头，它有三个旋转的巴掌。这些巴掌上面还有牙齿在钻头旋转时用以切削、剥落岩石（图4.35）。一些三牙轮钻头上用碳化钨硬钮代替牙齿。还有

Drill Bits
- *Cable Tools*
- *Rotary Bits*
  - *Opener Bits*
  - *Core Bits*
  - *Drag Bits*
  - *PDC Bits*
  - *Tricone Roller Bits*
  - *Diamond Bits*

Figure 4.32　Drill Bits

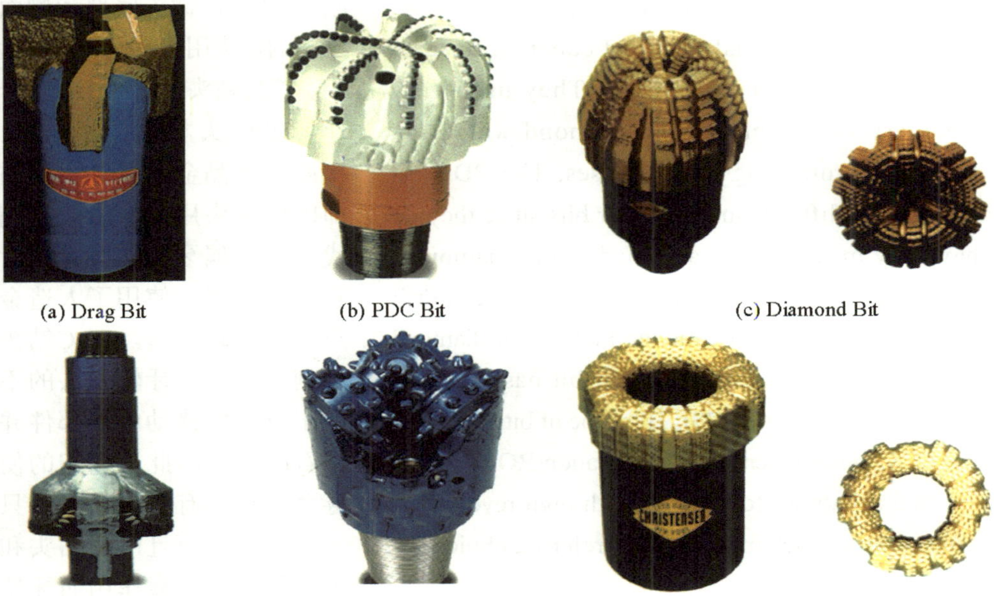

(a) Drag Bit　(b) PDC Bit　(c) Diamond Bit

(d) Opener Bit　(e) Roll Cone Bit　(f) Core Bit

Figure 4.33　The Rotating System
(Source: http://ffden-2.phys.uaf.edu)

(a) Cable Tools (b) Diamond Bits (c) Tricone Bits

Figure 4.34 Drill Bits
(Source: Overview of the Oil Industry, Richard Lau)

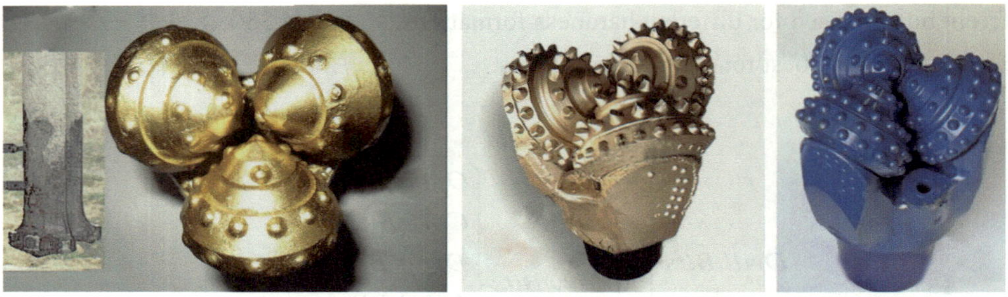

(a) Rotary Tool Bit (b) Tricone Rotary Bit (c) Tricone Roller Bit

Figure 4.35 Cable Tool Bits & Rotary Tool Bits
(Source: Overview of the Oil Industry, Richard Lau)

A new improvement in the material used for drill bits. The polycrystalline diamond compact bit or PDC bit is fitted with industrial diamond cutters instead of hardened metal teeth(Figure 4.36). They used very hard polycrystalline man made diamond wafers on the face of a protruding bronze bases. The PDC bit or diamond bit differ from the roller bits since they do not have any moving parts and made from diamonds. They last longer, one bit can sometimes be used to drill an entire well. At the same time the PDC bit or diamond bit has a sharing action while a roll bit has crushing action(Figure 4.37, Figure 4.38). This type of bit is popular because of its much better rate of penetration(ROP), longer lifetime and suitability for drilling with high revolutions per minute(rpm), which makes it the preferred choice for turbine drilling.

金刚石钻头、镶齿钻头，用金刚石镶嵌在钻头底部和侧面。不同的钻头用于不同硬度的地层中。镶齿钻头最适用于石灰岩。

在钻头材料上有很大的提高，聚晶金刚石复合片钻头即PDC钻头用人造工业金刚石代替硬金属牙齿（图4.36）。它的切削刃上使用了人造金刚石和烧结碳化钨。PDC钻头和金刚石钻头与牙轮钻头的不同在于没有可移动的零部件并由钻石缔造。因此，它们的使用寿命更长，有时钻一口井只用一个钻头。而且PDC钻头和金刚石钻头有研磨作用而牙轮钻头只有压碎作用（图4.37、图4.38）。PDC钻头和金刚石钻

The selection of bit type depends on the composition and hardness of the formation to be drilled and the planned drilling parameters.

头有更快的钻速、更长的寿命、更有效而更受欢迎，特别是在地下涡轮钻进时。

钻头类型的选择在于地层硬度和成分，也和钻井计划有关。

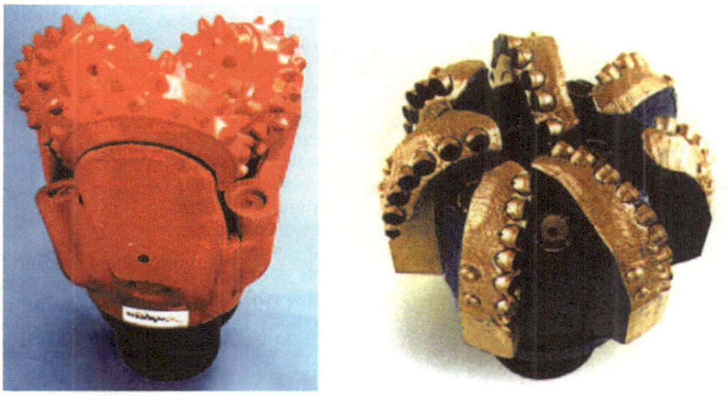

(a) Roller Cone Bit (b) PDC Bit

Figure 4.36 Roller Cone Bit and PDC Bit
(Source: Hydrocarbon Exploration and Production)

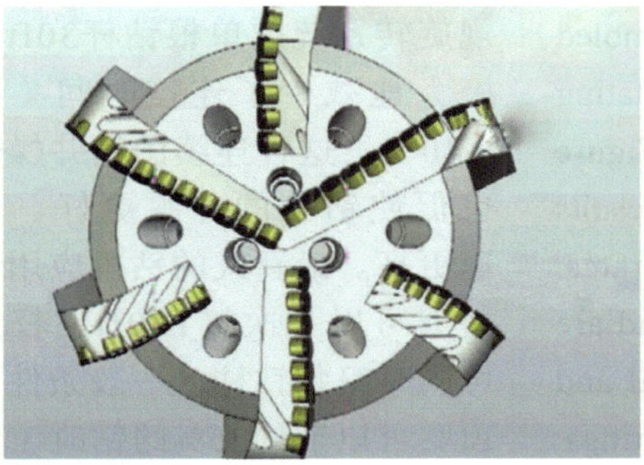

Figure 4.37 PDC Bit
(Source: Overview of the Oil Industry, Richard Lau)

Figure 4.38 Bits Teeth Wear Out
(Source: Overview of the Oil Industry, Richard Lau)

4.5.1.2 Top Drive Rotary System

Another new type of rotating system is called the top drive(Figure 4.39). Modern rigs the kelly and its parts have been replaced by the top drive also known as the power swivel. Using hydraulic or electric motors suspended above the drill pipe enables top drives to rotate and pump continuously while drilling or during the removal of drill pipe from the hole. Most offshore units and an increasing number of land rigs use top drives.

4.5.1.2 顶驱旋转钻井

另一种旋转系统叫做顶驱旋转（图4.39）。现代钻机用动力水龙头取代方钻杆和相关部件。用高悬的水力和电力马达驱动钻杆旋转。顶驱使钻杆和泥浆泵连续旋转不管是不是在接单根还是在钻进过程中。大部分的海上钻机和越来越多的陆上钻机开始使用顶驱系统。

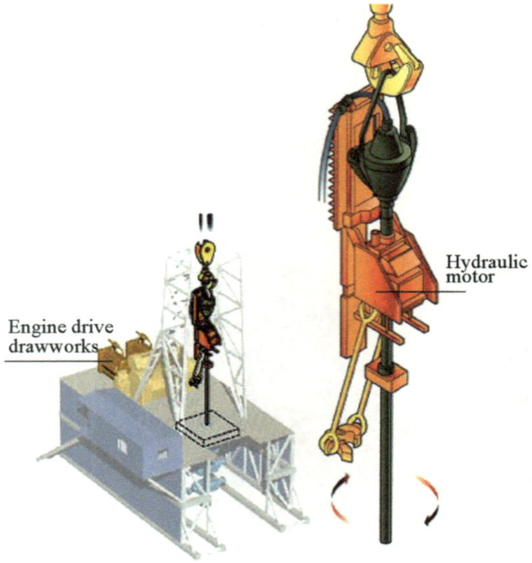

Figure 4.39 The Top Drive Rotating Drilling System
(Source: http://www.drillingfunds.com/rotating.html)

Instead of having a rotary table in the rig floor the drive mechanism for the drill string is mounted on guide rails and moves up and down inside the derrick. Thisallows drilling in segments of 90 ft of pre-assembled pipe at a time rather than 30 ft in conventional rotating system, significantly reducing connection time(Figure 4.40). The more continuous drilling process also results in better hole conditions and faster penetration rates. The latest rigs allow running of 120ft sections and are equipped with two derricks–one drilling the well and the other used to concurrently pre-assemble drill strings. Typically a top drive can reduce drilling time by about 25%(Figure 4.41).

顶驱系统取代转盘提升钻柱并在井架内上上下下。每次一柱即3根钻杆90ft而不是常规旋转系统的单根钻杆30ft，极大地减少了连接时间（图4.40）。这种连续的钻井操作也促使钻出的井眼更好、钻速更高。最新式的钻机应用双井架可以一次接120ft钻杆并可同时组装钻柱。一般条件下顶驱可以节约25%的钻时（图4.41）。

The manual labour on the rig floor has largely been replaced by a hydraulic system which picks up pipe from the rack, transports it up to the rig floor and inserts it into the drill string. The process is controlled by the rig crew manning a workstation housed beside the rig floor(Figure 4.42).

钻机平台上的大量人工操作由水力系统取代，提起钻杆、运送到平台上、连接钻柱等工作现在已由钻井工人坐在平台控制室完成（图4.42）。

Top Drive Advantages:

◆ Reduces Drilling Time;
◆ Rotates Drill String Efficiently;
◆ Handles Pipe Efficiently;

顶驱的优势有：

◆ 节省钻井时间；
◆ 钻柱旋转高效；
◆ 有效接钻杆；

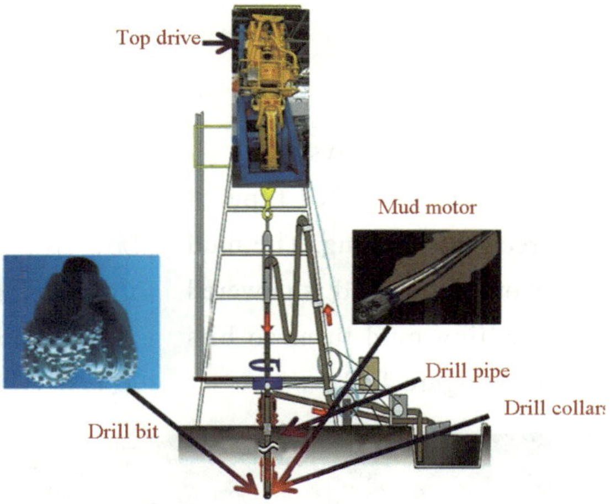

Figure 4.40 Top Drive Rotary System (Source: Overview of the Oil Industry, Richard Lau)

Figure 4.41 Top Drive System (Source: Hydrocarbon Exploration and Production)

Figure 4.42 The Rig Crew Controlling the Automated Pipe System Visible Outside the Control Room (Source: Hydrocarbon Exploration and Production)

- ◆ Variable Rotating Power;
- ◆ Allows Rotation at Any Time;
- ◆ Rapid Response to Well Kicks.

- ◆ 可调节旋转动力；
- ◆ 允许任何时间旋转；
- ◆ 快速反应处理井涌。

4.5.1.3 Down Hole Mud Motor Rotary System

The down hole mud motor is the 3rd type of rotating system, used in directional drilling. The mud motor is mounted directly about the bit and is powered by the drilling mud. The drilling mud turns to bits without rotating the drill strings(Figure 4.43).

4.5.1.3 井下马达旋转钻井

井下马达是第三种旋转系统，用于定向井钻井。井下马达直接安装在钻头上由钻井液驱动。钻井液驱动钻头而不需旋转钻柱（图4.43）。

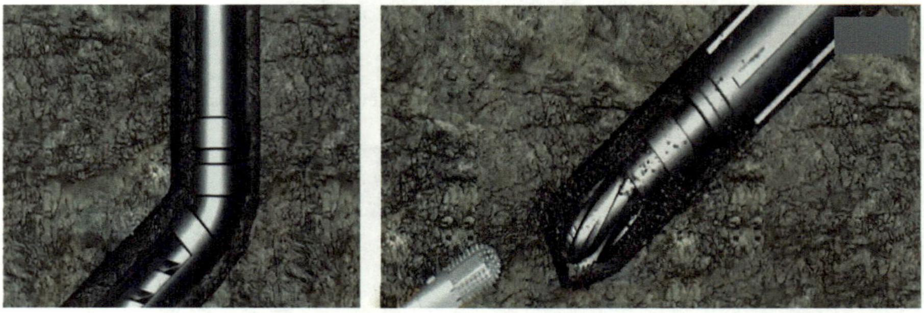

Figure 4.43 Down Hole Mud Motor Rotary System
(Source: Overview of the Oil Industry, Richard Lau)

4.5.2 Hoisting System

The Hoisting System is used to raise and lower and suspend equipment in the well. The drilling line (wire rope) is usually braided steel cable about $1^1/_8$ inches in diameter. It is wound around a reel in the draw works. The engines are connected to the draw works and let the drilling line in or out. The derrick or mast is the steel tower. If the tower comes on a tractor-trailer and is jacked up, it is a mast. If the tower is erected on the site, it is derrick. The drilling lines go over a pulley, called the crown block, at the top of the derrick, and then down to another pulley, called the traveling block.

Below the traveling block is a hook to which equipment can be attached. As the drilling line is reeled in or out of the draw works, the traveling block rises and falls in the derrick. This raises and lowers the equipment in the well.

Rigs are often rated as to how deep they can drill.

4.5.2 提升系统

提升系统用于提起、降落和悬挂钻井设备。钻井用钢丝绳一般是直径为$1^1/_8$in的辫织钢绳，缠绕在绞车上。引擎联动绞车使钢丝绳绞进、绞出。井架一般为钢制塔式，可以是车载的也可以现场搭建。钢丝绳绕过的上滑轮叫天车，绕过的下滑轮叫游车。

游车下是大钩，用以勾住钻井设备。钢丝绳绕在绞车上，通过绞进或绞出提升和下降游车，从而起降钻井设备。

钻机通常以钻井深度来评级。越深的井需要更强大的

This is because deeper wells require more powerful engines and stronger derricks. A deep well could have a drill string (pipe) weighing 500000 lb suspended down the hole. The engines must be able to raise that equipment out of the hole and the derrick must be strong enough to support that weight(Figure 4.44~Figure 4.46).

引擎和更强的井架。钻深井的钻柱重量可达500000lb。引擎必须能提升这么重的钻柱，同时井架必须能够承载这个重量（图4.44~图4.46）。

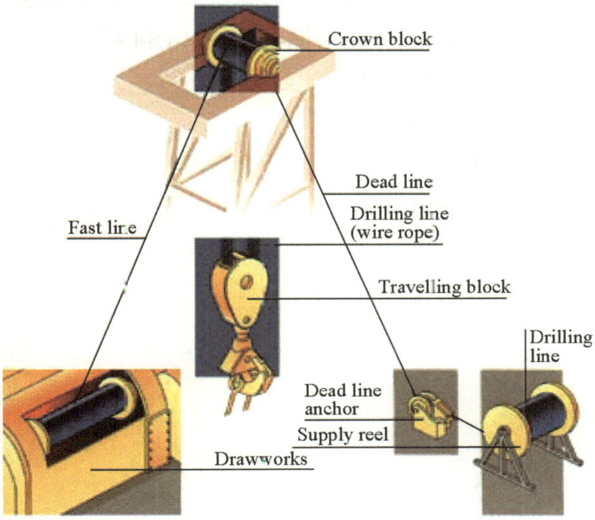

Figure 4.44 The Hoisting System in Drilling Rig
(Source: http://www.drillingfunds.com/hoisting.html)

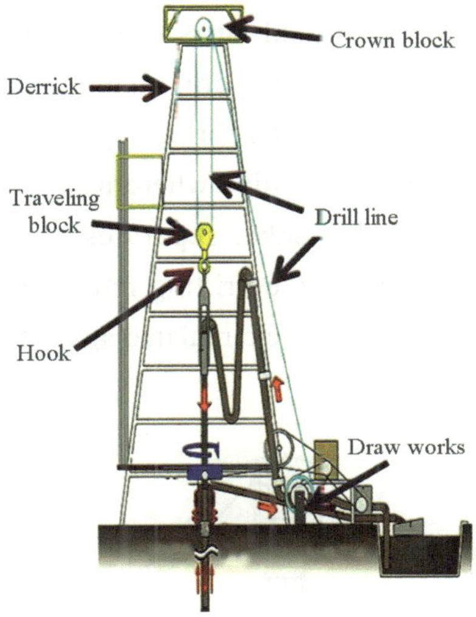

Figure 4.45 Hoisting System

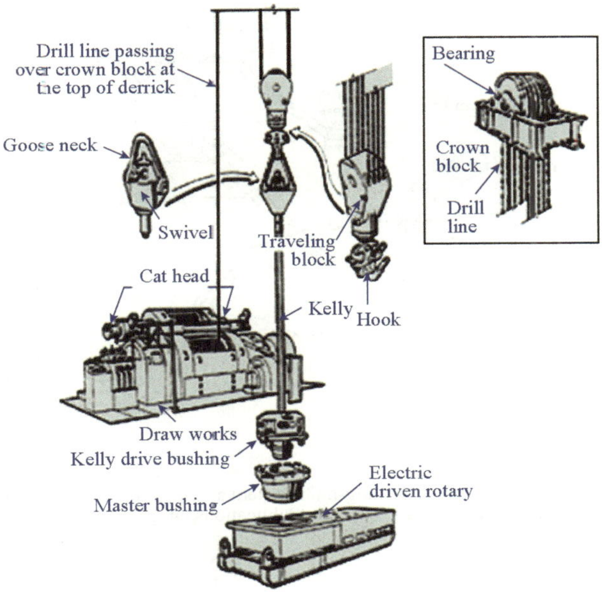

Figure 4.46 The Diagram of Hoisting System
(Source: http://ffden-2.phys.uaf.edu)

4.5.3 Circulating System

The drill string is assembled and lowered into the conductor pipe. Drilling fluid, better known as drilling mud, is circulated through the kelly and the drill string by means of pipes and flexible hose connecting the drilling fluid or mud pumps and swivel device attached to the upper end of the kelly. The mud pump draws fluid from mud tanks or pits. The drilling mud passes through the kelly, drill pipe, drill collars, and drill bit. It is returned to the surface by means of the well bore and the conductor pipe where it is directed to shale shaker. The shale shaker separates the drilling cuttings and solids from the drilling mud, which is returned to the mud tanks to be circulated again(Figure 4.47).

4.5.3 循环系统

钻杆沿导管下放安装。钻井液即泥浆通过泥浆泵和水龙头泵入方钻杆，流入钻杆、钻铤和钻头。再经环空返回到地面振动筛，振动筛筛去泥浆中的岩屑固体后，流入泥浆池沉淀再循环使用（图4.47）。

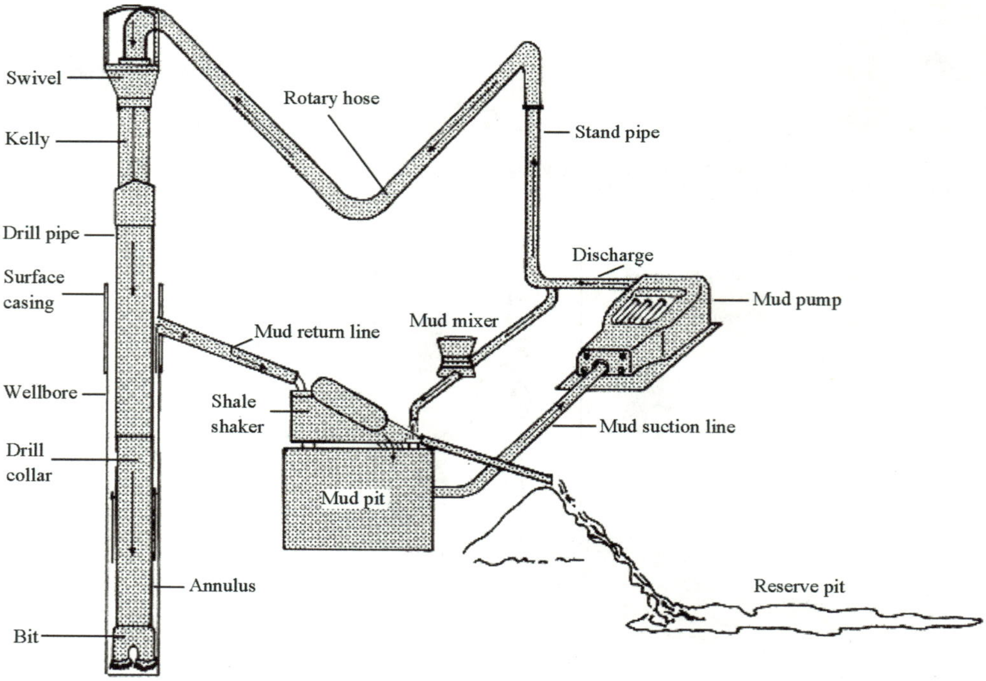

Figure 4.47 Diagram Illustrating the Drilling-fluid System and the Flow of Fluids through the System (Source: http://www.kgs.ku.edu/Publications/Oil/primer12.html)

Earlier on when we described the cutting action of the drill bit, we learned about the drilling fluid or mud. Pumps, called mud hogs, forcing the drilling mud through the hollow rotating drill string and jet out through holes in the drilling bit on the bottom of the well. The mud system circulates drilling mud in the hole. The drilling mud picks rock chips (cuttings) from the bottom of the well. It flows up the well in the space between the rotating drilling string and well walls (annulus). So the mud cools the bit and also removes the cuttings by carrying them up the annulus, outside the drill pipe. At the top of the well, the mud flows through the blowout preventers and runs over a number of moving screens, the shale shakers, which remove the cutting for disposal. The fine particles which pass through the screens are then removed by desanders and desilters, the latter usually hydrocyclones.Having been cleaned, the mud is transferred into and stored in steel mud pits beside the rig or the mud tanks, which are large treatment and storage units at the surface. From there a powerful pump brings the mud up through a pipe (stand pipe) and through a hose connected to the swivel (rotary hose)forcing it down the hole inside the drill string. Eventually the cleaned mud will exitagain through the bit nozzles. This is circulating system, it is the backbone of the rotary drilling(Figure 4.48).

在早前的钻井过程中，我们提到钻井液也就是泥浆。泥浆泵促使钻井液通过中空钻柱从钻头喷嘴喷出。泥浆在整个井筒中循环。泥浆能携带岩屑从井底井环空空间循环出井口。而且，泥浆能冷却钻头。当泥浆经防喷器出井后，会途经多个筛网的振动筛，从而移走岩屑。细小的岩屑通过振动筛后由除砂和除泥装备移走，最后是旋液分离器清除岩屑。清除岩屑的泥浆被转移到钢制存储罐或泥浆池中。泥浆泵再将泥浆泵入立管，经过水龙带和水龙头进入钻柱，再到井底，从钻头水眼喷出。这就是循环系统，是旋转钻井的核心（图4.48）。

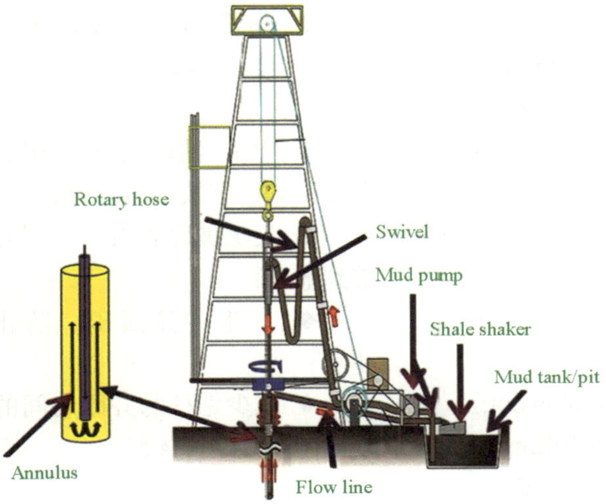

Figure 4.48 Circulating System
Source: Overview of the Oil Industry, Richard Lau)

Circulating the drilling fluid serves several purposes. The drilling fluid removes cutting from the bottom of well. As the drilling fluid flows across the bit, it cleans cutting from the teeth. The drilling fluid cools the bit from heat generated by the friction of drilling. In very soft sediment, such as coastal plain, the jetting action of the drilling mud squirting out of the bit on the bottom of the well helps cut the well hole. The drilling fluid also controls pressures in the well and prevents blowouts. At the bottom of the well, there are two fluid pressures. Pressure on fluids in the rock tries to cause the oil or water to flow into the well. Pressure exerted by the weight of the drilling fluid tries to force the drilling mud into the surrounding rocks. If the pressure on the fluid in subsurface rock is greater than the pressure of the drilling mud, the water, gas, or oil will flow out of the rock into the well. This often causes the sides of the well to cave in and trapping the equipment. In extreme cases, it causes a blowout. In order to control subsurface fluid pressure, the weight of the drilling mud fluid is adjusted to exert a greater pressure on the bottom of the well. This is called overbalance drilling. And the drilling mud fluid is then forced into the surrounding rocks. The rocks act as a filter and the solid mud particles cake to the sides of the well as the fluids enter the rock. This filter of mud cake is very hard. Once the filter cake has formed, the sides of the well are stabilized and subsurface fluids cannot enter the well.

We sum up circulation functions as following:

◆ To lubricate and cool rotary blades, helps to greatly extend the bits useful life;

◆ To flush cutting and clean the hole allowing the bit to penetrate more efficiently;

◆ To collect the loose rock and dirt then bring them back to the surface;

◆ To exert back pressure to prevent blowout;

◆ To reduce friction between the drilling string and the side of the hole(Figure 4.49~ Figure 4.52).

循环泥浆有多重作用，从井底清除岩屑、冷却和清理钻头。在软地层中，如海岸平原喷出的泥浆有助于钻进。泥浆还可控制井筒压力防止井喷。在井底有两种流体压力。地层压力促使油、水流向井筒，而井筒中泥浆压力促使泥浆流入井周岩层。假如地下地层压力大于泥浆压力，油、气、水将流入井筒。这常造成井壁塌陷、卡钻等事故，严重时会造成井喷。为控制地下流体压力，加入重泥浆在井底产生回压，这叫过平衡钻井。这也使泥浆泄漏到井周岩层中。周边岩石像滤纸一样将泥浆中固态颗粒物滤出留下从而在井壁附近形成一层泥饼，液体进入地层。一般来说，这种泥饼是相对坚固的。一旦形成泥饼，井壁就被加固，还阻挡地下流体进入井筒。

循环系统的功用有以下几方面：

◆ 润滑和冷却旋转牙轮叶片，极大地延长钻头有效寿命；

◆ 冲洗岩屑、清理井眼，允许钻头更有效地破碎岩石；

◆ 收集破碎岩石、岩屑并带到地表；

◆ 产生井筒回压，防止井喷；

◆ 减少钻柱与井眼之间的摩擦（图4.49~图4.52）。

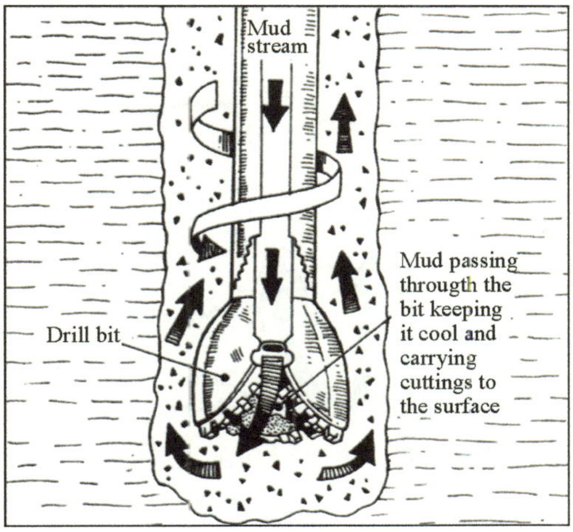

Figure 4.49 Mud Passing through the Bit

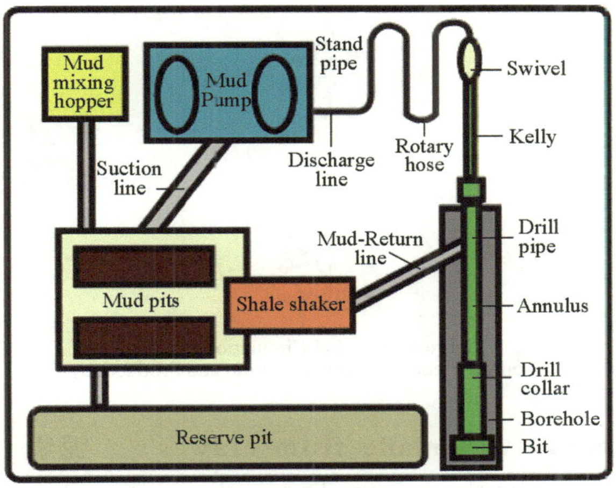

Figure 4.50 Drilling Fluid Circulation
(Source: http://www.kgs.ku.edu/Publications/Oil/primer12.html)

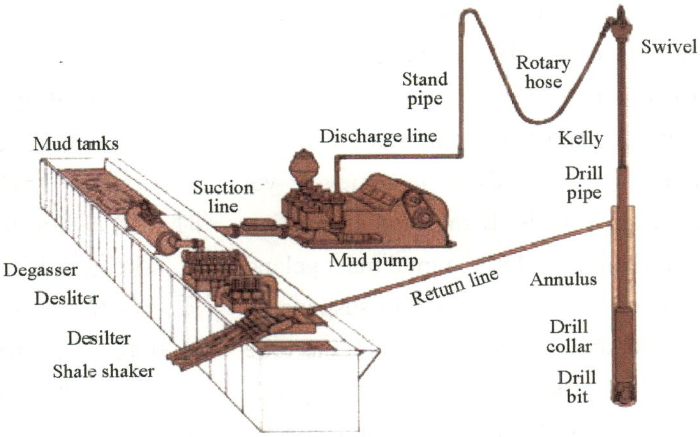

Figure 4.51 Circulation System
(Source:https://www.osha.gov/SLTC/etools/oilandgas/drilling/mud_system.html)

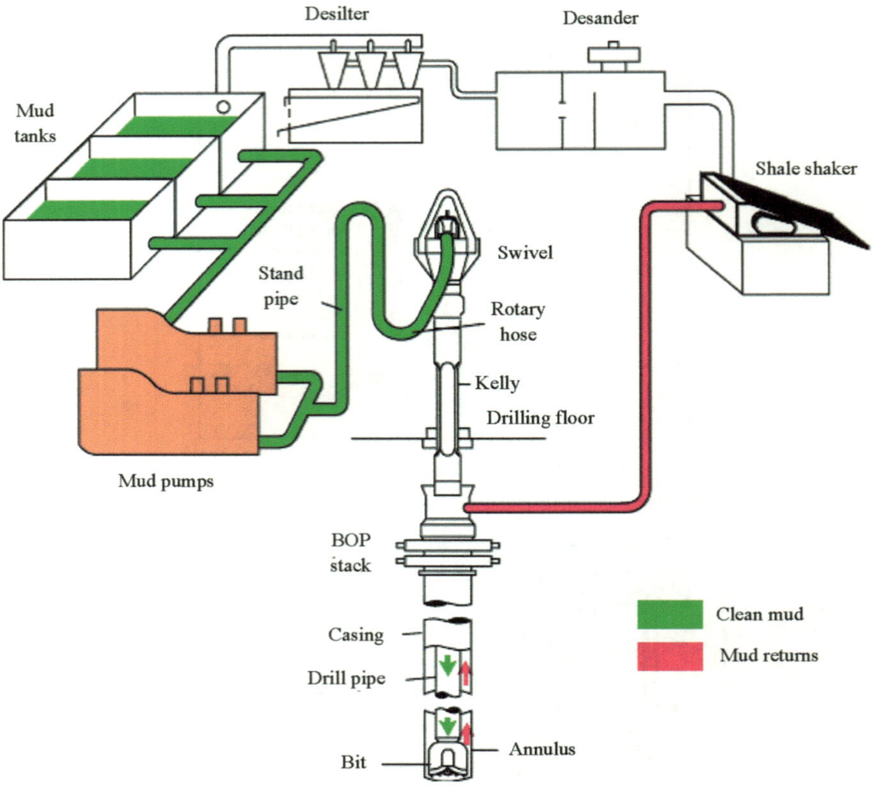

Figure 4.52 Mud Circulation System
(Source: Hydrocarbon Exploration and Production)

Originally, ‘mud’ was made from clay mixed with water——a simple system.Today the preparation and treatment of drilling fluid has reached a sophistication which requires specialist knowledge. The reason for this becomes clear if we consider the properties expected(Figure 4.53).

In order to effectively lift the cuttings out of the hole a certain fluid viscosity needs to be achieved; yet the mud must remain pumpable. If the mud circulation stops,for instance to change the bit, the mud must geland any material suspended init must remain in suspension to avoid settling out at the bottom of the hole. It has tobe stable under high temperatures and pressures as well as at surface conditions. Mud chemicals should not be removed by the mud-cleaning process. Drilling fluids

泥浆一般由黏土混合水形成。今天，配制泥浆是一项复杂工作，需要专业技术人员完成（图4.53）。

为有效举升井眼中的岩屑，钻井液必须有一定的黏度，而钻井液又不能太稠，否则泵不动。当钻井液循环停止时，比如更换钻头时，钻井液必须胶化使其中岩屑之类的物质悬浮在钻井液中，而不是沉入井底。钻井液在高温高压情况下也要像在地面情况下稳定。钻井液中的化学成分不能随地面钻井液清理过程而被去

Figure 4.53 Drill Muds and Special Fluids
(Source: http://www.ntpbt.com/en/services/)

haveto be capable of carrying weighing material such as barites in order to control the formation pressures. They have to be compatible with the formations being drilled,for example they should prevent the swelling of formation clay and not permanently damage the reservoir zone. Last but not least, since these fluids are pumped,transported and disposed in large quantities, they should be environmentally friendly and cheap!

Often drilling fluids are made up using water and are called water-based muds(WBM). Another frequently employed system is based on oil, oil-based mud(OBM). The advantage of OBM is better lubrication of the drill string, compatibility withclay or salt formations and they give a much higher ROP. Diesel fuel was usuallyused for the preparation of OBM. During operations, large quantities ofcontaminated cuttings

除。钻井液还得能携带加重物质如重晶石等，以便能控制地层压力。钻井液还得与地层流体兼容，比如钻井液得防止地层黏土膨胀，不能永久性地伤害储层。最后，要求钻井液有大批量的可泵性、可运输性、可废弃性，对环境友好而又便宜。

大多数钻井液由水配制，称水基钻井液（WBM）。时常也用到由油配制的钻井液，称为油基钻井液（OBM）。油基钻井液的优势是能更好地润滑钻柱，在黏土层和盐层得到更高的采收率。柴油常用于油基钻

were formerly disposed of onto the seabed. This practice is no longer considered environmentally acceptable. If OBM or any other hazardous fluid is contained in the cuttings, a closed-loop mud system will be required. Thecuttings will be either decontaminated in a dedicated onshore plant or re-injected asa slurry into a suitable formation. New mud compositions and systems are continuously being developed, for instance synthetic drilling fluidswhich rival theperformance of OBM but are environmentally benign [e.g. synthetic oil-based mud(SOBM)].

井液的配制。在钻井过程中，大量污染的岩屑之前处理到海底。但现在认为对环境不利。油基钻井液或其他有害液体侵染岩屑，要求进行封闭钻井液循环。污染的岩屑要求在陆上工厂进行净化或混合成泥浆再注入合适的地层中。现在新的钻井液成分和系统不断发展，例如，合成钻井液其性能可比拟油基钻井液，但对环境却很友好。

A common drill mud is made of bentonite clay. Heavier drilling mud can be made by adding barite ($BaSO_4$). Various chemicals are also used in different situations. Drilling fluid is described by its weight. The heavier the drilling fluid, the greater pressure exerts on the bottom of the well.

另一钻井液常用物质是膨润土。较重的钻井液可以加重晶石（$BaSO_4$）。不同的化学剂适合不同的钻井情况。钻井液以其重量衡定。越重的钻井液产生越大的井底回压。

Air or pneumatic drilling is used in many shallow wells or in cases in which the drilling fluid would damage a subsurface reservoir rock. Air, gas and mist can be used as drilling fluid which often are used in nearbalance or underbalance drilling. Air is forced down the drill string similar to circulating drilling mud. The air mixes with water, which is always present on the bottom of the well, and form foam. The foam picks cuttings off the bottom of the well and returns through the annulus to the surface. The foam and cuttings blow out the side of the rig in a blooey line. Air drilling is faster and less expensive than mud drilling but has the disadvantages that it cannot control subsurface pressures and sometimes air mixes with subsurface gas, causing an explosive mixture.

空气钻井和充气钻井用于许多浅井，以防止钻井液对地层岩石的伤害。空气、气体、雾都可用作钻井液进行近平衡或欠平衡钻井。压缩空气进入钻柱如同循环钻井液。空气与井底时常存在地层水混合，形成泡沫，此泡沫携带岩屑脱离井底，通过环空空间上返到地面。泡沫和岩屑由钻机旁的放空管线排出。空气钻井比常规钻井液钻井更快更便宜，它的缺点是不能控制井下压力，有时空气与井下气体混合导致爆炸性混合气。

The choice of drilling fluid has a major impact on the evaluation and production of a well. After drilling a well, we often investigate the interaction between drilling fluids, logging operations and the potential damage to well productivity caused bymud invasion into the formation(Figure 4.54).

钻井液选择对评估和生产有很大关系。钻完井后，我们时常会调查钻井液、测井作业过程泥浆侵入地层对井生产能力的损害（图4.54）。

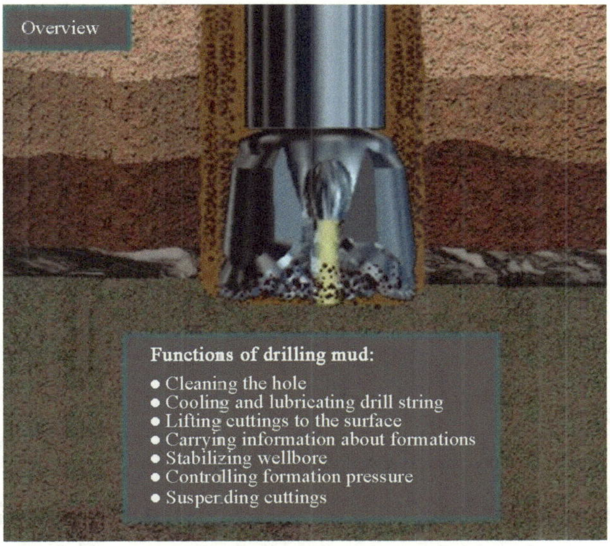

Figure 4.54 Functions of Drilling Mud
(Source: Drilling Fluid Functions)

4.5.4 Power System

The engines (or prime movers) supply the power to the rig(Figure 4.55). Most engines are diesel and they are often used to generate electrical power. Depending on the size of the rig and the drilling depth, there are two to four engines. On a big rig, the engines develop 3000 hp or more horsepower. The power is used primary to turn the drill string and raise and lower equipment in the well(Figure 4.56).

4.5.4 动力系统

引擎给钻机提供动力（图4.55）。绝大多数的引擎是柴油机，常用于产生电能。依据钻机和钻井深度的不同，一般有2~4组引擎。在大钻机上，引擎提供3000hp或更多的马力。动力主要用于提升和下落钻柱（图4.56）。

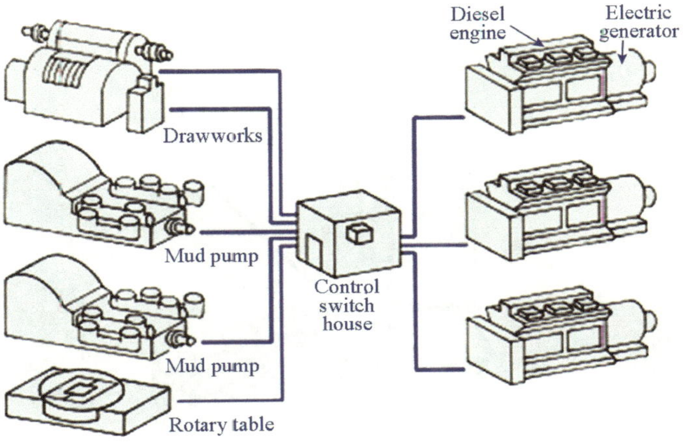

Figure 4.55 The Power System
(Source: http://ffden-2.phys.uaf.edu)

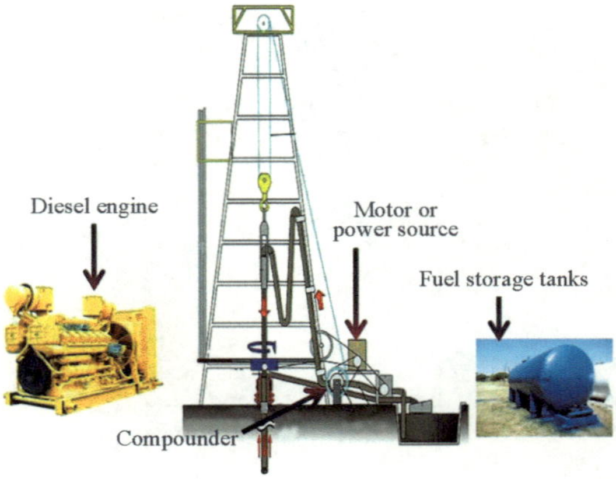

Figure 4.56　Power System
(Source: Overview of the Oil Industry, Richard Lau)

4.5.5　Pressure Control System

The last is Pressure Control System, it is principal component is Blowout Preventers (BOPs). BOP is also an important safety feature on every modern rig. BOPs are designed to close off the well. They are attached to the top of the well below the derrick floor(Figure 4.57). Two types are often used on the same well. One type is designed to close around the drill pipe to shut off the annulus. The other type closes the well by shearing off the drill pipe with rams.

4.5.5　压力控制系统

最后是压力控制系统，其主要部件是防喷器。防喷器也是每个现代钻机重要的安全必备。设计防喷器的目的是为了关井。防喷器安装在钻井平台下的井口处（图4.57）。应用最广泛的主要有两种防喷器，一种是抱死钻柱关闭环空关井；另一种是用闸门剪断钻柱关井。

Figure 4.57　Pressure control System
(Source: Overview of the Oil Industry, Richard Lau)

As discussed earlier on, one of the purposes of the drilling mud is to provide a hydrostatic head of fluid to counter balance the pore pressure of fluids in permeable formations(Figure 4.58). However, for a variety of reasons the well may ‘kick’, that is formation fluids are may enter the wellbore, upsetting the balance of the system, pushing mud out of the hole, and exposing the upper part of the hole and equipment to the higher pressures of the deep subsurface. If left uncontrolled, this can lead to a blowout, a situation where formation fluids flow to the surface in an uncontrolled manner.

如前所述，泥浆的一个作用是提高水头压力以平衡地层流体压力（图4.58）。但因各种原因，可能会出现井涌现象，即地层流体流入井筒破坏压力平衡，将泥浆挤出井口，地下高压下井涌会破坏井口装置。一旦失去控制，就会导致井喷，地层流体以不可控的方式流向地面。

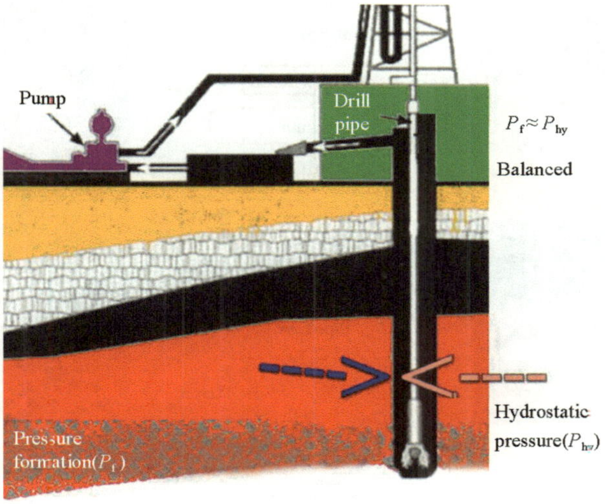

Figure 4.58 Balanced Pressure System
(Source: Overview of the Oil Industry, Richard Lau)

The BOPs are a series of powerful sealing elements designed to close off the annular space between the pipe and the hole through which the mud normally returns to the surface. By closing off this route, the well can be ‘shut in’ and the mud and/or formation fluids are forced to flow through a controllable choke, or adjustable valve. This choke allows the drilling crew to control the pressure that reaches the surface and to follow the necessary steps for ‘killing’ the well, that is restoring a balanced system. Figure 4.59 shows a typical set of BOPs. The annular preventer has a rubber sealing

防喷器是一系列用以关闭钻柱和井筒间环空的密封组件。打开防喷器后，井将被关闭，泥浆和地层流体将被迫从控制阀或调节阀流出。而钻井操作者可以通过控制阀控制流体到地表的压力直至最后一步的压井，以平衡流体压力。图4.59展示典型的防喷器组件，环形防喷器有一个橡胶密封部件，在水力压力下能膨胀密封

element that is hydraulically inflated to fit tightly around any size of pipe in the hole. Ram type preventers either grip the pipe with rubber-lined steel pipe rams, block the hole with blind rams when no pipe is inplace or cut the pipe with powerful hydraulic shear rams to seal off the hole.

BOPs are opened and closed by hydraulic fluid stored under a pressure of 3000 psi in an accumulator, often referred to as a 'Koomy' unit.

井眼钻柱。闸门类防喷器用水力剪切闸板密封井眼。

防喷器的开关靠Koomy储能器，当压力达到3000psi时实现开或关。

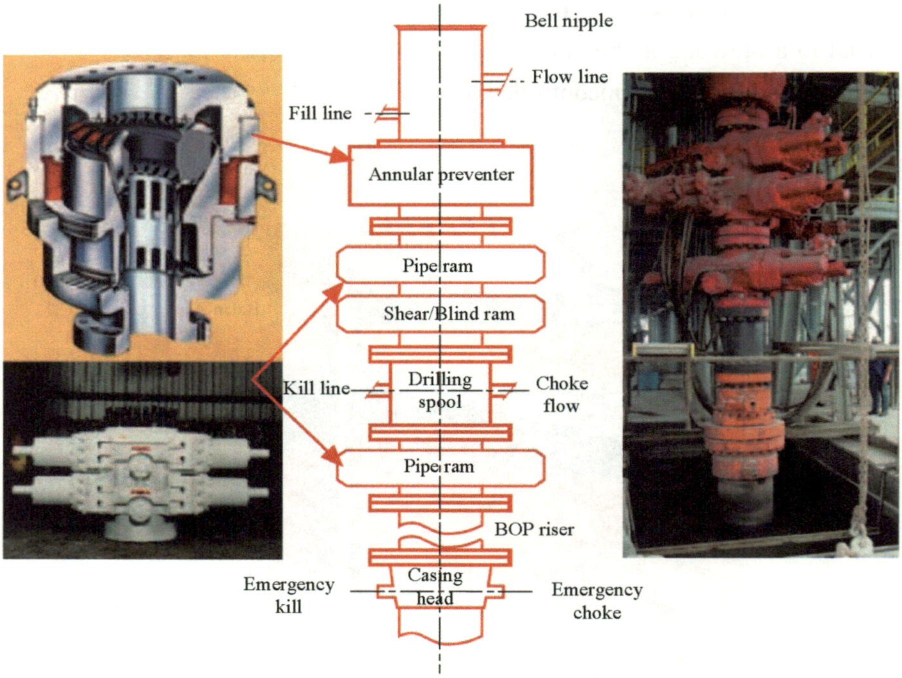

Figure 4.59 Blowout Preventers
(Source: Hydrocarbon Exploration and Production)

4.6 Routine Drilling Procedures

All of the efforts of the petroleum landmen, geologists and geophysicists, and everyone else involved in activities up to this point culminate in a decision by the management of the petroleum company to either drill or not drill a well. If the exploration activities conducted thus far have not found encouraging indications for the accumulation of petroleum in the area being explored, the area will likely be dropped. If exploration activities have been encouraging, however, a decision will likely

4.6 常规钻井程序

石油工人、地质学家、地球物理学家等人的所有努力都集中在钻井还是不钻井这个问题上。假如之前的勘探调查还没有发现工业油气，这块勘探区很可能会被放弃。如果发现有可能存在油气藏，石油公司将作出决定是否打勘探井。这项决定关乎着成百上千万美元

be made to drill an exploratory well. Such a decision may require the commitment of hundreds of thousands or even millions of dollars for the drilling program. Drilling a well is commonly called making-hole. In this section we will discuss the activities related to the drilling of an exploratory well.

的钻井投资。本节将详述打勘探井的过程。

4.6.1 Rigging Up

If the well is very shallow or medium depth, the entire equipment comes on the back of a truck or trailer. This is a truck-mounted or potable rig. The mast is jacked up, and secured for drilling. If the well is deep, the equipment is so massive that it comes on several tractor-trailers and has to be assembled on the site.

Rigging up is placing and assembling the various parts of equipment that make up the rig, and preparing the rig for drilling(Figure 4.60).

4.6.1 安装钻机

假如打浅井或中深井，钻机在卡车或拖拉机背后。这叫做车载钻机。立起井架为钻井做准备。假如是打深井，钻机可能巨大而笨重，钻机由卡车或拖拉机拖运过来现场组装。

安装钻机指安置、组装钻机等钻井设备，为钻井做准备工作（图4.60）。

Figure 4.60 Rigging Up
(Source: https://www.osha.gov/SLTC/etools/oilandgas/drilling/rigging_up.html)

4.6.2 Spudding In

When the engines, pumps, and rotating and hoisting equipment are aligned and connected, the drill pipe and drill collar are laid out, then the drilling contractor is ready to begin drilling operations (spud the well).

For the very first section of the borehole, a base from which to commence drilling is required. In a land location, this will be a cemented ‘cellar’ in which a conductor or stove pipe will be piled prior to the rig moving in. The cellar will later accommodate the ‘Christmas tree’ (an arrangement of seals and valves to control production), once the well has been completed and the rig has moved off location(Figure 4.61).

4.6.2 开钻

当引擎、泥浆泵、旋转设备和提升设备安置、连接好，钻杆、钻铤都摆放好，钻井方将开始钻井。

最初的井眼在钻井安装前开钻，钻到表土层以下的基岩，水泥固井后形成钻井的地基用于安装导管和采油树。钻井完成后，钻机将搬移井场（图4.61）。

Figure 4.61 Cellar with Christmas Tree on a Land Location
(Source: Hydrocarbon Exploration and Production)

Once the drill bit has drilled below the conductor the well is said to have been spudded. Spudding in, or to spud a well, means to begin drilling operations(Figure 4.62).

一旦钻头钻到导管下，就被认为开钻，正式开始钻井。开钻，即开始钻井，表明钻井作业正式开始（图4.62）。

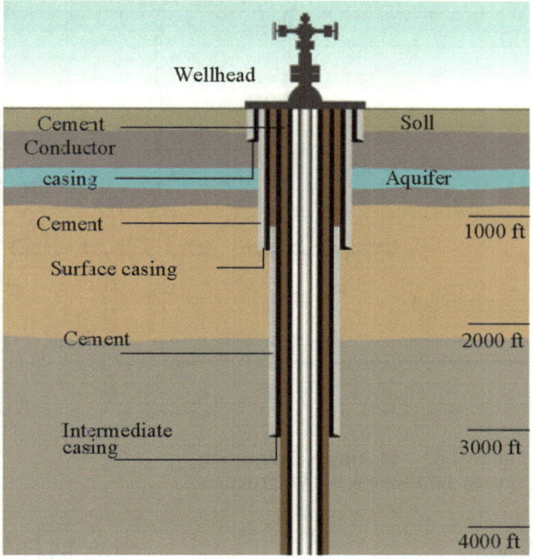

Figure 4.62 Well Casing
(Source: http://frackwire.com/wp-content/)

4.6.3 Drilling Ahead

Most of the rig's time is occupied by drilling ahead, so called turning to the right or making a hole. The bit is on bottom drilling, and this continues until the kelly joint is drilled down (upper coupling approaching the rig floor). Another joint of drill pipe is then added, called making a connection, and drilling resumes. When a top drive is used, the entire stand (90ft) is drilled down, and an entire new stand then added.

While drilling ahead, the driller watches the weight indicator closely, slacking off on the brake to lower the drill string and keep the proper weight on the bit. He is often alone on the rig floor while the roughnecks are occupied elsewhere with routine maintenance duties.

4.6.3 钻进

大部分钻井时间用于地层钻进，因此，叫打钻或打洞。钻头在井底连续钻进，直到方钻杆基本下降到钻台下面（上接箍接近钻台）。这时，需要在钻柱上接一根新的钻杆，称为接单根。然后继续向下钻进。当进行顶驱钻井时，可以在钻到一根立柱长（90ft）时，连接新的立柱。

在钻进过程中，司钻必须密切观察指重表，适当放松刹车保持钻压。在这一过程中，通常只有司钻一人在钻台上操作，其他场地工则在进行日常维护工作。

4.6.4 Making a Connection

Steel drill pipe comes in 30ft sections that are threaded on both ends. Each section of drill pipe is called a joint. The kelly must always be located on the top of the drill pipe. After drilling 30ft, the kelly must be raised and another joint of pipe added below the kelly(Figure 4.63).

4.6.4 接单根

方钻杆之下是钻杆，钢制钻杆一般为30ft一节，靠两端丝扣拧紧。方钻杆一直处于钻杆顶端，下钻30ft以后，提起方钻杆再加一根钻杆，这叫接单根（图4.63）。

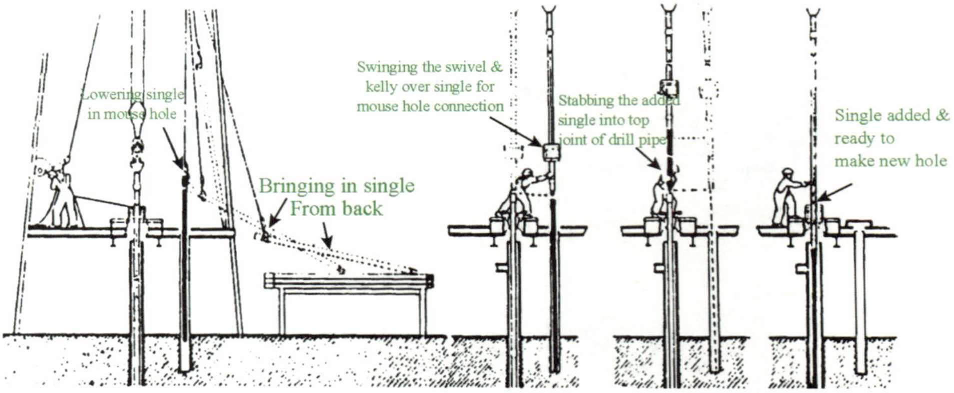

Figure 4.63 Making a Connection
(Source: http://www.netwasgroup.us)

When the Kelly has drilled down:

当方钻杆钻下后，将进行如下程序：

◆ The driller stops the rotary, raises the drillstring off bottom, and shuts down the mud pump.

◆ 停止钻台旋转，提升钻柱，停止钻井泵；

◆ The floorhands set the slips and driller slacks off (lowers) the drillstring until it is suspended from the slips.

◆钻工打卡瓦，司钻放管柱直到卡瓦坐住并悬挂钻柱重量；

◆ Using tongs, the floorhands break out the Kelly from the drill string.

◆ 用液压钳卸下方钻杆；

◆ The Kelly assembly (which includes the Kelly, Kelly bushing, swivel, and rotary hose) is swung over to the mousehole, stabbed into the joint of drillpipe standing in the mouse hole, and the joint is made up.

◆ 方钻杆总成（包括方钻杆、方钻杆补心、水龙头和水龙带）被荡至小鼠洞上，接入到小鼠洞中的一根钻杆上，下一步进行接单根操作；

◆ The Kelly assembly with the new joint attached is pick up, swung over to the drillstring, stabbed, and made up.

◆ 方钻杆总成连同新接上的单根钻杆被连续接到钻柱上；

◆ The string is picked up, the slips pulled, the mud pumps started, the bit run to bottom and rotation resumed.

◆ 提升钻柱，取下卡瓦，启动钻井泵，下放钻头，继续旋转钻进。

4.6.5 Making a Trip

4.6.5 起下钻

Bits wear out after 8 to 200 hours of rotation, with an average bit wearing out after 24 hours. A worn bit can be detected by the noise on the derrick floor that the rotating drill pipe makes and by a decrease in rate of drill penetration. Making a trip is necessary for changing the bit. All the pipes are pulled out of the hole and stacked

钻头的平均寿命为24h，一般旋转8~200h后需要更换钻头。钻头的磨损情况可以通过钻井平台上的噪声音和钻速发现。更换钻头时必须要起下钻，需要把所有的钻柱提出并

in the derrick. The bit is then changed and the pipe is put back into the hole. This takes rig time and costs money. The deeper the well, the longer the trip takes. Making a trip is also called tripping. Major improvements in the durability of bits and the formulation of drilling fluids have greatly reduced the number of trips required to drill a well. Many shallow wells today are drilled without a bit change.

堆放在井架旁，换上新钻头后再下放到井内。这样既费时又费钱。井越深，花的时间越多。钻头耐磨损程度和泥浆性能的提高能减少钻井费用，今天很多浅井都是一根钻头钻完而不需要换钻头。

Ripping out means pull out the drilling string of the hole:

◆ The slips are set, the Kelly is broken out, the assembly is set back in the rathole, and the swivel is unlatched from the hook;

◆ The elevators attached to the hook are lowered and latched onto the drill sting below the box that is "looking up";

◆ The string is picked up the length of a stand (usually three joints) and the slips reset;

◆ The floorhands break out the joint, push the pin end of the stand off to the side, and set it down on wooden sills;

◆ The derrickman unlatches the elevators and leans the stand back against the fingerboard. The elevators are run back down to pull another stand and the process repeated until the entire string is racked in the derrick.

起钻指将钻柱从井中起出，包括：

◆ 打卡瓦，卸下方钻杆总成到大鼠洞中，将水龙头从大钩上取下；

◆ 吊卡接在大钩下面，卡在最上一根钻杆的接头下方；

◆ 向上提升一根立柱的距离，打卡瓦；

◆ 卸下立柱，将带外螺纹的立柱下端推向一旁，放置在木板上；

◆ 卸下吊卡，将立柱靠放在指梁上。接着下放吊卡继续起管，直到全部钻杆起出，卸下并码好。

Tripping in or run in the hole means the process is reversed to go back into the hole. The derrickman latches the elevators on the stands, the stands are picked up, stabbed, and made up into the string.

下钻指将管柱重新下放到井眼中。钻工用吊卡卡住立柱，吊起、对接、上扣。

The handing of each stand is a complex operation involving many sequential steps, and the sequence needs to be repeated over and over. A rhythm develops, requiring split-second timing between the driller, the floor hands, and the derrick man. It's actually safer to work quickly in this smooth, deft, manner with complete concentration than at a slower pace, allowing the mind to wander(Figure 4.64, Figure 4.65).

对每根钻柱的操作都要经过若干步骤，在作业过程中，这些步骤需要一遍一遍的重复。这一操作需要在操作人员中间形成快速准确的节奏，这样不仅有效率而且可以增加安全性（图4.64、图4.65）。

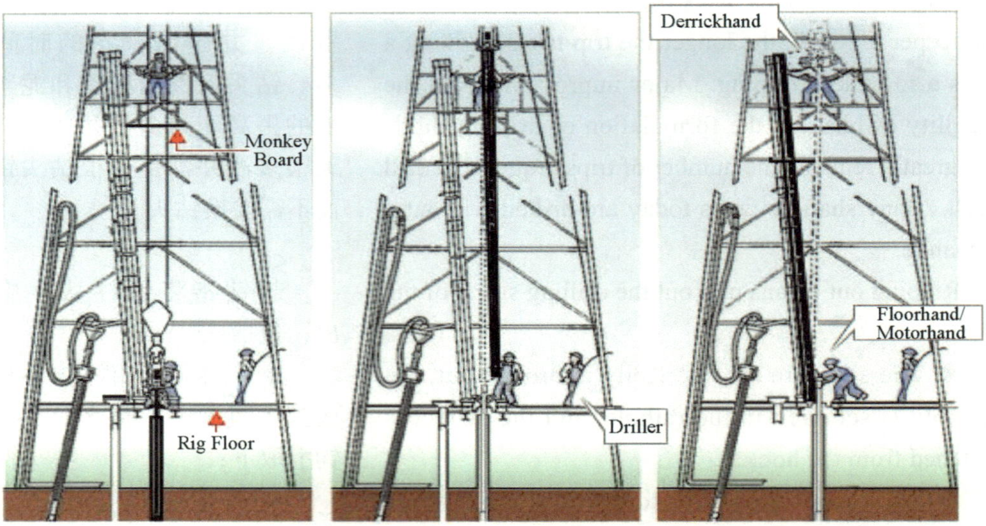

Figure 4.64 Drilling Rig Crews
(Source: http://www.rigtech.ca/live/Green_Hands)

Figure 4.65 Making a Trip
(Source: http://cmsimg.stargazette.com/)

If the string breaks or gets stuck in the hole, a specialist is called in to help the crew go fishing with special tools. No one wants to lose an expensive bit and bottom-hole assemble, but the blocked hole is the real problem. As a last resort, the crew drills a curved section called a sidetrack to bypass the debris(Figure 4.66、Figure 4.67).

假如钻柱折断或发生卡钻，修井专家会被叫来打捞落鱼。没人想要损失昂贵的钻头和井底设备，但是堵住的井眼确实是一个大麻烦。最后一招是从旁边钻一口井绕过障碍（图4.66、图4.67）。

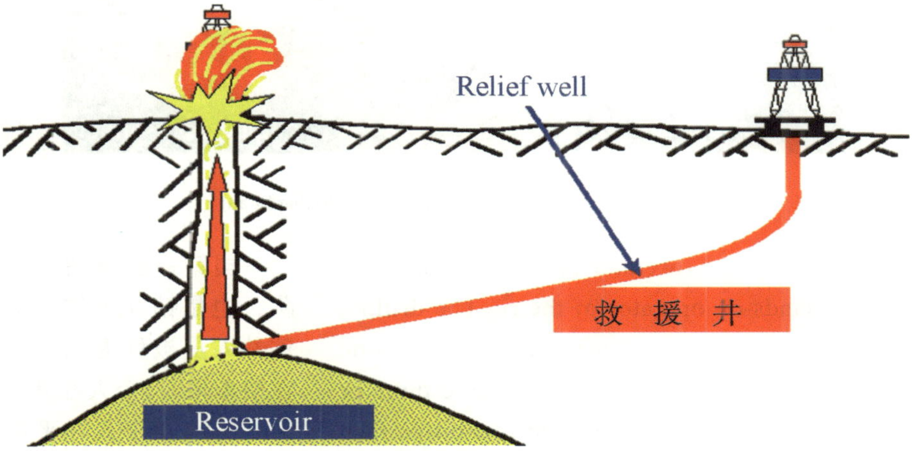

Figure 4.66 A Relief Well

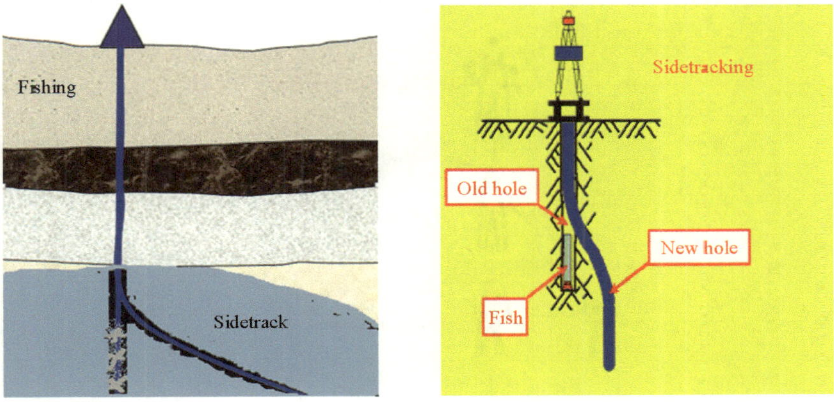

Figure 4.67 A Sidetrack to Bypass the Debris

4.6.6 Drilling Crews

The supervisor is known as the rig manager or “toolpush”. Crews include driller, derrickman/ mudmen, motormen (who maintain motors), floorhands (roughnecks) and leasehands (laborers)(Figure 4.68).

4.6.6 钻井作业人员

钻井监理一般为井场经理。井场上包括司钻、井架工、电工、平台工和井场工等（图4.68）。

Figure 4.68 Drilling Crew
(Source: Overview of the Oil Industry, Richard Lau)

The driller is the foreman of the drilling crew and also the hands-on operator of the rig.The drilling crew are called roughnecks.During a roundtrip, one roughneck function as the derrickman, while three or four function as floor hands; Between roundtrips they have varied duties checking and maintaining all the rig equipment(Figure 4.69).

司钻一般是钻井人员的领队，也是钻井操作的第一人。钻井人员被称为钻工，平台工在接单根时1个人在井架上接，另外3个或4个人在平台上装。在钻井作业中，负责检查和维护整个钻机运行（图4.69）。

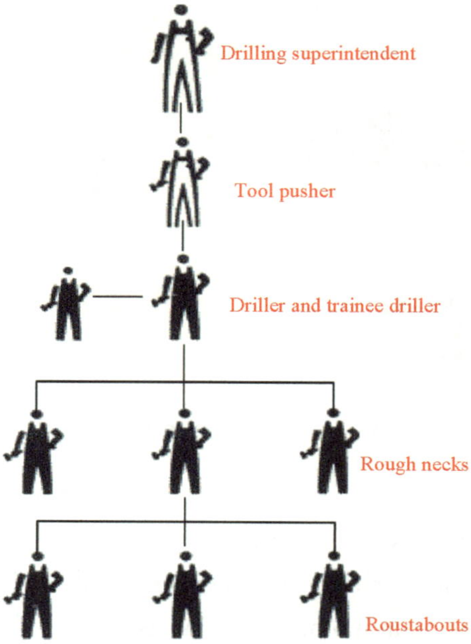

Figure 4.69 Drilling Crew (Source: http://thediagram.com/4_1/drilling_crew)

In addition to the drilling crews, drilling operations require a number of specialists for mud engineering, logging, fishing, etc., not to forget maintenancecrews, cooks and cleaning staff. A well site geologist monitors well cutting, examines cores and identifies potential reservoirs(Figure 4.70).

除了钻井相关人员外，井场上还有各种专家，泥浆工程师、测井工程师、打捞落鱼专家，还有其他维护人员及厨师、清理人员等。驻井地质师监测岩屑、检查岩心和识别潜在油气藏（图4.70）。

Figure 4.70 Drilling Rig Crews
(Source:http://www.rigtech.ca/live/Green_Hands/Drilling_Rig_Crews)

All drilling activities will be carried out around the clock, seven days a week, and the crews usually works in 8 or 12 hours shifts for two weeks and then have a week off. It is common to have some 40~90 people on site,depending on the type of rig and the location. The operation is managed on site by a 'Company Man' who represents the operator and a 'Rig Manager' who is the drilling contractor's representative.

所有的钻井活动都实行倒班制，每周工作7天，每班工作8个小时或12个小时，每两周换班，休息一周。井场上一般为40~90人，根据钻机和井场位置不同而不同。井场上的操作由甲方代表和井队的乙方代表管理。

4.7 Non-Vertical Wells Drilling

4.7 非直井钻井

4.7.1 Directional Drilling

Directional drilling is usually done with a rotary steerable system. A downhole steering and control unit is located in the near-bit assembly. A set of small electronically controlled rotating stabiliser pads (actuators) exert a continuous directional force onto a drive shaft which orients the drill bit into the desired direction(Figure 4.71) . The drill string is rotated at the same time, allowing hole cleaning. A control unit near the bit ensures that the hole angle is not increased

4.7.1 定向井钻井

定向井一般由底驱系统钻得。底驱系统在近钻头部位有方向控制件。一套小型电控旋转稳定器持续施加作用力导航方向（图4.71）。同时，钻柱旋转清理井筒。近钻头的控制件保证井眼角度不增加和降低过快，以免形成狗腿导致多余的扭转力和拉伸力。钻头附近的旋转导向系统配合测井工具

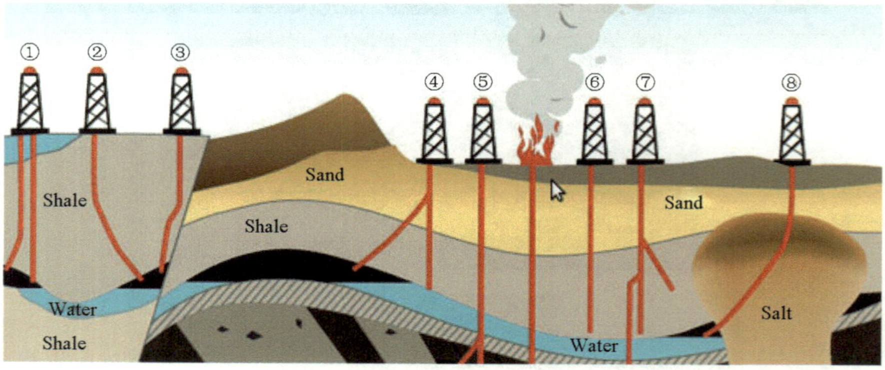

Figure 4.71 Examples of Directional Drilling
(Source:drilling-group4-directionaldrilling-1.wikispaces.com)

①–Multiple Wells Pffshore; ②–Inaccesible Location; ③–Straightening & Sidetracking; ④–Shoreling Drilling; ⑤–Stratigraphic Traps; ⑥–Salt Dome ⑦–Fault Control; ⑧–Relief Well;

or decreased rapidly creating ‘dog legs’ which will result in excessive torque and drag. The rotary steerable system is combined with logging tools in the drill string close to the bit, allowing a continuous optimisation of the well trajectory(Figure 4.72).

Mud turbines and mud motors are also used for directional drilling(Figure 4.73).

Advances in drilling and completion technology today allow us to construct complicated wells along 3D trajectories. In addition to vertical wells, directional drilling allows us to build, maintain or drop hole angle and to turn the drill bit into different directions. Thus, we are able to optimise the wellpath in terms of reservoir quality, production or injection requirements.

允许钻得最优的井眼轨迹（图4.72）。

泥浆涡轮和泥浆泵也用于定向钻井（图4.73）。

今天，在钻井和完井方面的进步帮助我们打造三维立体井眼轨迹。相比直井，定向井让我们增加、保持或降低井斜角，使钻头沿不同的方向钻进，从而优化油藏质量、产量和注入方式。有时候是地表（如绕过障碍）需要特殊的井眼轨迹，有时候是地下（如薄气层、断层、透镜体油

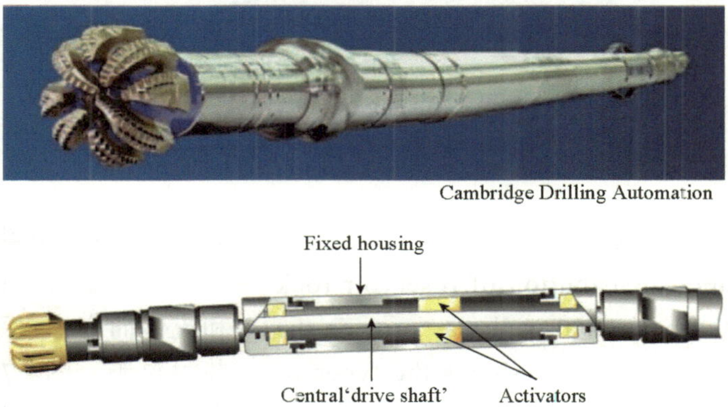

Figure 4.72 Rotary Steerable System
(Source: Hydrocarbon Exploration and Production)

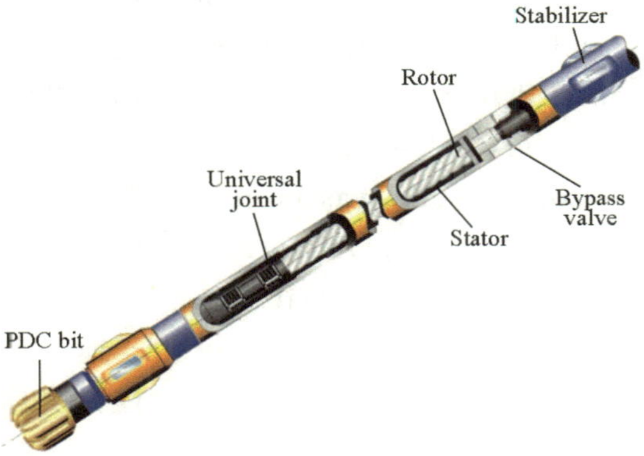

Figure 4.73 Mud Turbine
(Source: Hydrocarbon Exploration and Production)

Sometimes constraints at the surface (e.g. built-up areas) or subsurface (e.g. shallow gas, faults, lenticular reservoirs) may require a particular well trajectory to be followed(Figure 4.74).

藏）需要特殊的井眼轨迹（图4.74）。

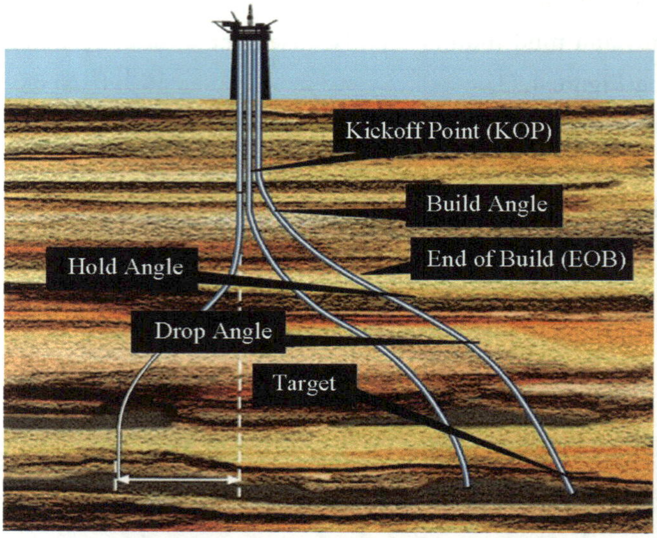

Figure 4.74 Directional-drilling Well
(Source: http://tntenergy.ca/directional-drilling/)

4.7.2 Horizontal Drilling

Given the lateral distribution of reservoir rock or reservoir fluids, a horizontal well may provide the optimum trajectory. Figure 4.75 shows the types of horizontal wells being drilled. The build-up rate of angle is the main distinction from a drilling point of view. Medium radius wells are preferred since they can be drilled, logged and completed with fairly standard equipment. The horizontal drilling target can be controlled within a vertical window of less than 2 m.

The success of horizontal wells was largely dependent on the development of tools which relay the subsurface position of the drill bit in real time to the drill floor. Improvements in this technology have greatly improved the accuracy with which well trajectories can be targeted. MWD is achieved by the insertion of a sonde into the drill string close to the bit.

4.7.2 水平井钻井

对于横向分布的油藏储层和流体而言，水平井是最优的井眼轨迹。图4.75展示了不同类型的水平井。从钻井上看，水平井类型的不同主要在于它们增斜角的不同。中等半径的井应用最广，因为它们可以用常规的钻进工具完成钻井、测井和完井。水平井的钻井靶点可以控制在小于2m的垂直窗内。

水平井的成功主要归结于地下实时定位工具将钻头的位置及时传给钻井平台。此项技术的提高极大提升了钻头导向的准确性。随钻测量是将探

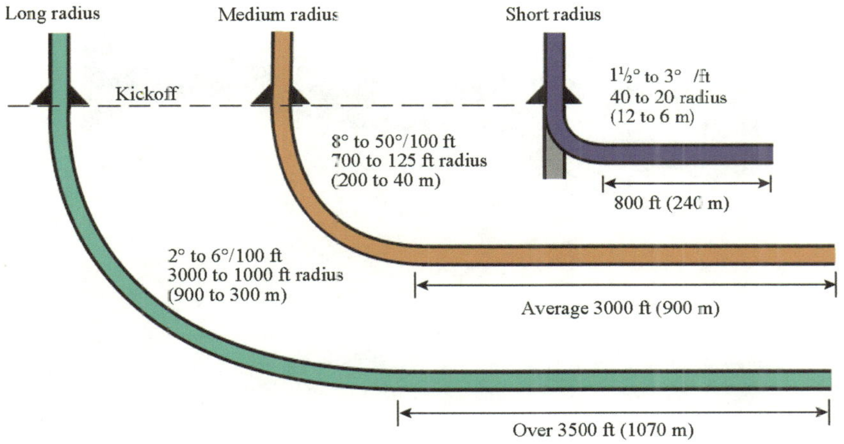

Figure 4.75 Horizontal Well Radii
(Source: Hydrocarbon Exploration and Production)

Initially providing only directional data, the tools have been improved to the point where petrophysical data gathering [gamma ray (GR), resistivity, density and porosity] can be carried out whilst drilling(Figure 4.76, Figure 4.77).

头感应器插入钻头后部。刚开始，探头只能提高方向数据，现在探头已经提升到边钻边收集地层物理数据（伽马射线、电阻率、密度和孔隙度）的程度（图4.76、图4.77）。

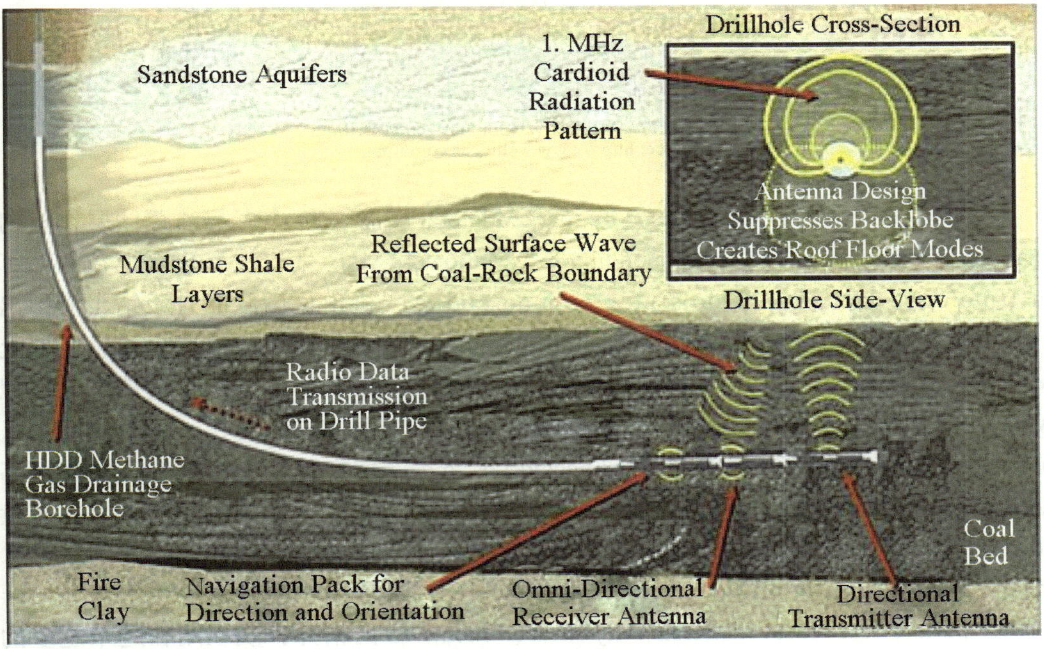

Figure 4.76 Measurements While Drilling for Horizontal Drilling, Navigating, and Structure Detection
(Source: http://www.netl.doe.gov/technologies/oil-gas/Petroleum/)

Figure 4.77　MWD/LWD for Horizontal Drilling
(Source: http://www.wellideas.com/images)

Most reservoirs are characterised by marked lateral changes in reservoir quality corresponding to variations in lithology. Computing tools now commercially available allow the modelling of expected formation responses ‘ahead of the bit’. This is possible in areas where a data set of the formations to be drilled has been acquired in previous wells. The expected GR and density response is then simulated and compared to the corresponding signature picked up by the tool. Thus, in theory, it is possible to direct the bit towards the high-quality parts of the reservoir. Resistivity measurements enable the driller to steer the bit above a hydrocarbon

绝大多数油藏储层在横向上产生岩性变化。现在的计算机模型已经能先于钻头模拟出下一地层的情况。在感应器上接受的伽马和密度信号与模拟反应进行对比，随时更新。从而在理论上指导钻头钻进更高质量的油藏区块。电阻率测量帮助司钻导向，确保钻在油水界面之上。这种技术叫地质导向钻井，现在用在越来越多的油田开发优化中。地质导向钻

water contact (HCWC), a technique used, for example, to produce thin oil rims. These techniques, known as geosteering, are increasingly being applied to field development optimisation. Geosteering also relies on the availability of high-quality seismic and possibly detailed palaeontological sampling.

The benefit of horizontal wells is that they optimize the exchange surface between the well and the reservoir rock. Since the drainhole is much longer in the reservoir rock, horizontal wells can produce up to ten times more than vertical wells(Figure 4.78, Figure 4.79).

井依赖高质量的地震资料和详细的古生物岩心分析。

水平井最大的好处在于增加了储层与井之间的泄油面积，因为在储层岩石中泄油面积增大、增长，从而提高了产量，水平井产量一般是直井产量的10倍或更多（图4.78、图4.79）。

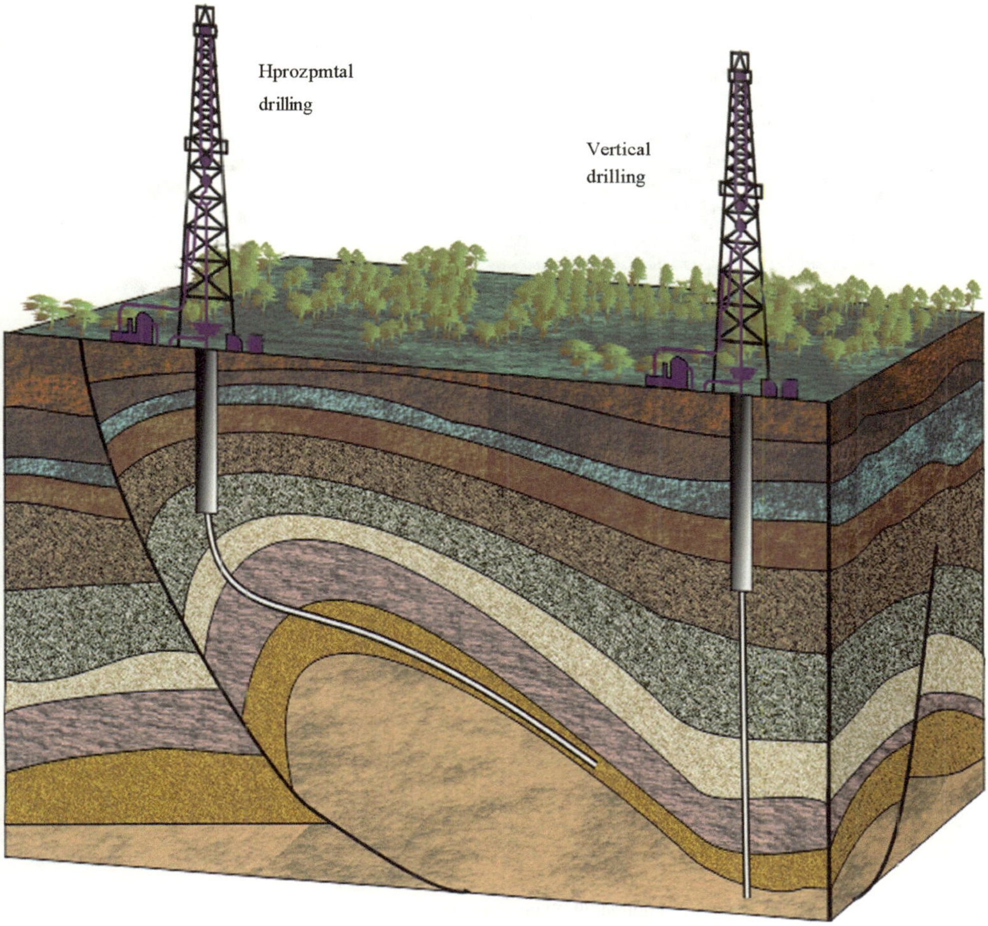

Figure 4.78 Vertical Well and Horizontal Well
(Source: http://www.intragaz.com/en/geophysics_drilling.html)

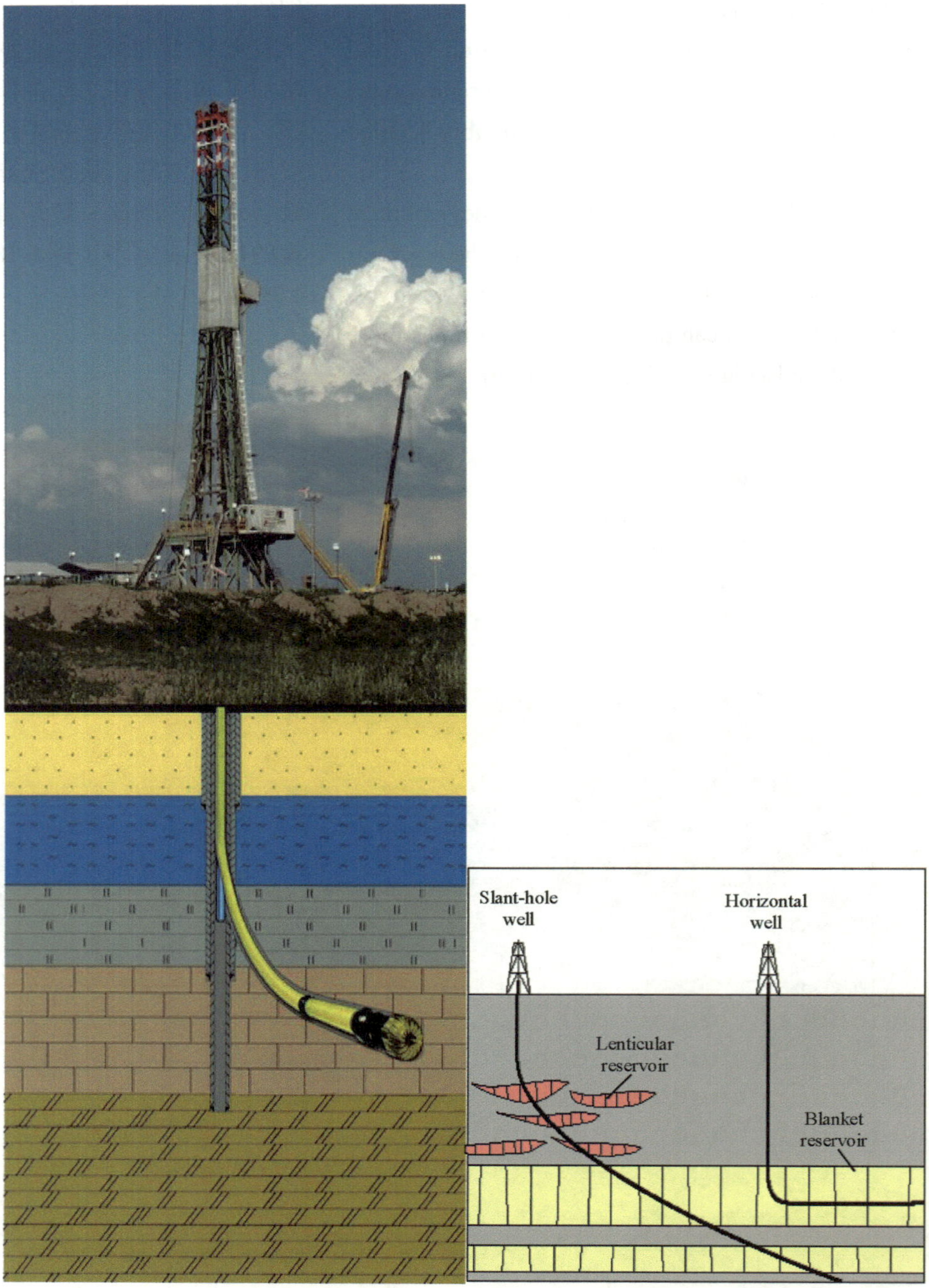

Figure4.79 Slanted and Horizontal Well
(Source: http://www.naturalgas.org/images/)

4.7.3 Multilateral Wells

Drilling a number of holes, branching from a central borehole, is an attractive option in the following cases:

◆ where reservoir productivity is low but can be significantly improved by increasing the reservoir surface area exposed to the well (Figure 4.80);

4.7.3 多支井

多分支井指从主轴井眼分支出多条井道，以适应以下情形：

◆ 当油藏生产能量低时，多分支井能提高产量（图4.80）；

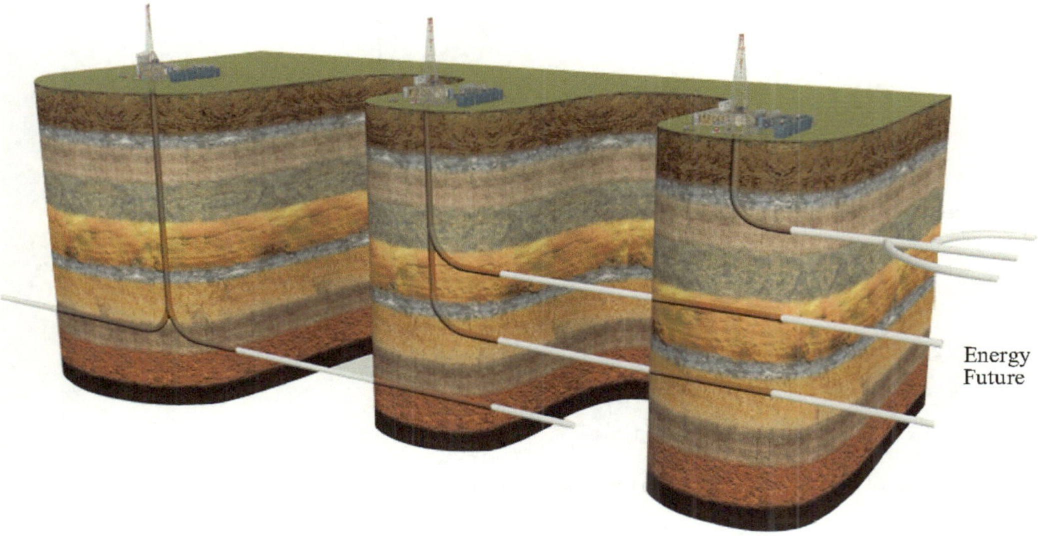

Figure 4.80 Drilling of Multilateral Wells
(Source: http://www.energy-future.com)

◆ in reservoirs which are lenticular;

◆ where reservoir layers are vertically segregated by permeability barriers.

Whilst drilling and, in particular, completing is more complex, multilateral wells have the advantage that they only require one borehole from the surface. If there are no spare conductors on the platform, this is an attractive option. In mature fields, multilaterals are often best suited to drain remaining pockets of hydrocarbons. In subsea developments, multilaterals can offer a substantial cost advantage over conventional wells(Figure 4.81, Figure 4.82).

◆ 油藏是透镜体油藏；

◆ 油层被不渗透障碍垂直分隔。

多分支井在钻井和完井方面都更复杂，但它的优势在于能从井底多渠道采油。假如在地面上没有多余位置安装导管，这是一个不错的选择。在一些老油田，多分支井最适宜采口袋油。在海上，多分支井比传统井更有回报（图4.81、图4.82）。

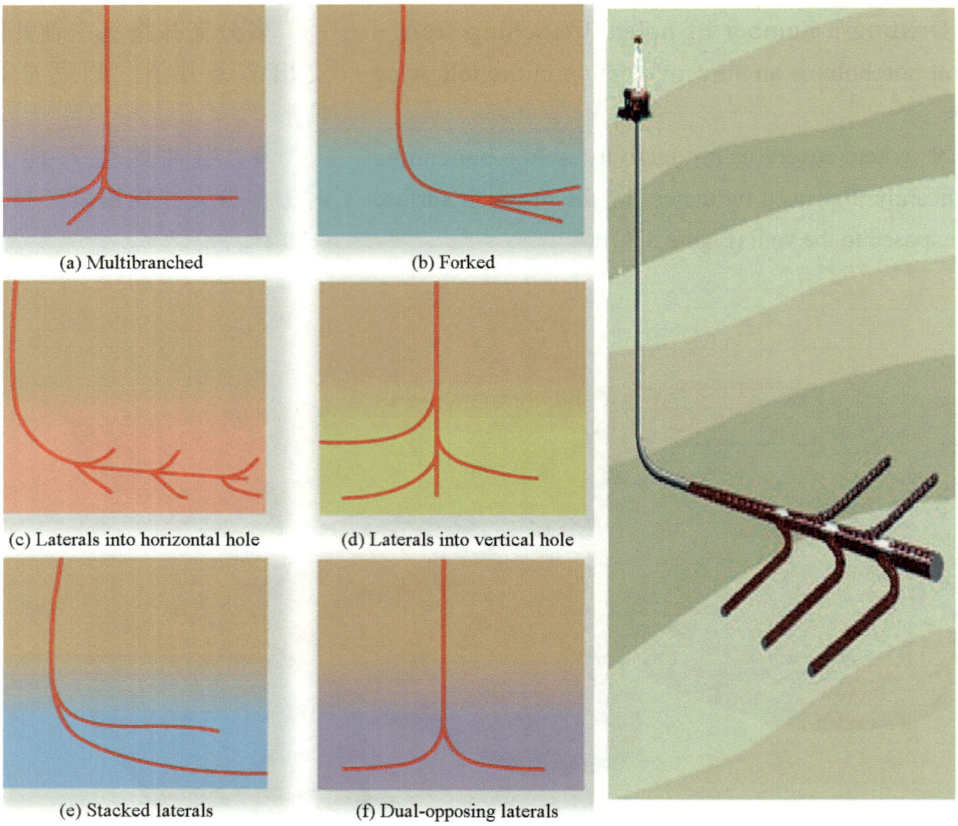

Figure 4.81 Multilateral Well
(Source: http://www.ccgc.cn/art/2012/12/28/art_541_19165.html)

Figure 4.82 Model of Multilateral Well to Increase Reservoir Contact
(Source: Hydrocarbon Exploration and Production)

4.7.4 Extended Reach Drilling

An extended reach well is loosely defined as having a horizontal displacement of at least twice the vertical depth. With current technology, a ratio of over 4 (horizontal, displacement/vertical depth) can be achieved.

Extended reach drilling (ERD) wells are technically more difficult to drill and because of the degree of engineering required for each well, the term ‘designer well’ is frequently used.

ERD will be considered:

◆ where surface restrictions exist;

◆ where marginal accumulations are located several miles from existing platforms/clusters;

◆ where ERD allows a reduction in the number of platforms required.

The high deviation (often up to 85°) and the long horizontal displacement expose the drill string to extreme drag and torque. Hole cleaning (cutting removal) and cementing of casing is more difficult due to the increased effect of gravity forces compared to low-angle wells. Thus, ERD wells usually require heavier and better equipped rigs compared to standard wells, and take longer to drill. Top drive systems are routinely employed in combination with rotary steerable systems.

Not surprisingly, costs are several times higher than conventional wells. Nevertheless, overall project economics may favour ERD over other development options. For example, BP developed the offshore part of the Wytch Farm Oilfield (which is located under Poole Harbour in Dorset, UK) from an onshore location. The wells targeted the reservoir at a vertical depth of 1500m with a lateral displacement of over 11,000m (Figure 4.83). The alternative was to build a drilling location on an artificial island in Poole Bay. ERD probably saved a considerable amount of money and advanced first oil by several years.

Offshore, subsea satellite development may be a viable alternative to ERD wells.

4.7.4 大位移井钻井

大位移井的简单定义是水平段至少比垂直段长2倍以上的井。目前的技术甚至可以达到4倍以上。

大位移钻井技术更难，需要专业的井身设计者。

大位移井可用在：

◆ 地表有障碍存在；

◆ 位于钻井平台远处的边际油田；

◆ 钻井平台不足的情形下。

大斜度（通常超过85°）长水平段使钻柱承受极大的拉伸和扭转力。相比小曲率半径的水平井重力影响增大，井筒清理（排屑）和注水泥非常困难。因此，相比标准井，大位移井通常需要更大、更强的钻机，更长的时间钻井。顶驱和地质导向系统也常用于大位移钻井。

不足为奇，大位移井比常规井耗时几倍以上。然而，从项目总的经济来看大位移井却是开发油田的好选择。例如，BP公司开发的海上油田Wytch Farm（位于英国Dorset，Poole港）。从陆上打大位移井延伸11000m到达垂深1500m的目标油藏（图4.83）比在Poole湾建人工岛钻井省钱多了，更早收回投资。

在海上，可用卫星定位帮助大位移钻井。

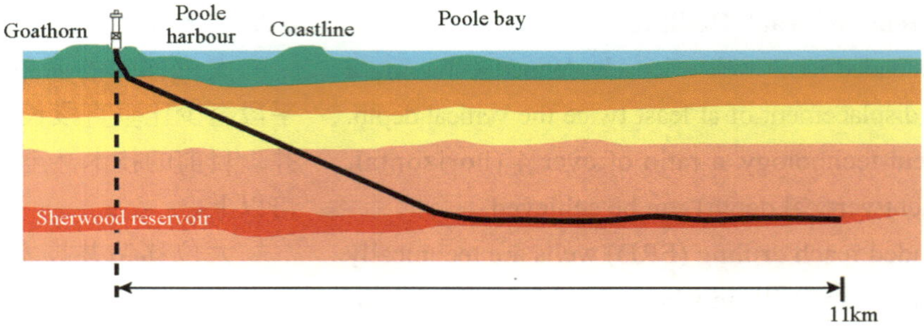

Figure 4.83 Extended Reach Drilling (BP, Wytch Farm).
(Source: Hydrocarbon Exploration and Production)

4.7.5 Coiled Tubing Drilling

A special version of slim hole drilling has emerged as a viable alternative: coiled tubing drilling (CTD) (Figure 4.84). Whilst standard drilling operations are carried out using joints of drill pipe, CTD employs a seamless tubular made of high-grade steel. The diameter varies between $1^3/_4$ and $3^1/_2$ in. Rather than being segmented, the drill string is reeled onto a large-diameter drum.

4.7.5 连续油管钻井

相对常规钻井现在有一种特殊的可行性选择——小井眼钻井方式，叫连续油管钻井（图4.84）。常规的钻井操作需要连接一根根钻杆，而连续油管钻井使用一根高钢级无缝油管连续钻进，中途无需接单根作业。连续油管直径为 $1^3/_4$in或$3^1/_2$in，缠绕在大直径鼓盘上，而不像常规钻杆分段堆放。

Figure 4.84 Coiled Tubing Drilling Unit
(Source: Hydrocarbon Exploration and Production)

Compared with conventional techniques using full-scale rigs and standard drill pipe, CTDcan significantly reduce environmental impact. CTD is mainly a reentry drilling service that enables operators to more effectively find hydrocarbon pockets still untapped in the reservoir. Reentry wells reduce the area at the surface impacted by drilling, but also reduce the volumes of steel, cement, and drill cuttings created in the drilling process(Figure 4.85).

相比传统钻井用全尺寸钻机和标准钻杆，连续油管钻井能显著减少环境破坏。连续油管钻井能更有效地发现油藏未开发的口袋油。此外，连续油管钻井能减少钻井对地表的影响，也减少了钻井进过程中套管用量、注水泥量和岩屑量（图4.85）。

Figure 4.85 Coiled Tubing Drilling
(Source: http://www.kingroyalty.com/)

Coiled-tubing technologies replace the traditional rigid, jointed drill pipe with a long, flexible coiled pipe string. This can reduce the cost of drilling, as well as provide a smaller environmental footprint. In particular, less drilling mud is required, which reduces the use of nonrenewable resources and minimizes the potential for unplanned releases. Coiled tubing also employs slim

连续油管用一个长长的可弯曲钻杆代替传统僵硬的需不断接单根的钻杆。这样能减少钻井成本和环境破坏。特别是减少钻井液的用量，从而减少不可再生资源的用量和计划外排放量。连续油管用小井眼

hole drilling techniques to achieve very cost-effective drilling and less impact on the environment. In addition, because drilling operations are not interrupted for pipe connections, CTD can use "closed" mud circulation techniques, which reduce the risk of spills and blowouts.

钻井技术更有效地钻井，对环境影响也更少。另外，因为钻井操作不被接单根作业打断，因此可以使用封闭泥浆循环技术，从而减少泥浆外溢和外喷的风险。

The advantages of CTD are several:

◆ nearly no pipe handling;

◆ better well control allows at-balance or even underbalanced drilling, resulting in higher penetration rates and reduced potential for formation damage;

◆ less environmental impact;

◆ lower cost for site preparation, lower day rates, lower mobilisation and demobilisation costs;

◆ easier completion by using the computerised tomography (CT) as a production string.

连续油管钻井的优点有以下几个：

◆ 不用添加钻杆，中途无需停下来进行接单根作业；

◆ 在近平衡甚至欠平衡钻井过程中更好地控制井压力从而有更高的钻速，更小的地层伤害；

◆ 对环境影响更小；

◆ 降低经常准备费用，减少钻井时间，降低人员开支；

◆ 连续油管作为生产管柱利于电脑成像完井。

However, CTD is limited to slim holes, and the reliability of some of the drill string components such as downhole motors is still improving. Presently, the cost of building a new customised CTD rig limits the wider application of this technology.

然而，连续油管还仅限于小直径井眼，受限于井下马达性能，钻井可靠性有待加强。目前，特制一套新的连续油管的花费制约着该技术的应用发展。

FORMATION EVALUATION
5 地层评价

Drilling a well not only provides an opportunity to produce oil or gas, it is also an opportunity to collect information about the formations being penetrated. In fact, expendable holes are often drilled solely to gather information with no intent to ever complete them. From the gather information we will decide to abandon or complete the well. So we need formation evaluation(Figure 5.1, Figure 5.2).

钻井不仅提供一个生产油气的通道，也为收集钻遇地层的相关数据提供一个机会。实际上，有大量价值不菲的井眼仅仅是为了收集地下数据而根本没打算完井。所以，需要开展地层评价（图5.1、图5.2）。

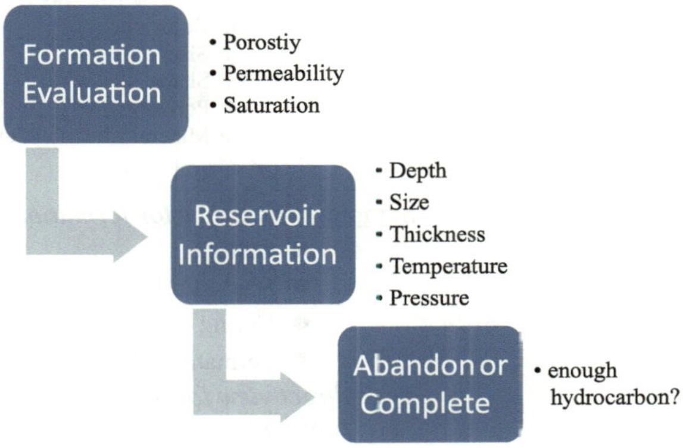

Figure 5.1 Why We Need Formation Evaluation

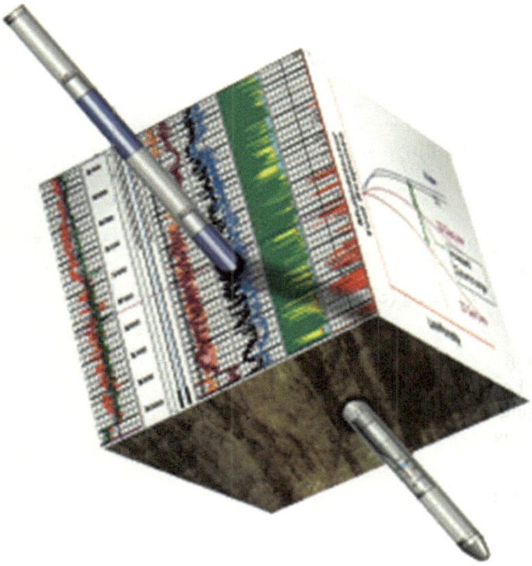

Figure 5.2 Formation Evaluation
(Source: http://www.halliburton.com/en-US/ps/)

Measure: *indirect, partial interpretive view for formation characteristics*

- Driller
  - ✓ Drill Time
  - ✓ Drill Rate
  - ✓ Rate of Penetration (ROP)
  - ✓ Weight on Bit (WOB)
  - ✓ Hookload
  - ✓ Torque in Drill String
  - ✓ Rotary Speed (rpm)
  - ✓ Pump Pressure and Rate
- Mud Logger
  - ✓ Mud Characteristics
    - ✧ Mud weight in and out of the hole
    - ✧ Volume of mud in the tanks
    - ✧ Viscosity
    - ✧ Weight
    - ✧ Salinity
  - ✓ Oil and Gas Show
  - ✓ Gas Logging
  - ✓ Fluorescence Logging
  - ✓ Sample Description
- Site Geologist
  - ✓ Rock Characteristic (Lithology)/Stratigraphic Column
    - ✧ Color
    - ✧ Size
    - ✧ Shape
    - ✧ Porosity
    - ✧ Mineral Identification
  - ✓ Correlation

(Driller, Mud Logger, Site Geologist → Drilling Operation Log (Mud Log))

Sample: *direct view for formation characteristics*

- Cutting
- Mud Return
- Coring
- Formation Fluid

Test: *direct view for formation characteristics, fluid pressure, fluid volumes*

- Drill Stem Test (DST)
- Wireline Formation Test (WFT)
- Openhole Well Log
  - ✓ E-Logging
    - ✧ Electric Logging (Resistivity)
    - ✧ Micrologging
    - ✧ Induction Electric Logging (SP)
  - ✓ Acoustic Logging
  - ✓ Radioactivity Logging
    - ✧ Gamma Ray Logging
    - ✧ Neutro Logging
    - ✧ Density Logging
  - ✓ Caliper Logging
- Initial Potential (IP) Test

(DST, WFT, Openhole Well Log → Open Hole; Initial Potential (IP) Test → Cased Hole)

The data is gathered at different stages(Figure 5.3). Data from cores, drill-stem tests, and mud logs is obtained while the hole is actually being drilled. Open-hole well logs are run after total depth is reached but before the casing is run. Reservoir sampling takes place after the well is completed and equipped, but before it is produced. The 24-hour initial potential test at the start of production is the final piece of data from a new well.

地下信息是分不同阶段收集的（图5.3）。在钻进过程中，同时收集岩心和中途试井、泥浆测井的数据。裸眼测井是当钻达最终深度而又没有下套管时进行的。油藏流体取样是在完井和装配后，正式生产前进行的。24h初始产能试井是新井开始生产前最后一项测试。

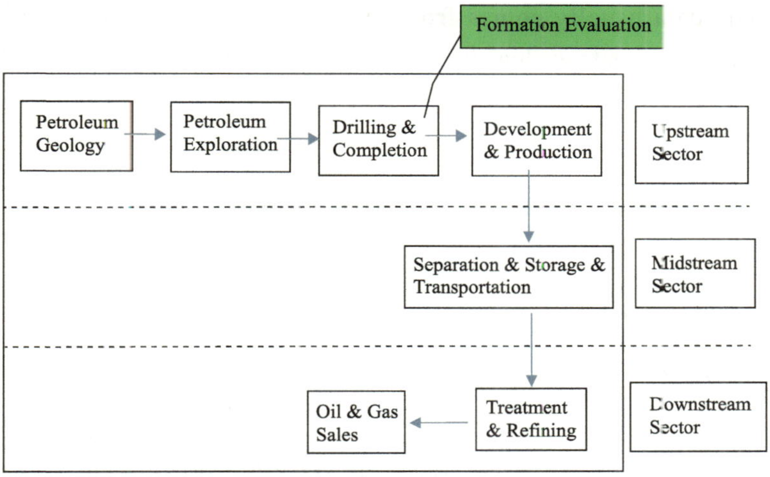

Figure 5.3 We Need Evaluate the Well before Completion

Since there are costs associated with each type of data gathered, not all the possible data is gathered from each well. A new-field wildcat, for example, justified gathering a great deal more data than does an in-fill development well in a mature field.

每一项数据收集都是有成本核算的，因此不能每口井都收集全所有数据。例如，打野猫井时收集的数据比加密开发井时收集的数据多得多。

5.1 Drilling Operations Log

5.1 钻时录井

The drilling operation log is a continuous, foot-by-foot record of the hole as it is being drilled. All pertinent data gathered from the cuttings, mud returns, and drilling operations are plotted against depth on the same strip-chart so that they can be interpreted simultaneously. The log is maintained by the operating company' s wellsite

钻井记录是钻进中连续、一英尺一英尺地记录钻井过程。从岩屑、上返泥浆和钻井记录中汇集关键参数，在带状图上画出深度信息以方便后期解释。钻井记录一般由甲方的

geologist, if one has been assigned. Otherwise, the contract mud logger does it.

The operations log helps the wellsite geologist track the well’s progress down through the stratigraphic column. The process is much like that of a navigator continuously plotting an airplane’s position. It’s essential that the geologist anticipates when the bit is approaching each zone of interest so that it can be thoroughly evaluated. This is particularly demanding with rank wildcats where little is known about the stratigraphy.The data usually comes from the three input: driller, mud logger, site geologist(Figure 5.4).

驻场地质师完成。否则，由乙方录井师完成。

钻井记录帮助驻场地质师跟踪钻井进展，钻遇构造岩层。其过程像绘制连续飞机定位导航。钻井记录对于地质师预测钻头钻遇地层很有用，需要详细连续记录。特别是在野猫井钻井时，地下地层信息未知，钻井记录更加重要。井场数据一般来源于司钻、泥浆师和驻场地质师（图5.4）。

Formation Evaluation
- *Drilling Operation Log (Mud Logging)*
  - *Drill Time*
  - *Mud Charateristic*
  - *Oil and Gas Shows*
  - *Gas Logging*
  - *Fluorescence Logging*
- *Coring*
- *DST*
- *WFT*
- *Openhole Logging (Wireline Logging)*
  - *E–Logging*
    - *Electric Logging*
    - *Micrologging*
    - *Induction Electric Logging*
  - *Acoustic Logging*
  - *Radioactivity Logging*
    - *Gamma Ray Logging*
    - *Neutro Logging*
    - *Density Logging*
  - *Caliper Logging*
- *IP*

Figure 5.4 Formation Evaluation Methods

5.1.1 Drilling Data

The driller and assistant driller will man the drilling controls on the rig floor from where instrumentation will enable them to monitor and control the drilling parameters, like the drilling time, the speed of kelly’s descent (drilling rate), running count of the number of stands of pipe that have gone down into the hole, and so on:

◆ hookload;

5.1.1 钻井参数

司钻或副司钻在钻井平台上通过多种参数监测来控制钻井，这些参数包括钻井时间、钻井速度、多少根钻柱钻进入井等，还包括：

◆ 大钩载荷；

◆ 钻杆扭矩；

◆ 钻头载荷；

- ◆ torque in drill string;
- ◆ weight on bit (WOB);
- ◆ rotary speed (rpm);
- ◆ pump pressure and rate;
- ◆ ROP (in min/ft).

- ◆ 转速；
- ◆ 泵压和泵速；
- ◆ 钻速。

5.1.2 Mud Logging

The mud logging(Figure 5.5) is an important record of drilling activity prepared by the petroleum geologist' s assistant during the drilling of a well (the assistant is called a mud logger). The mud log contains the entire geological record of the drilled hole, along with other important parameters such as drill time, mud characteristics, and oil and gas shows. If the logger sees any oil in the well samples, it will be very clearly noted on the mud log. Any indication of oil or gas is called a show, and the shows are one of the important things the mud logger will detect. Mud logger looks for oil and gas shows using florescent under ultraviolet and gas chromatogragh.

5.1.2 泥浆录井

泥浆录井是钻进过程中的一项很重要的工作（图5.5），由石油地质师（录井师）在钻井过程中取得。泥浆录井包含钻进过程中所有的地质信息和其他相关参数，比如钻时、泥浆特性和油气显示。假如录井师在钻遇过程中发现油样显示，他将清晰明确他在录井日志中标明。任何形式的油、气显示都是录井师重点观察的。泥浆录井师利用紫外线和气相色谱下的荧光寻找油气显示。

Figure 5.5 Mud Logging
(Source: http://en.wikipedia.org/wiki/Mud_logging)

Here is an old-style mud log, drawn by hand(Figure 5.6). As the well cuttings come up from the bottom of the hole, they are examined and identified. The mud log is updated each day of drilling. The mud log and the electric logs (which will be run after the holes is drilled) are the two most important items the PG needs to evaluate the potential well.

下面是老式的泥浆录井记录，由手绘完成（图5.6）。当岩屑从井底上返到地面时，它们会被详细地检查和识别。泥浆录井日志每日更新。泥浆录井和电测录井（可在完钻后进行）是石油工程师评估油井的最重要的两项参数。

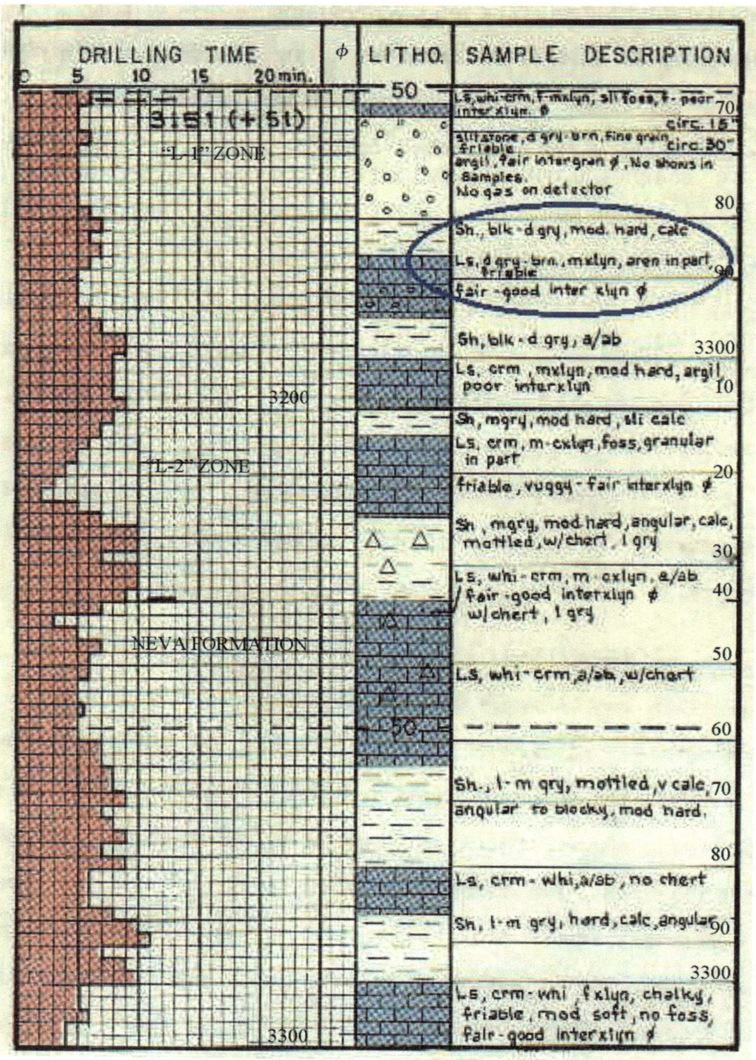

Figure 5.6 An Old-style Mud Log
(Source: http://www.geomore.com/the-mud-log/)

On the right side of the mud log, the mud logger or site geologist records his interpretation of the rocks drilled. There are two types of interpretation shown, a drawing down the center of the log (abbreviated litho), and a sample description.

Let's talk first about the section called "litho" (lithology, or rock type). The logger makes a drawing down the middle of the page. He uses standard symbols and colors to illustrate the different kinds of rocks, textures, fossils, and bedding. Limestones are shown as blue blocks on the log, shales are shown as short black horizontal lines, and sandstones/siltstones are shown with a dotted pattern. He looks at the small well cuttings under a microscope to figure out what the rock types are. It takes a lot of experience and know-how to figure out what rocks are being drilled just by looking at the tiny well samples.

The sample description is a more detailed summary of what the logger sees. Abbreviations are used because so much detail must be recorded. For example, look at the words inside the blue circle. It's a little hard to read, but it says "shale, black to dark-gray, medium hard, calcareous (containing limestone, or calcium carbonate).

Right below the shale is a limestone described as dark-gray to brown, medium-crystalline (medium grained), arenaceous (containing quartz sand), friable (breaks apart like sand), with fair to good inter-crystalline porosity.

The mud logger has also picked a couple of formation tops, and placed them on the left side of the log. These are the L-1 and L-2 Zones, and the Neva Formation. Mud loggers are very often the most experienced workers for interpreting well samples in a given area, and they are often able to pick up obscure tops or markers the geologist is unfamiliar with.

Other data that will be posted on the log are all the chemical properties of the mud (viscosity, weight,

在泥浆录井图的右边，录井师或驻场地质师解释钻遇岩层。有两类需要解释，一类是岩性确定；另一类是岩样描述。

首先说岩性如何确定。录井师在页面中部用标准的象征符和颜色代表某类岩石、纹理结构、化石和层理绘出来。比如灰岩在录井图上由蓝色块状图表示，页岩用黑色水平短线表示；砂岩/泥岩用点状图表示。录井师在显微镜下仔细分辨微小的岩屑确定其岩性。这需要大量的经验和实践才能做到以微小井内岩屑确定岩石成分。

岩样描述指录井师对所见岩样进行较为详细的描述。一般用缩写记录更多的细节。例如，图中蓝色圆框所写的，有点难以看懂，其实是“页岩，黑到深灰，中等硬度，钙质（包含灰岩、碳酸钙）”的意思。

页岩右边是灰岩描述为深灰到棕色，中等结晶（中等粒度），含砂质的（包含石英砂），易碎的（容易碎成散砂），中等到好的内结晶孔隙结构。

录井师也要识别地层顶面结构，写在录井图的左边。比如图中的L-1和L-2层为NEVA地层。录井师一般为该区域最有经验的岩样分析师，他们经常能识别地质师都不熟悉的模糊顶面结构和标志层。

别的数据如泥浆的化学性（黏度、重度、盐度等）、下钻、起钻、气体燃烧、泥浆池

salinity, etc.), drill string trips that are made in or out of the hole, records of gas flares, oil in the mud pit, or any other circumstances that might be noticed while the well is drilling.

的油气显示，或其他钻井时的特征也需要记录在录井图上。

It is especially helpful when the mud logger notices some action being taken at the rig that he does not believe to be correct, and notes it on the log (and calls the geologist). For example, the geologist may have ordered that drilling samples be taken every 10 feet, but the mud logger notices that samples are collected randomly, or not at all. The mud logger has to look out for sample "stacking." Stacking is a term used when the sample collector (generally the lowest status drilling hand, or "worm") skips a bunch of samples, then collects several samples at once and labels them with different intervals.

当录井师注意到某些钻井作业不正常时记录在录井图上（或通知地质师）是非常有用的。例如，地质师要求每钻10ft取岩样,但录井师注意到岩样收集的随机性或收不到岩样。录井师就必须去寻找岩样重叠。岩样重叠是一个术语用于指岩样收集者（一般为较低等级的钻井副手，叫"worm"）筛一批岩样，收集几个样本，标识清楚后，再筛一批再收集。

On the left side of the log, drilling time is recorded. The rig keeps track of the drilling time and a lot of other parameters using a recording clock device known as a geolograph. The geolograph makes a precise record of the time it takes the drill bit to drill the rock. Drill-time is recorded in increments of 2 feet. The numbers across the top indicate how long it took to drill that two feet, in minutes per foot. For instance, the depth from 3200′ to 3202′ (1 block), drilled at 7 minutes per foot, or a total of 14 minutes for the two feet. Drill time is incredibly useful in picking out zones that may be productive, or correlating the electric logs to the mud log.

在录井图左边，钻时也需要记录。钻机运转的相关参数和钻进时间用地质记录仪记录下来。地质记录仪记录下钻头钻进岩石的准确用时，钻时指每钻进2ft所用的时间。多少钻时指花了多少时间钻进了2ft，即每分钟多少英尺。例如，以7min 1ft计算，从3200~3202ft，总共花了14min钻进2ft。钻时在识别层系特别是生产层系时很有用，这时泥浆录井比电测井更有用。

Making a chart of drill time is very important, because it gives the PG one more hint to help figure out what rocks are being drilled. Remember that the well has not been logged with wireline tools at this point. Notice that in this well, the shales tend to drill "slow" (5~10 minutes per foot), while the limestones drill "fast" (2~6 minutes per foot). This is because the soft shales are ground into a thick, sticky mud by the drill bit, and "ball up" on the end of the bit. This

记录下钻时非常重要，它让石油工程师能猜到下面钻遇的岩石类型。记住此时还不能进行电测井。注意到钻遇页岩时钻进很慢（5~10min钻进1ft），而钻遇灰岩时相对较快（2~6min钻进1ft）。这是因为钻遇泥页岩时,在井底形成又厚

causes the bit to cut through the rock slower, which slows the penetration rate.

Usually the slowest-drilling formations will be extremely hard, well-cemented, non-porous sandstones or limestones. For these types of formations, a special (and expensive) bit called a PDC bit, is sometimes used.

Porous sandstones and limestones don't ball up, so they usually drill really fast. In fact, one of the things a petroleum geologist likes to see is a good "drilling break" in his target sand. This means the sand drills very fast (3 minutes per foot or less), indicating it is soft or porous.

The mud log will be used in combination with the electric logs run on the well to make a decision about whether to complete the hole (try to produce oil or gas from it) or plug the well. Mud logs are an extremely important tool to the petroleum geologist.

The site geologist collects and washes the cutting, then examines them under a microscope, analyzes their characteristic and rock type(Figure 5.7). To determine exactly where the bit is in the stratigraphic column, the geologist tries to spot the

又黏的泥层包裹在钻头上，降低了钻头的破岩速度。

通常钻遇最慢的地层为极坚硬、强胶结、无孔隙的砂岩层或灰岩层。针对这种地层需要用到特殊的钻头，PDC钻头。

多孔的砂岩和灰岩不起泥包层，因此钻进相对较快。实际上，石油地质学家希望看到一种好的钻井中断，即钻遇目标砂层，钻进速度非常快（每3min或更短时间就钻进1ft），显示是软砂岩或多孔砂岩。

泥浆录井结合电测井能确定该井下一步是开发生产还是关井。泥浆录井对石油地质学家而言是非常重要的工具。

井场地质师收集清理岩屑，用显微镜观察岩屑的特征和岩性（图5.7）。确定钻头是否钻遇地层柱状图，地质师识

Figure 5.7 Inside a Mud Logging Cabin
(Source: http://en.wikipedia.org/wiki/Mud_logging)

very first cutting from each new zone drilled. Lithology, porosity,oil and gas show and hydrocarbon odor will be described(Figure 5.8). This is a difficult task where experience counts greatly.

别出钻遇新层的第一批微小岩屑。岩性、孔隙性、油气显示和碳氢化合物气味都要描述记录（图5.8）。这是一项艰巨的任务需要大量实践经验。

Figure 5.8 Sample of Drill Cuttings of Shale While Drilling an Oil Well in Louisiana. For Reference, the Sand Grain and Red Shale are Approximately 2 mm. in dia (Source: http://en.wikipedia.org/wiki/Mud_logging)

5.1.3 MWD/LWD

Drilling operations which employ a measurement while drilling (MWD) system will also provide the drilling engineers with formation parameters,downhole pressure and directional data in real time(Figure 5.9, Figure 5.10).

Logging While Drilling (LWD) is a technique of conveying well logging tools into the well borehole downhole as part of the bottom hole assembly (BHA) (Figure 5.11).

5.1.3 随钻测量/随钻测井

随钻测量是用传感器等测量仪器在钻进过程中实时测量地层参数、地层压力、钻头方向（图5.9、图5.10）。

随钻测井是将测井工具安装在钻头上，将测井作为钻井的一部分（图5.11）。

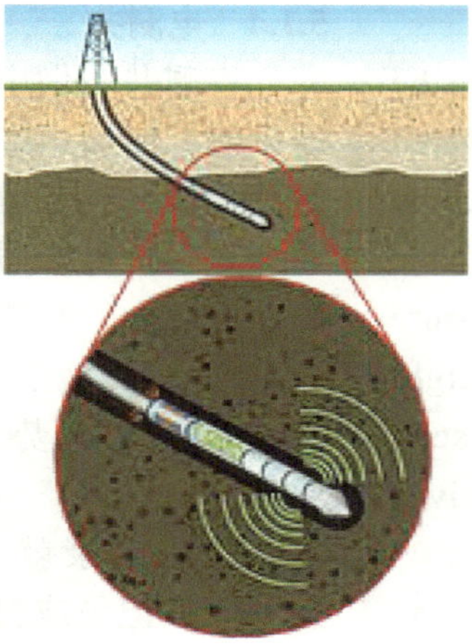

Figure 5.9 Measurement While Drilling
(Source: http://aircraft.saftbatteries.com/
MarketSegments/Oilgas/tabid/)

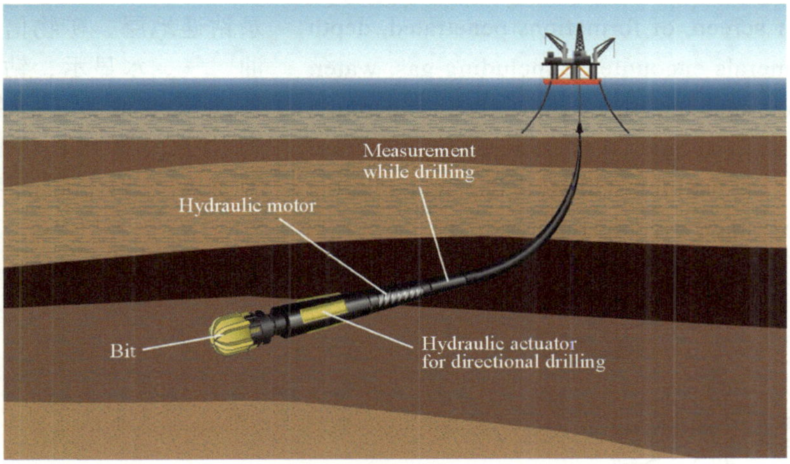

Figure 5.10 Offshore Measurement While Drilling (MWD)
(Source: http://www.rwe.com/web/cms/mediablob/en/)

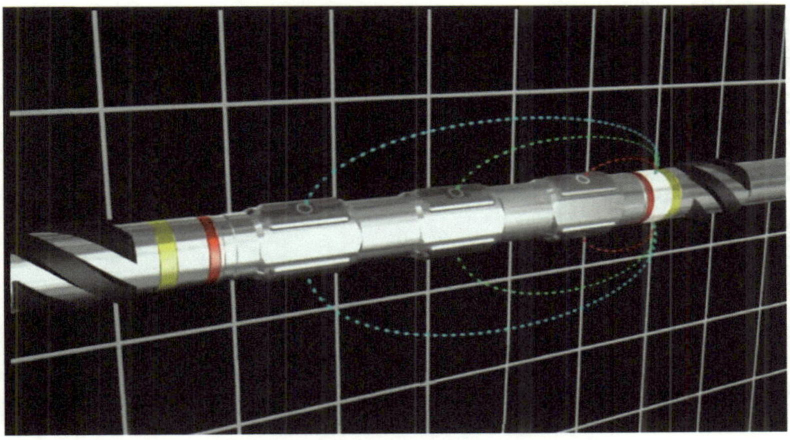

Figure 5.11 Logging While Drilling (LWD)
(Source: Overview of the Oil Industry, Richard Lau)

5.1.4 Wireline Logging

Historically, well logging has also come to be known as wirelining since the logging tool are lowered all the way to bottom and then reeled slowly back up by means of a cable——the wireline(Figure 5.12). As the tools come back up the hole, they are able to measure the properties of the formation they pass. It might display physical properties such as resistivity, self-potential, and gamma ray intensity or velocity at given depths.

This record, known as a well log. And logs may be records, either printed out in long strips or displayed on a computer screen, of formations penetrated, depth records of minerals encountered including gas, water, or oil, the size and length of pipe used, and much more information about a well. The exact structure and content of the well log will vary, depending on the purpose for creating the record(Figure 5.13).

5.1.4 电测井

将测井叫做电测井是因为原来测井时将电缆缓慢下放到井底再绕轴提起测井仪器，从而测量每一处经过地层的性质（图5.12）。它能显示地下每一深度下的地层物理性质，如电阻率、电势、伽马线和声波。

这些记录就是测井。测井可能被记录或打印在长条带卷轴和展示在计算机屏上，记录钻遇地层、矿物信息，包括油、气、水显示，钻杆尺寸、长度等多种关于井的信息。测井的内容和具体形式可能因测井目的不同而不同（图5.13）。

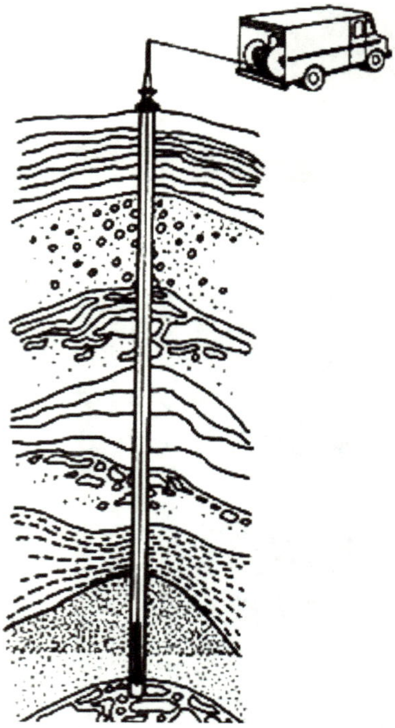

Figure 5.12 Wireline Truck
(Source: http://en.wikipedia.org/wiki/Well_logging)

Figure 5.13 Logging Tool Cutaway View
(Source: http://www.geomore.com/electric-logs/)

Today, well logging has larged its extent, there are more than 100 different types of logs available. Main methods include electric logging, micrologging, induction electric logging, acoustic logging, radioactivity logging, gamma ray logging, neutron logging, density logging and mud logging. Actually, well logging is one of most common and most important methods for stratigraphic comparison, for tectonic analysis, and for any evaluation(Figure 5.14~Figure 5.19).

今天，测井已经扩大其内涵，要超过100多种测井方式可供应用。主要的测井方式包括电测井、微电极测井、感应电导测井、声波测井、放射性测井、伽马测井、中子测井、密度测井、泥浆录井等。实际上，测井是为地层对比、岩层分析和各种评估提供依据的最主要、最常见的方式（图5.14 ~ 图5.19）。

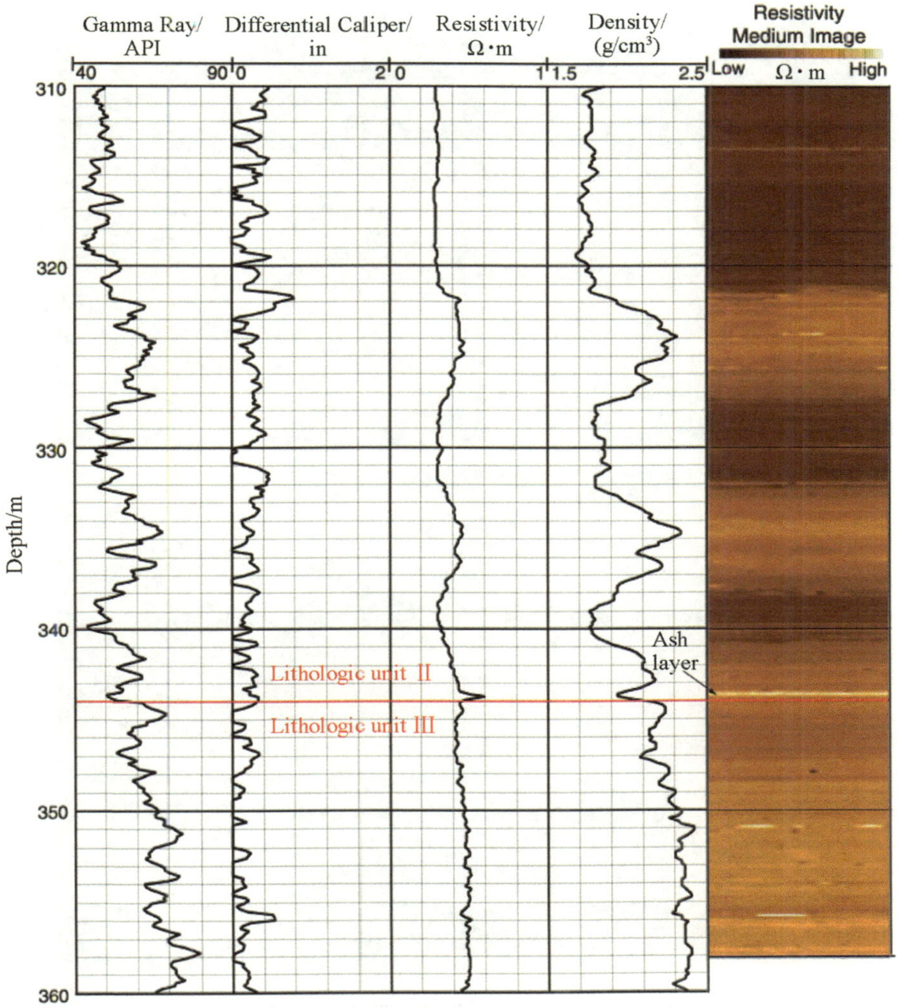

Figure 5.14 Wireline Log Consisting of Gamma, Caliper, Density and Resistivity Logs
(Source: http /www-odp.tamu.edu/publications)

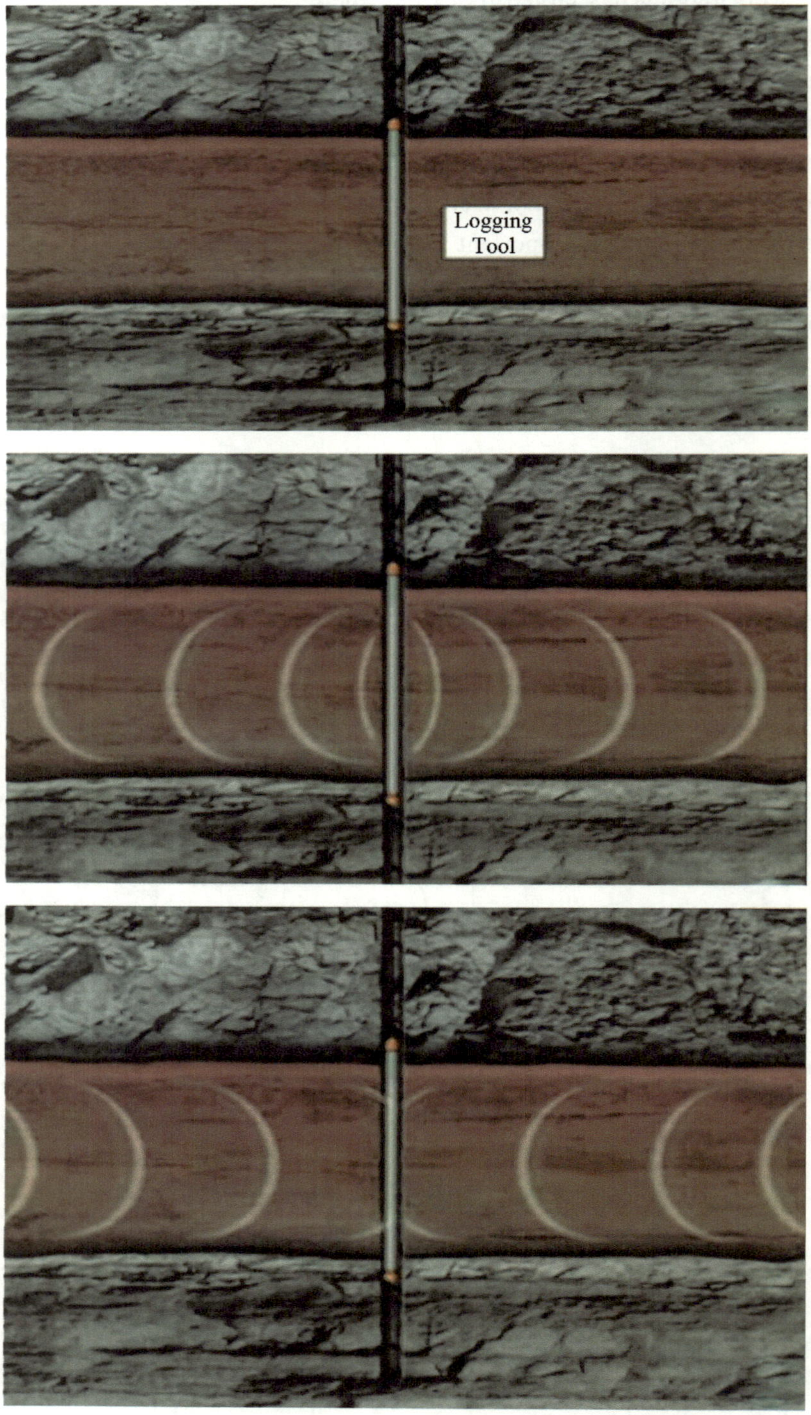

Figure 5.15 Logging Tool Lowered into Hole by Wireline
(Source: www.eei-inc.com)

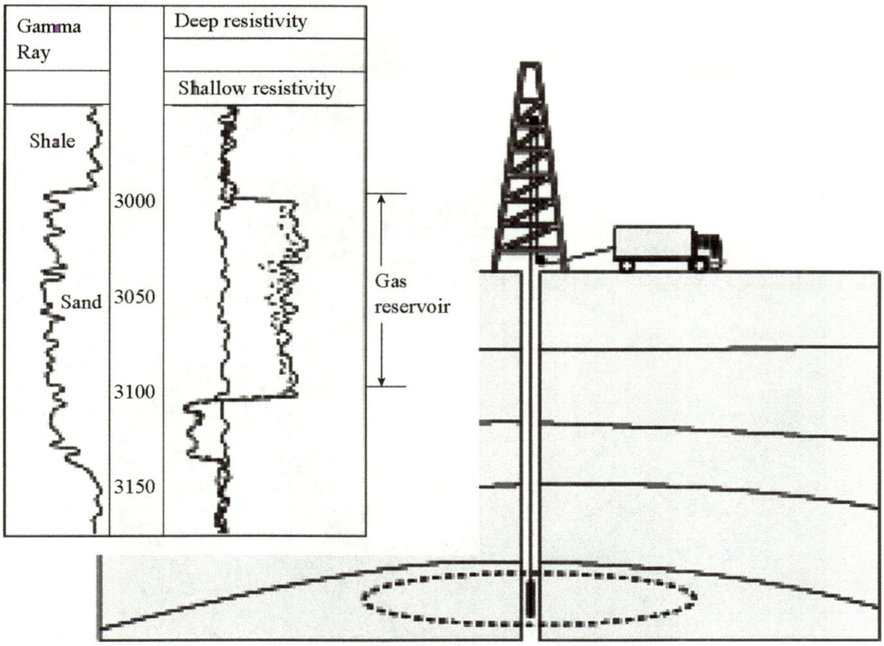

Figure 5.16 Gas Logging
(Source: http://www.natgas.info/html/gasexploration.html)

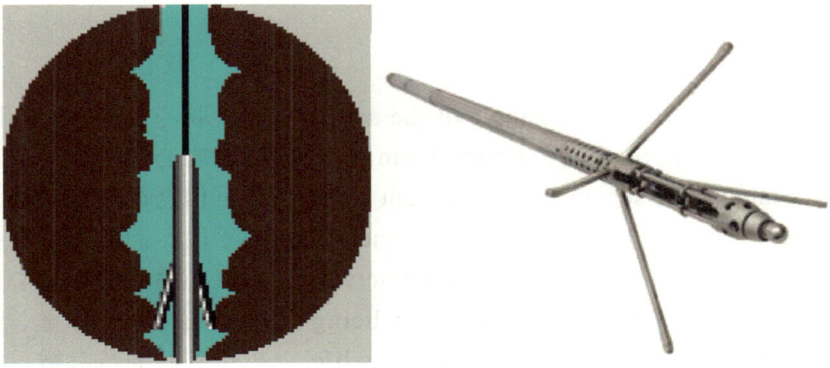

Figure 5.17 Caliper Logging
(Source: http://ny.water.usgs.gov/projects/bgag/intro.text.html)

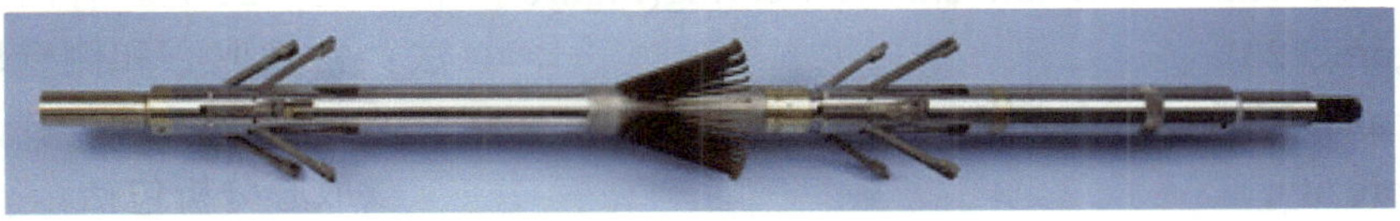

Figure 5.18 Caliper Logging Tool
(Source: http://www.gfm.ru/en/production/downhole/73skpd/
& http://www.bakerhughes.com/products-and-services/evaluation/)

Figure 5.19　Fluorescence Logging
(Source:http://www.chinookconsulting.ca/Services/Wellsite-Geology.html)

5.2　Coring

Coring operations capture a sample of the actual subsurface rocks(Figure 5.20). Other formation evaluation techniques are somewhat indirect, but cores are very direct. They can be picked up, examined in detail, smelled, weighed, and analyzed for key reservoir parameters. A core has lasting value, often being subjected to additional analysis late in a field' s life when secondary or tertiary reservoir-drive projects are being considered(Figure 5.21). There are two types of coring:conventional coring and sidewall coring(Figure 5.22, Figure 5.23)

Core analysis routinely include the following:

- ◆ Porosity;
- ◆ Fluid saturations;
- ◆ Permeability.

5.2　取心

取心指取得地下岩层的一段岩心（图5.20）。其他的地下信息评估方法或许是间接的手段，而取心是直接手段。用取心钻头钻取岩心后带到地面仔细检查，称重、嗅闻、分析得到其油藏参数。岩心具有持续价值，当油藏二次或三次开发时，对岩心做进一步的分析（图5.21）。有两种取心方法：常规取心和侧壁取心（图5.22、图5.23）。

取心分析包括：

- ◆ 孔隙度分析；
- ◆ 流体饱和度分析；
- ◆ 渗透率分析。

Figure 5.20 Running Down the Coring Tool
(Source: TOTAL)

Figure 5.21 Collecting Coring
(Source: Overview of the Oil Industry, Richard Lau)

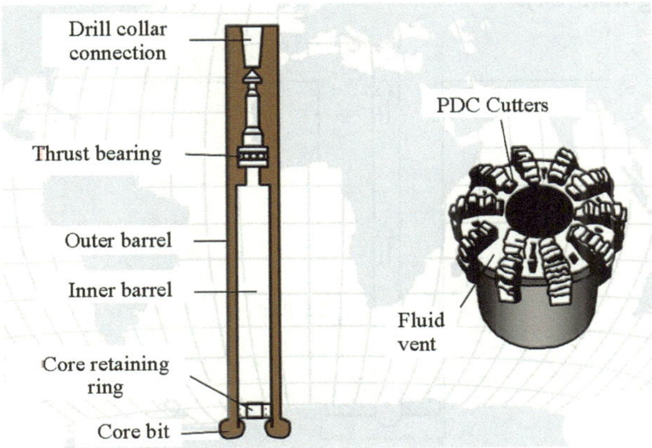

Figure 5.22 Conventional Coring: Coring Assembly on Bottom of Drill String and Coring Bit
(Source: https://www.spec2000.net/09-corepore.htm)

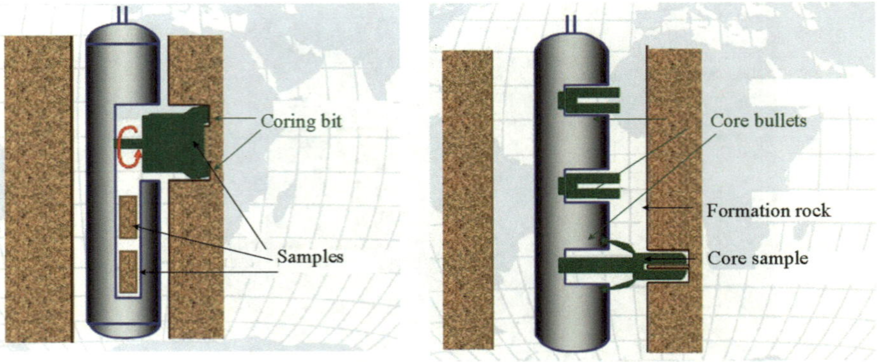

Figure 5.23 Sidewall Core Gun with Steel Bullets
(Source: https://www.spec2000.net/09-corepore.htm)

5.3 Drill–Stem Test (DST)

Since a drilling rig is usually on the site during exploration well testing, this rig is used to run and retrieve the test string. Moreover, drill pipe may be used as a portion of the production tubing for the temporary completion. For these reasons, a single-well test conducted on an expandable exploration well is usually referred to as a drill-stem test (DST). A drill-stem test can be further classified as an open-hole DST or a cased-hole DST. In an open-hole DST, the portion of the hole that penetrates the target zone is left uncased. The test string is equipped with open-hole packers that can be set so as to isolate the zone to be tested. The zone is produced by opening valves in the test string adjacent to the test interval(Figure 5.24).

5.3 中途测试

钻勘探井时，测井通常与钻机运转同时进行。钻杆当作生产油管以临时完井。探井测试或单井测试一般做这种中途试井。中途试井可进一步细分为裸眼中途试井和套管内中途试井。在裸眼中途试井中，部分产层被射穿而不下套管，用封隔器封隔上下层而独立出测试层。该产层放喷测产能（图5.24）。

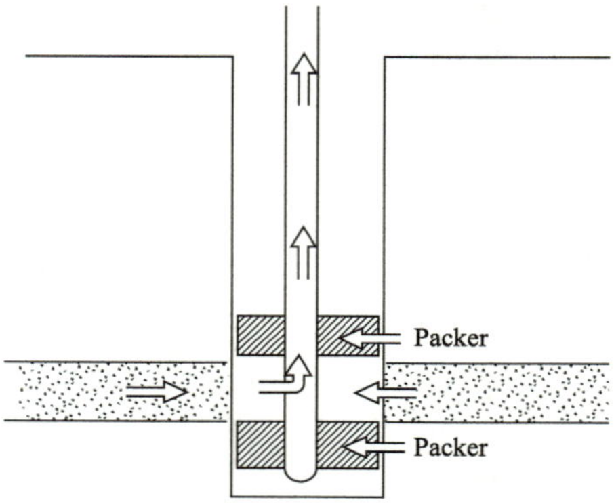

Figure 5.24 A Drill Stem Test
(Source: http://pages.geo.wvu.edu/~jtoro/Petroleum/Review%203.html)

5.4 Initial Potential (IP) Test

Obtaining an actual flow test on a potential reservoir has unique significance. Although the well logs might indicate that a reservoir has the necessary permeability and saturations to produce commercially, the logs are only indirect indications and can be misleading. Until oil or gas is actually produced, an operator can never be completely certain of a well's productivity.

Some very costly development mistakes have been made using well log data only, the productivity of gas reservoir is particularly difficult to assess without extended production testing. Productivity testing is so important that the Society of Petroleum Engineers (SPE) requires that the commercial producibility of reserves designated as proved be supported by actual production or formation tests.

IP test is run to determine its initial productivity.

The test period is usually 24 hours, but shorter tests are often taken and erected to a 24-hour basis. The produced barrels of oil, barrels of water and MCF of gas (or metric equivalents) are reported along with the tubing pressure and choke size if the well is flowing. If the well is artificially lifted, the specifications of the lift equipment are included.

A great deal of attention is paid to the IP test since it often is the first dependable indication of a well's productivity, and productivity largely determines profitability.

The IP is usually the signal to remove the well from the drill report. It marks the well's change from drilling status to production or other status.

5.4 IP测试

获得工业性油流靠测井是不够的，测井只能间接显示油藏的渗透率和饱和度，很可能产生误导。只有油气实际开始生产时，我们才能了解地下井的生产能力。

只采用测井数据开发油田造成过非常大的生产损失，特别是气藏的生产能力如果不生产就无法了解。生产测试是如此重要，以至于石油学会要求用实际生产测试计算可采储量值。

初始产能测试确定井的最初生产能力。

初始产能测试一般为24h，更短时间的测试也可以，最后换算成24h制就可以了。测试中生产的油、水、气量及相应的油管压力、阀门大小都要记录。假如该井采用人工举升，则举升设备的相关内容也要记录。

初始产能测试至关重要，因为它决定井的生产能力，而生产能力又决定井的产量和盈利。

初始产能测试以后，该井就由钻井方移交到生产方或其他方。

5.5 Abandonment & Reclamation

If tests indicate the well is a dry hole, not capable of producing commercial quantities of oil and gas, the drilling crew plugs the wellbore with cement and cleans up the site. A similar procure is followed if a producing

5.5 废井&复原

假如产能测试显示该井为干井，无商业价值的油、气产量，钻井方将用水泥填埋井眼并清理井场。如果生产井不能

well is no longer economical to operate.

继续生产有经济价值的油、气也将被填埋和清理。

Because of stringent environmental regulations, the cost of abandonment and cleanup is significant factor in an oil company’s planning. The company remains responsible for the site until a reclamation certificate is issued by government authorities. It can take years to complete the reclamation.

由于严格的环境法规，这部分废弃和清理费将纳入油田公司的核算中。油田公司将对井场生态负责直至政府当局验收。这将花费很多年来复原生态。

6 WELL COMPLETION & CEMENTING
完井与固井

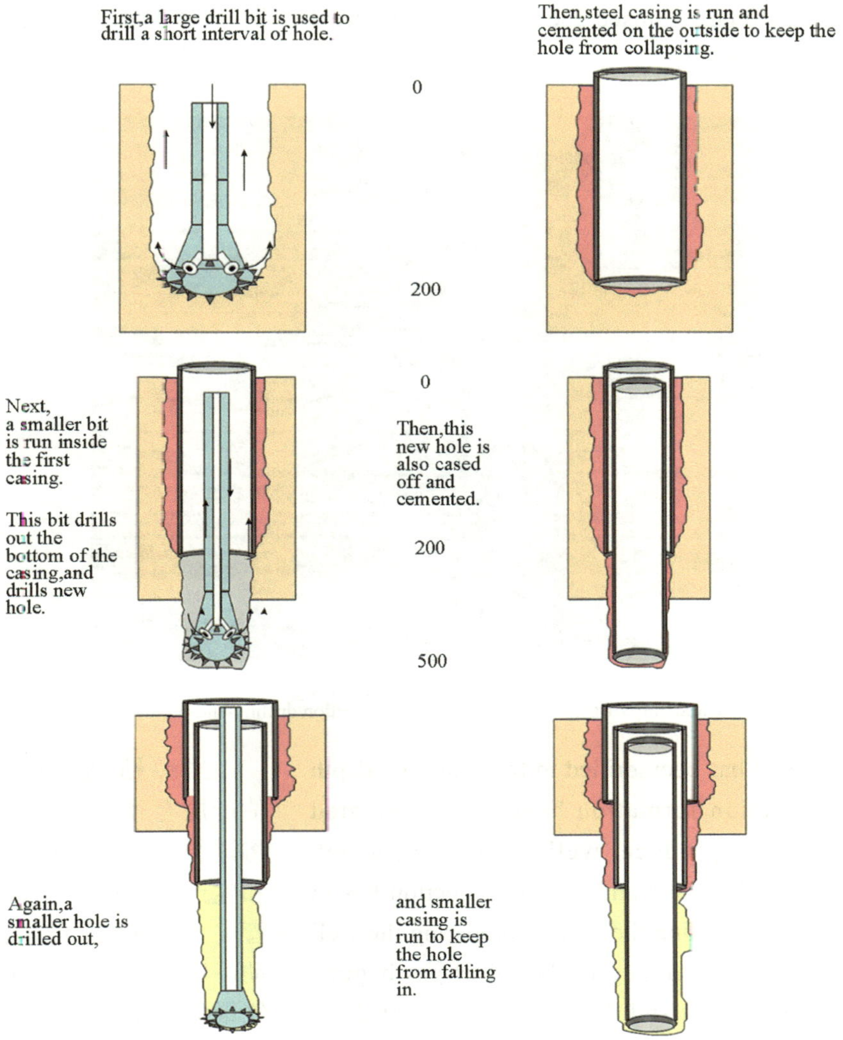

Figure 6.1 Well Operations
(Source: http://www.ppdm.org/wiki/index.php/Well_Operations_Reference_guide)

6.1 Downhole Strings

Imagine that a reservoir exists at a depth of 2500 m. We could attempt to drill one straight hole all the way down to that depth. That attempt would end either with the hole collapsing around the drill bit, with the loss of drilling fluid into formations with low pressure or in the worst case with the uncontrolled flow of gas or oil from the reservoir into unprotected shallow formations or to the surface (blowout). Hence, from time to time, the borehole needs to be stabilised and the drilling progress safeguarded(Figure 6.2).

6.1 井下管柱

想象一下，油藏存在于2500m的地下，我们试图从地面钻直井到油藏，但钻头附近的井眼随时可能坍塌，损失钻井液泄漏到地层中，更糟的是失控的油、气可能从油藏流入不受保护的地表浅层或流出地面发生井喷。因此，井眼需要时不时地加固，为钻井护航（图6.2）。

Figure 6.2 Drilling Casing
(Source: http://fracfocus.ca/groundwater-protection/drilling-and-production)

After a well has been drilled to the projected depth and the production formation have been evaluated as economical to produce, well completion which connects the finished hole and actual production has to be done, including setting the casing, preparing the well for production, and bringing in the oil or gas begins. Completion equipment and the methods employed are quite varied, and the decisions for an individual well are usually based on the type of oil or gas accumulations involved, the requirements that may develop during the

当井钻到目标深度，评估产层有经济产量，完井工作将进行。完井即连通钻井井眼和生产通道的工作，包括下套管、生产准备和带油、气流动。完井的装备和方式基于油气聚集类型、井的寿命和经济性的不同而不同。低强度钢管甚至是二手钢管用于边际油藏，其他开支相应缩减。高压

life of the well, and the economic circumstances at the time the work is done. Low-pressure pipe, sometimes secondhand, will be employed if the oil accumulation has a marginal payout, and other expenditures will be scaled down accordingly. If high pressure is anticipated and well life is expected to be long, however, the best grade of pipe will be needed(Figure 6.3).

Many oil and gas wells require four concentric strings of large pipe: conductor pipe, surface casing, intermediate casing, and production casing(Figure 6.4).

The casing design will usually start with a 23 in conductor, then 18$^{5}/_{8}$in surface casing, 13$^{3}/_{8}$in intermediate casing above reservoir, 9$^{5}/_{8}$in production casing across reservoir section and possibly 7 in. production ‘liner’ over a deeper reservoir section. A liner is a casing string which

油藏和长开发井需要高强度钢管（图6.3）。

油井、气井一般有4层同心套管：导管、表层套管、中间套管和生产套管（图6.4）。

套管设计一般以23in的导管开始，接下来是18$^{5}/_{8}$ in的表面套管，13$^{3}/_{8}$ in的中间套管直达油藏上部，然后是9$^{5}/_{8}$ in的生产套管深入油藏，更深的油藏部分可能有7in的生产衬管。衬管是一节由一段封隔器挟持住前一层套管底部的套管段，衬管不用从地下延伸到地面，因此节约成本（图6.5）。

Downhole Strings
- Conductor
  - To prevent the hole from caving in
  - To prevent the rig foundation from collapsing
- Surface Casing
  - To provide protection for freshwater formations
  - To prevent the hole from caving in
  - To control fluids crossflow
- Intermediate Casing: To solve some troubles encountered in interval
- Production Casing: To prevent the oil-bearing from caving in
- Liner: To be used as production casing
- Tubing: To be used for production

Figure 6.3 Downhole Strings
(Source: English for Petroleum Engineering)

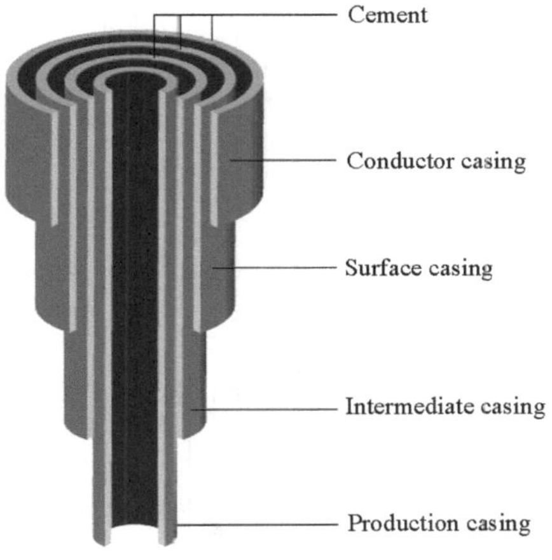

Figure 6.4 Typical Well Casing Diagram
(Source: http://www.encana.com)

is clamped with a packer into the bottom part of the previous casing; it does not extend all the way to the surface, and thus saves cost(Figure 6.5).

Casing joints are available in different grades, depending on the expected loads to which the string will be exposed during running, and the lifetime of the well.

The main criteria for casing selection are:

◆ Collapse load: originates from the hydrostatic pressure of drilling fluid, cements lurry outside the casing and later on by ‘moving formations’, for example salt.

◆ Burst load: this is the internal pressure the casing will be exposed to duringoperations.

◆ Tension load: caused by the string weight during running in; it will be highest atthe top joints.

◆ Corrosion service: carbon dioxide (CO_2) or hydrogen sulphide (H_2S) in formation fluids will cause rapid corrosion of standard carbon steel and there fore special steelmay be required.

◆ Buckling resistance: the load exerted on the casing if under compression.

The casing will also carry the BOPs described earlier.

Running casing is the process by which 40 ft sections of steel pipe are screwed together on the rig

套管接头有不同等级，依据钻杆预期载荷和井的服务寿命选择。

套管的主要选择标准如下：

◆ 极限载荷：由钻井液和水泥浆的静水压力在套管外产生，由移动层压力抵消，如盐层。

◆ 破裂载荷：作业过程中套管内部压力。

◆ 拉伸载荷：由下套管过程中套管自重产生，最大值出现套管顶部接头处。

◆ 反腐处理：地层流体中含有的二氧化碳(CO_2)或硫化氢(H_2S)会迅速腐蚀常规钢套管，因此需要特殊钢套管应对腐蚀流体地层。

◆ 抗弯性：在压缩载荷作用套管上要求套管有抗弯性。

套管还需承载之前所说的防喷器BOPs。

下套管是在钻台上将一段段40ft长的钢管拧起来下入井筒

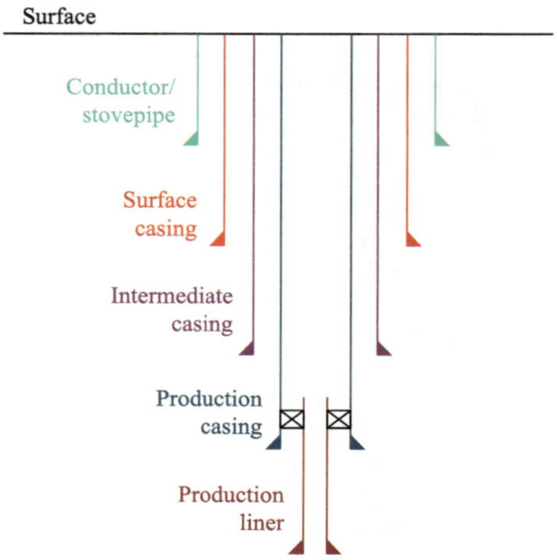

Figure 6.5 Casing Scheme
(Source: Hydrocarbon Exploration and Production)

floor and lowered into the hole.

的过程。

Spudding in usually begins in a medium or deep well with drilling a large-diameter but shallow hole called conductor hole. The conductor hole is lined with casing. The casing protects aquifers which may contain freshwater, provides a mounting place for the blowout preventer (BOP), and serves as the support for the production casing that will be placed in the well bore if the drilling program is successful. The conductor hole may be several hundred or several thousand feet deep.

中深井和深井开钻时用一个大直径的浅管，叫导管。导管里衬套管。套管保护水层特别是淡水层，也为防喷器提供安装位置，同时为后续的生产套管提供支撑。导管可以下到几百米或几千米深。

When the surface casing is set and cemented in place, it provides protection for freshwater formation. It also prevents loose shale and sand or gravel from falling into the hole and affords a means of controlling the flow of fluid from the well. Setting depths may vary from 500 ft to 5000 ft. Intermediate casing may be needed if troublesome zones are encountered below the surface casing and above the final depth of the well. The final casing for most wells is the production casing. The production formation is usually completely casing off, but in rare instances the production casing is set near or just on the top of potential pay zone.

当表层套管安装并水泥封固好后，它为淡水层提供保护，也防止疏松的页岩、砂岩和砾石落入井筒，还可提供一种控制井内流体流动的方法。表层套管一般下入500~5000ft。当在表层套管和最终井深前遇到难以控制的地层，这时需要用到中间套管。最后是生产套管，生产套管一般用水泥完全密封，也有少数井将生产套管下到产层顶部。

Another type of pipe that is not uncommon in wells over 10000 ft is called a liner. Liners are really just like casing——that is, they serve the same purpose——but they do not extend all the way to the surface. Instead, a liner hanger. A liner can function as production casing, in which case it is termed a production liner. Since it does not go to the surface, which is sometimes a considerable distance, the operator has a lower pipe cost(Figure 6.6).

另一种在10000ft以上井中不常见的管柱是衬管，衬管与套管功能一致，但并不延伸到地表，而是悬挂在井底。用作生产套管的衬管叫生产衬管。因为衬管不用延伸到地表，长度相对较短，所以成本更低（图6.6）。

The final string of pipe usually run in a production well is the tubing. Tubing is nearly always freely suspended in the well from the tubing head. In a flowing well its small diameter produces more efficient results than casing. Also, tubing is considerably easier to remove than casing when it becomes plugged or damaged. Tubing, when used in conjunction with a packer, keeps well fluids away from the casing because

最后一种安装在生产井中的管柱是油管。油管通过油管头自由地悬挂在井底。油管直径更小生产更有效。而且油管在损害后相比套管更容易移除。油管与封隔器一起连接，封隔器密封油管和环空保证流体流入套管中。井内流体腐蚀

Figure 6.6 Composites Liner&Epoxy Liner
(Source: http://www.compositesworld.com/articles/the-markets-oil-and-gas-2010 & http://www.compositesworld.com/articles/frp-liners-for-steel-piping)

the packer seals the space between the tubing and the casing. Well fluids corrode the casing and thus cause costly repair jobs later(Figure 6.7).

The packer consists of a pipe like device through which well fluids can flow, a rubber sealing element that forms a fluid-tight seal, and gripping elements (called slips) that hold the packer in the tubing-casing annulus

套管后会导致后期修井工作花费大量费用（图6.7）。

封隔器为一段中空管状设备，其中流体可在管内流动，管外橡胶膨胀后形成密封，封住产层段油管外的环空空间。由于封隔器密封了油管和套管

Figure 6.7 Seamless Steel N80 Oil Well Tubing Pipe
(Source: http://dongbaosy.en.alibaba.com)

just above the producing zone. Since the packer seals off the space between the tubing and the casing, formation fluids flowing into the well are forced into and up the tubing(Figure 6.8~Figure 6.12).

之间的空间，迫使地层流体流入油管（图6.8～图6.12）。

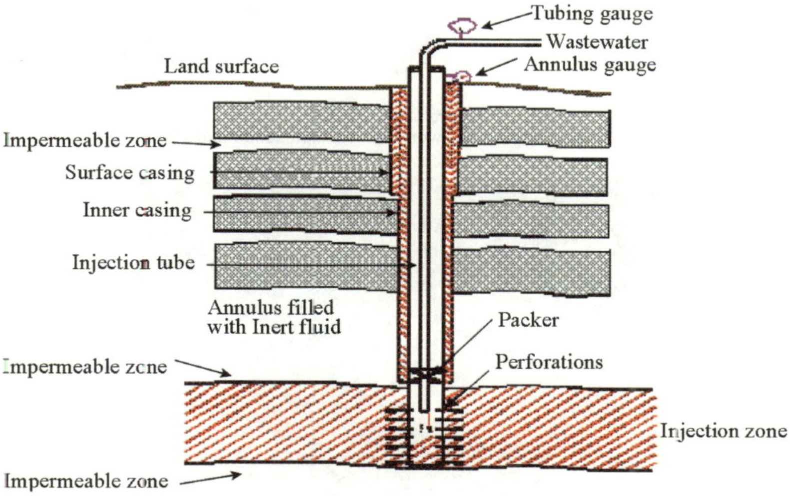

Figure 6.8 A Packer Seals off the Space between the Tubing and the Casing
(Source: http://web.deu.edu.tr/atiksu/ana58/deepwell.html)

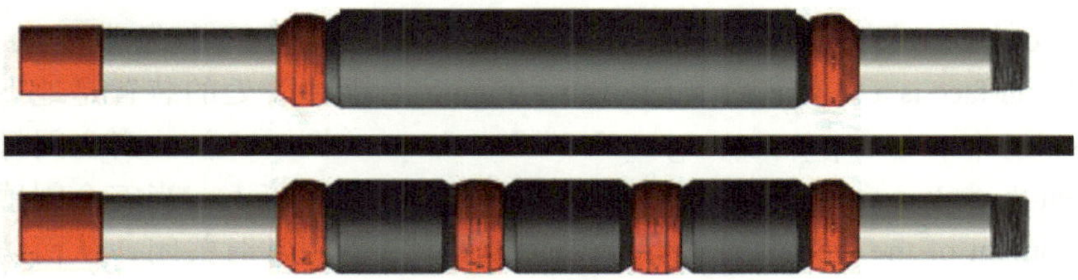

Figure 6.9 Swell Packers
(Source: http://cn.tradekey.com/product_view/id/1550856.htm)

Figure 6.10 Swell Packer
(Source: http://cn.tradekey.com/product_view/id/1550856.htm)

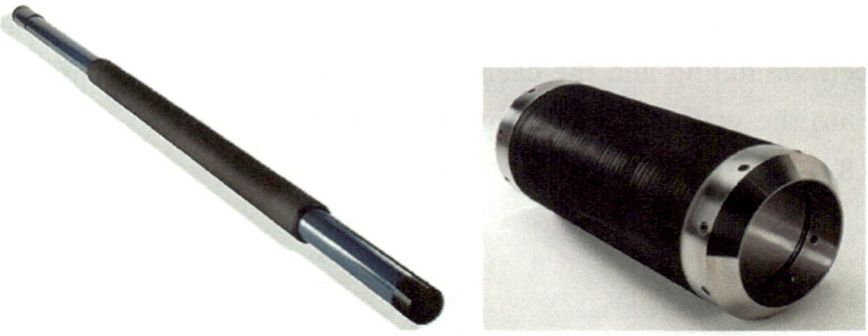

Figure 6.11 Packer
(Source: http://www.worldoil.com/uploadedimages/)

Figure 6.12 Swellable Packer
(Source: http://www.drillingcontractor.org/swellable-packers-qualified-for-valhall-11620)

Another device frequently installed in the tubing string near the surface is a subsurface safety valve (Figure 6.13). The valve remains open as long as fluid flow is normal. When the valve senses something amiss with the surface equipment of the well, it closes, preventing the flow of fluids(Figure 6.14).

另一个经常安装在近地面油管处的装置是地下安全阀（图6.13）。安全阀打开流体自由流入，当安全阀感知情况不正常时关闭，从而防止流体流入（图6.14）。

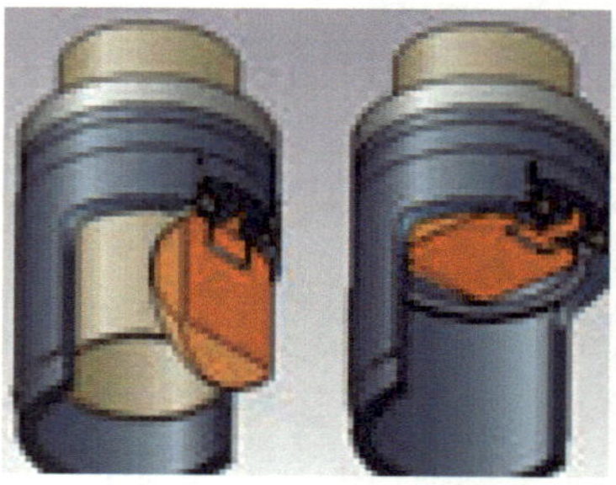

Figure 6.13 Subsurface Safety Valves
(Source: http://www.slb.com/resources/publication)

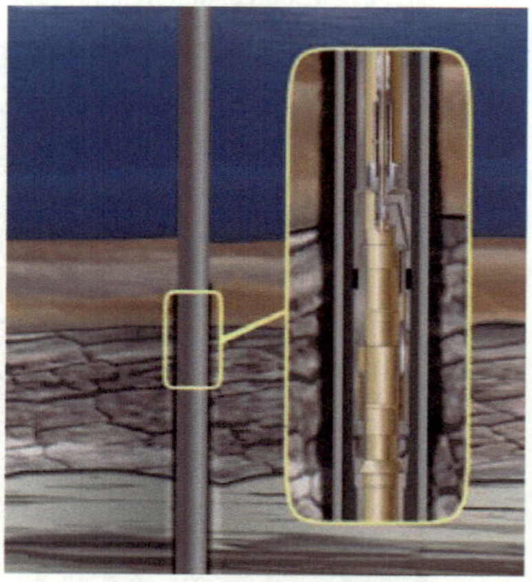

Figure 6.14 Downhole Subsurface Safety Valves
(Source: http://www.drillingcontractor.org)

6.2 Cementing

Cementing is the process of introducing cement to the annular space between the well-bore and casing. Most commonly, cementing is used to permanently shut off water penetration into the well(Figure 6.15, Figure 6.16).

6.2 注水泥

注水泥是将水泥注入井筒和套管之间的环空中，通常用于永久地封隔地层水流入井筒中（图6.15、图6.16）。

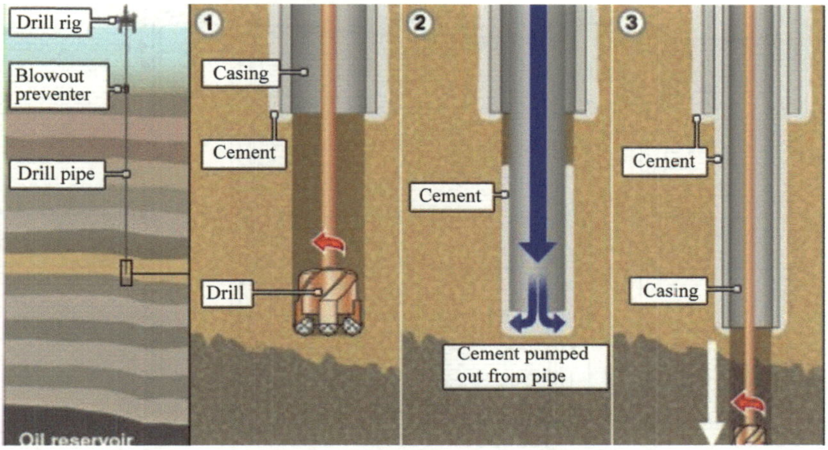

Figure 6.15　Cementing Casing and Drilling Out
(Source: http://blow-out.info/)

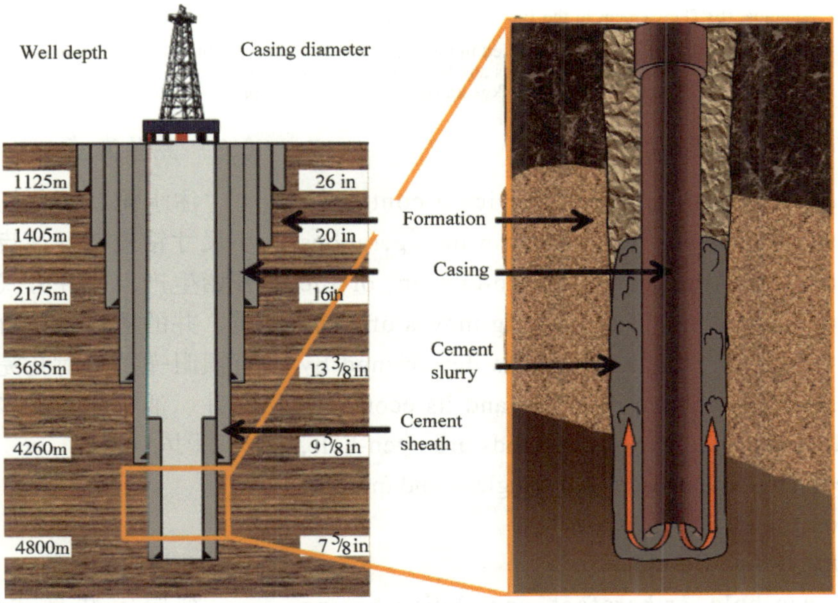

Figure 6.16　Schematic of a Cased and Cemented Oil Well (Left);
Principle of Placing the Cement Slurry between the Casing and the Formation (Right)
(Source: http://www.bauchemie-tum.de/master-framework/)

The cement supports the casing, seals off the annulus, isolates porous zones from each other, and protects the casing from corrosive formation fluids. Cements for use in oil or gas wells are manufactured and classified in relation to the ultimate depths to which they are to be pumped(Figure 6.17).

水泥支撑套管、封隔环空，隔绝各层孔隙带，保护套管不受腐蚀性流体侵扰。用于油井、气井水泥的制造和分类与最终泵入的深度相关（图6.17）。

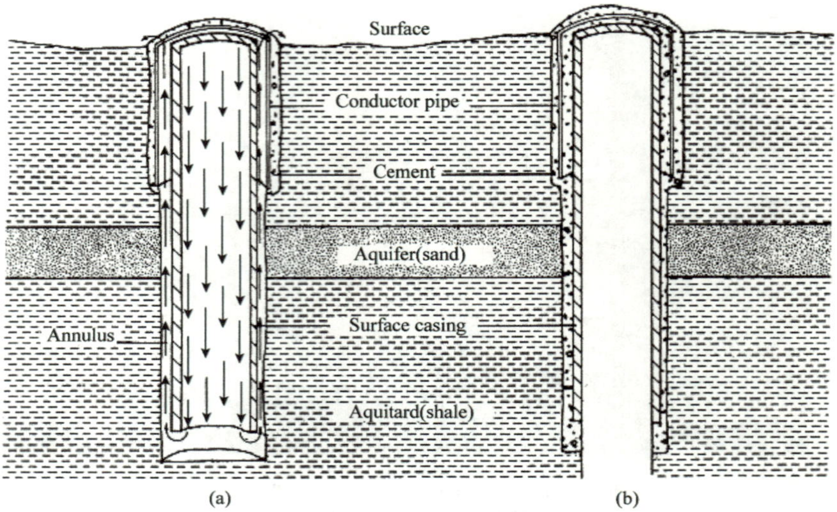

(a) The Conductor Pipe has been Cemented into Place. A Predetermined Amount of Casing Has Been Inserted into the Well Bore below the Deepest Freshwater Zone. Cement is Pumped Down the Inside of the Casing until Cement Flows to Surface Through the Aannulus. (b) The Cement in the Bottom of the Casing Has Been Drilled out so that Drilling can be Resumed.

Figure 6.17 A Casing is Installed in the Surface Hole to Prevent the Contamination of Freshwater Zones and to Support the Production Casing.
(Source: http://www.kgs.ku.edu/Publications/Oil/primer12.html)

6.3 Well Completion

After cementation, a well servicing contractor may move in a smaller rig to perform the operations necessary to put the well into production, or the work may be done before drilling rig moves off. The type of completion method used is determined by the characteristics of the reservoir and its economic potential. Among the various methods are open-hole, perforated, wire-wrappedscreen, tubingless, and multiple completion(Figure 6.18).

6.3 完井方式

注水泥后，井场服务方用小尺寸钻机钻开油流通道使井开始生产。完井方式由油藏性质、井的潜在经济价值决定。有裸眼完井、射孔完井、筛网完井、油管完井和多层完井方式（图6.18）。

6.3.1 Open-hole Completion

An open-hole, or barefoot, completion has no production casing or liner set opposite the producing

6.3.1 裸眼完井

裸眼完井是在产层不下生产套管或衬管，油藏流体直

formation. Instead, reservoir fluids flow unrestricted into the open wellbore. This type of completion, which is rarely used and in generally restricted to limestone reservoirs, is useful where only one productive zone and low-pressure formations exist. In an open-hole completion, casing is set just above the pay zone, and drilling proceeds into the productive zone as far as necessary to complete the well(Figure 6.19).

接无限制地流入井筒的完井方式。这种完井方式很少用到，并且一般只用于石灰岩的单层或低压产层。在裸眼完井中，钻井钻穿产层后下套管到产层之上（图6.19）。

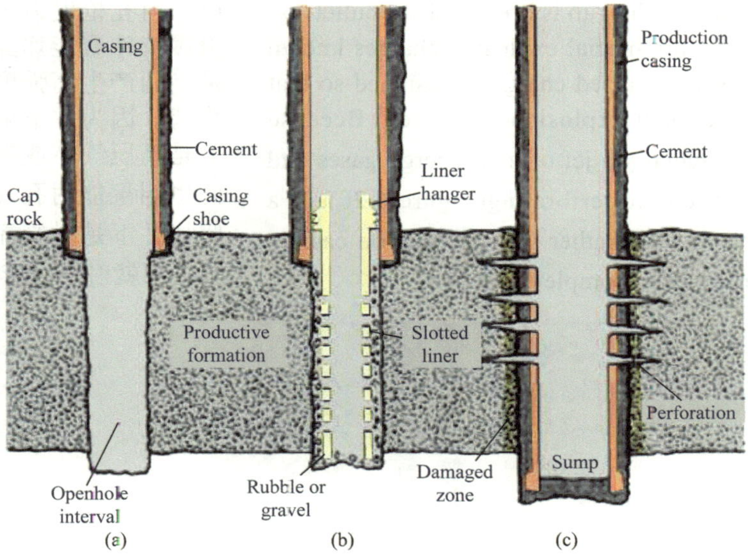

Figure 6.18 3 Types Completion
(Source: http://farm1.static.flickr.com)

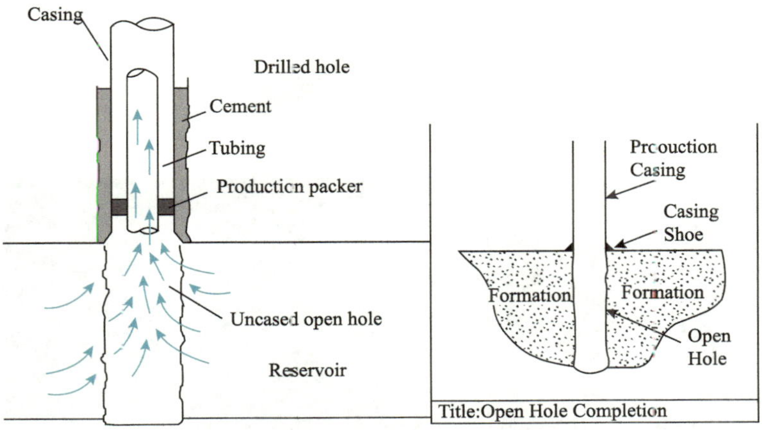

Figure 6.19 Barefoot Completion
[Source: http://www.accessscience.com/overflow.aspx&
http://en.wikipedia.org/wiki/Completion_(oil_and_gas_wells)]

6.3.2 Perforated Completion

The perforated completion——by far the most popular method of completing a well——requires a good cementing job and the proper perforating method. Perforating is the process of the piercing the casing wall and the cement to provide opening through which formation fluids may enter the wellbore. Perforating is accomplished by lowering a perforating gun down the production casing or the tubing until is opposite the zone to be produced. The gun is fired to shoot bullets or, more often, to set off special explosive charges known as shaped charges. A shaped charge is designed so that an intense, directional explosion is formed. Because the explosion is actually a jet of high-energy gases and particle, it is called jet perforating(Figure 6.20). If a production liner is used rather than production casing, the liner is perforated to complete the well.

6.3.2 射孔完井

射孔完井是目前最普遍应用的完井方式。它要求高质量的注水泥工作和准确的射孔方法。射孔是射穿套管壁和水泥环，提供一个使地层流体流入井筒的通道。射孔是通过在生产套管和油管中下入射孔枪正对产层射孔，油流出产层生产。射孔枪发射子弹，常常能引爆特殊聚能炸药。聚能炸药是专门产生高强度、定向性的炸药。因为井下射孔爆炸实际上是高压气体或颗粒喷射，因此也叫聚能射孔（图6.20）。假如井下是衬管而不是生产套管，那就射穿衬管以完井。

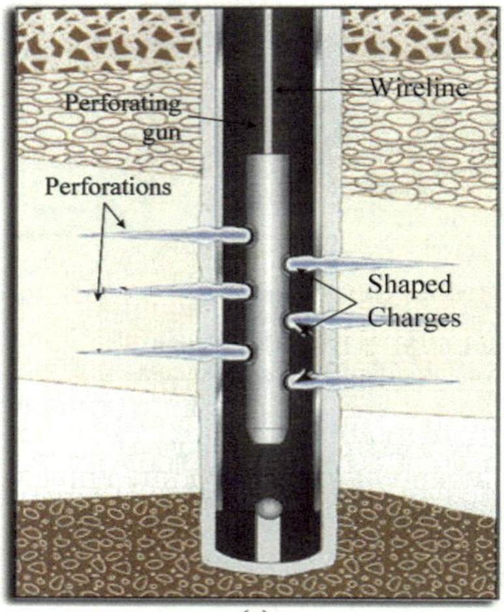

(a)

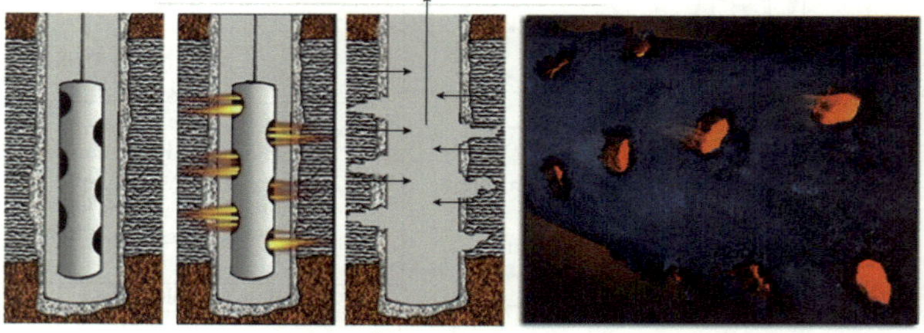

(b)

Figure 6.20 Shaped Charges Perforating & With Perforating Gun to Perforated Completion
(Source: http://www.blueridgegroup.com & http://www.oilinisrael.net/resources/)

6.3.3 Wire-wrapped Screen Completion

Another type of completion involves a wire-wrapped screen, which is a short length of pipe that has opening in its sides and is wrapped with a specially shaped wire. One method involves attaching the screen to the bottom of the tubing string and lowering it into a well that has been perforated(Figure 6.21, Figure 6.22).

6.3.3 筛管完井

另一种完井方式是筛管完井，它用到金属筛网，一种带有特殊开口的短管。这种筛管固定在油管底部，下入已射孔的井筒，其产层处已经完成射孔（图6.21、图6.22）。

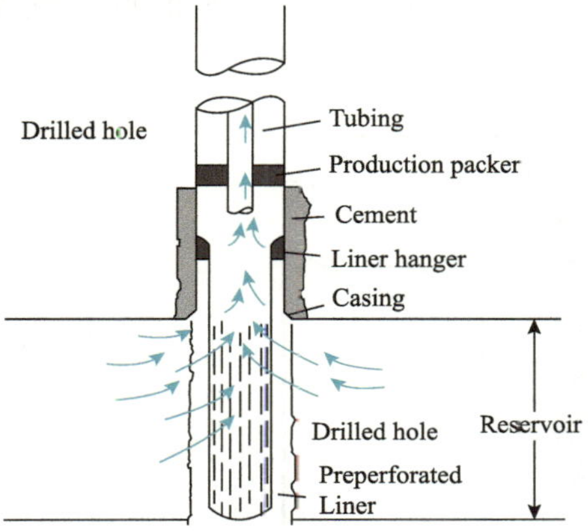

Figure 6.21 Liner-type Completion(Perforated Liner)
(Source: http://www.accessscience.com/)

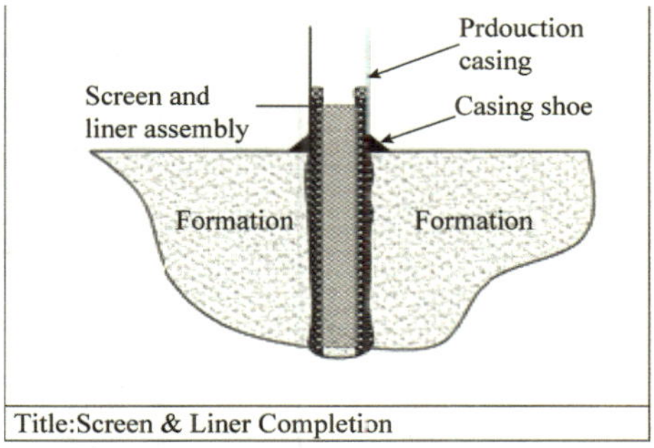

Figure 6.22 Screen and Liner Completion
[Source: http://en.wikipedia.org/wiki/Completion_(oil_and_gas_wells)]

Usually, wire-wrapped screens are run in conjunction with a gravel pack——that is, gravel is placed in the hole outside the screen(Figure 6.23). Well fluids flow through the gravel pack, through the wire-wrapped screen, and into the tubing. Wire-wrapped screen completions are most often utilized in cases where the producing zone or zones are likely to produce sand as well as oil and gas. The screen and gravel pack minimize the entry of sand into the wellbore, thus preventing problems caused by sand.

通常，筛管和砾石填充配合使用，砾石充填在筛管外部（图6.23）。地层流体经过砾石流经筛网最后流入油管。筛管完井绝大多数用在出砂或出油气产层，筛网和砾石最小程度地减少产层出砂，防止砂粒随油气流入井筒造成砂难。

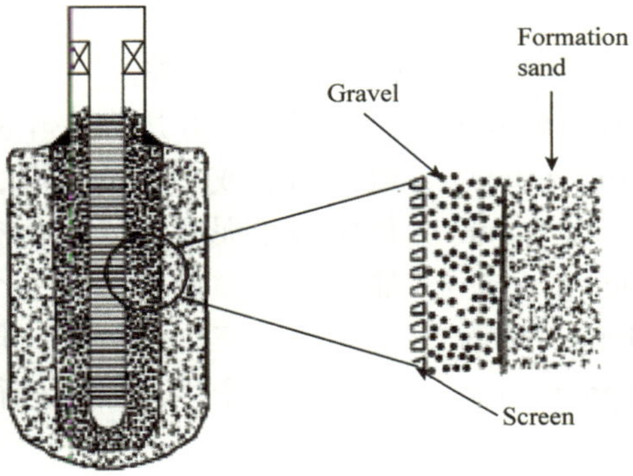

Figure 6.23 A schematic of the Function of a Slotted Liner or Screen in an Openhole Gravel Pack
(Source: http://petrowiki.org/Slotted_liners_and_wire_wrapped_screens)

6.3.4 Tubingless Completion

Although most wells are completed with tubing, a small-diameter well that uses small-diameter casing may be completed without the tubing. The small casing is cemented and perforated opposite the producing zones. Tubingless completions are used mostly in small gas reservoirs that produce few, if any, liquids and are low in pressure(Figure 6.24).

6.3.4 无油管完井

尽管绝大多数的井是油管完井，但部分小直径井也采用小直径的套管完井而无油管在其中。小直径的套管被水泥固结后在产层通过射孔射开生产。无油管完井大部分用在无水或无液产出的低压气井中（图6.24）。

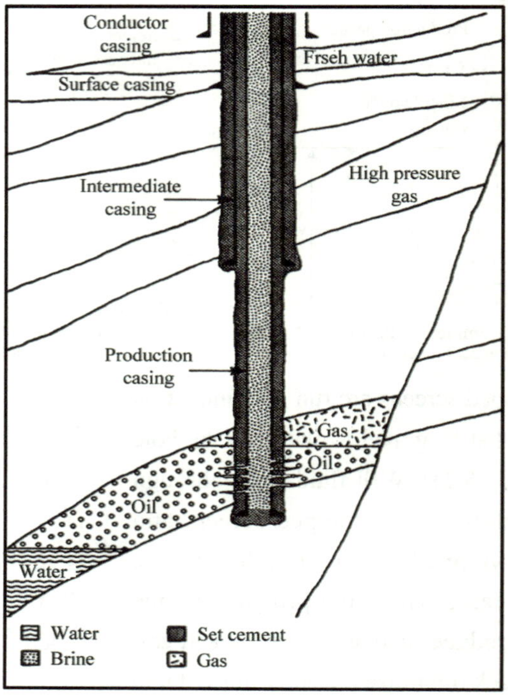

Figure 6.24 Primary Cementation
(Source: http://www.knowenergysolutions.com/)

6.3.5 Multiple Completion

Multiple completions can be used when one wellbore penetrates two or more producing zones. Usually, a separate tubing string with packers is run inside the production casing for each producing zone. For example, in a triple completion, three tubing strings and three packers can be utilized in a single production string(Figure 6.25).

6.3.5 多层完井

多层完井用在一个井筒中射开两层或多层的情况，通常每层用水泥塞封隔后由独立的油管生产。例如，在一个三层完井中，三个油管和三层水泥塞组成一个生产系统进行生产（图6.25）。

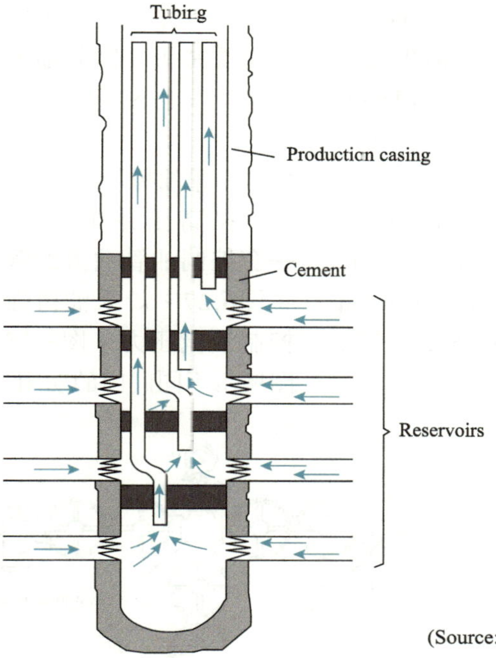

Figure 6.25 Multiple Completion
(Source: http://www.accessscience.com/overflow.aspx)

6.3.6 Wellheads

The wellhead is mounted on the surface casing. Each additional casing string and the tubing string are then hung from the wellhead as they are run. The wellhead also seals off the annular spaces between the strings(Figure 6.26).

6.3.6 井口装置

井头安装在表层套管上，每层套管和油管管柱悬挂在井头上并延伸下来。井头也用于密封各层管柱之间的环空空间（图6.26）。

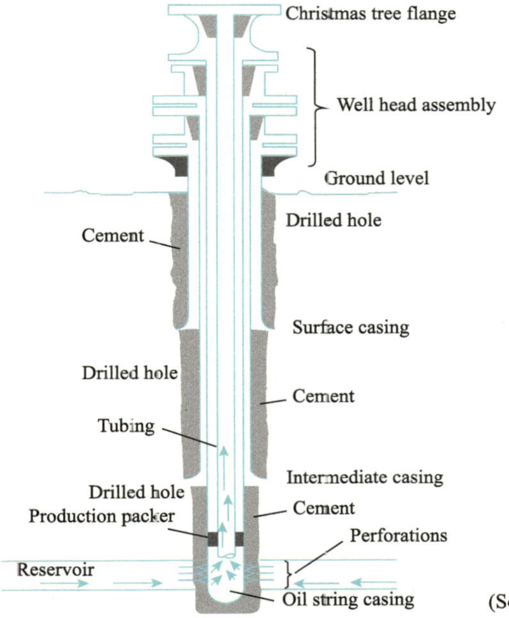

Figure 6.26 Wellhead Assembly
(Source: http://www.accessscience.com/)

Once the well is completed, operating personnel no longer need access to the wellhead. Onshore wellheads are therefore normally placed just below ground level, either buried or in cellars.

一旦完井后，操作人员将不需要井头，因此陆上井头一般被埋入地下。

6.3.7 Christmas Trees

In high-pressure flowing wells, the Christmas tree is the valve manifold that controls flow in the tubing(Figure 6.27).The tree is strongly constructed to contain full reservoir pressure(Figure 6.28). The functions of the various valves are follows:

6.3.7 采油树

在高压井中，采油树由一套控制油管油流的管阀组成（图6.27）。采油树强有力地控制油流压力（图6.28）。下图展示不同阀门的功用：

(a)

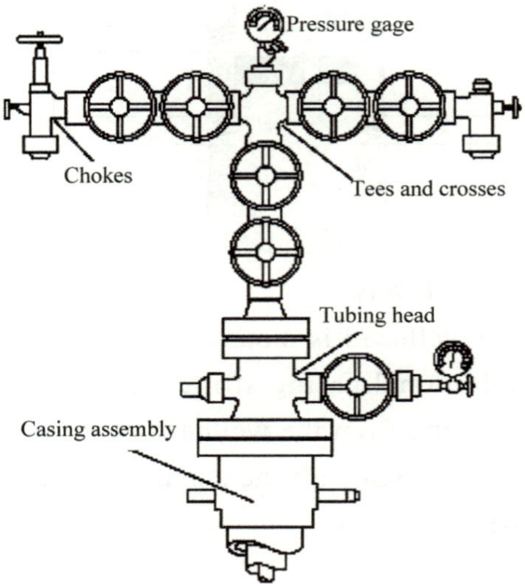

(b)

Figure 6.27 Christmas Tree Manifold
(Source: http://www.accessscience.com/)

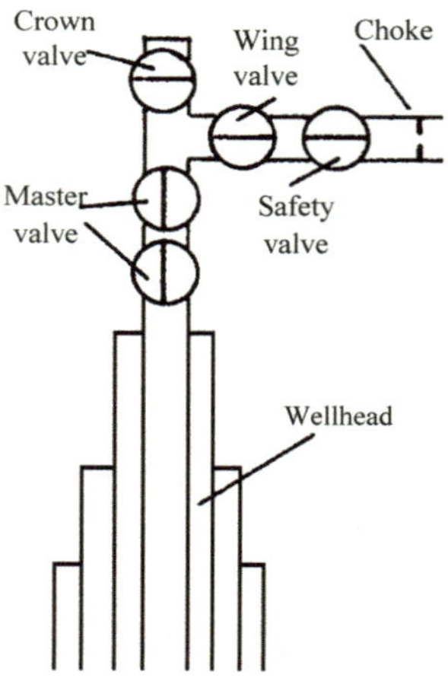

Figure 6.28 Wellhead and Christmas Tree

◆ The master valves are used to shut in the well. Many trees have dual master valves with only the top valve used while the bottom valve stays open. Should the top valve sustain wear and start leaking, the bottom valve is always in good condition to control the well while the top valve is repaired.

◆ 主阀门用来关井。许多采油树有两个主阀门，下面一个常开，上面一个供日常使用。如上面的磨损而不能关闭时，可以使用下面的主阀门。

◆ The crown (or lubricator) valve is used when a lubricator is attached to perform through-tubing well service.

◆ 防喷管阀门用来在过油管作业时对防喷管进行控制。

◆ The wing valve is normally used for routine opening and closing of the well.

◆ 生产阀门用来控制油井日常生产。

◆ The choke is an orifice that is varied in size to control the well's flow rate. It also protects downstream equipment by confining full well pressure to the tree.

◆ 大小不同的油嘴用来控制油井产量，并将压力控制在采油树内，保护下游设施不被过大压力损坏。

◆ The safety valve automatically shuts in the well if unsafe conditions occur, such as excessively high or low downstream pressure.

◆ 安全阀在井下压力异常时可以自动关闭来保障油井安全。

7 RESERVOIR PROPERTIES
油气藏性质

7.1 Properties of Reservoir Rock

In order to have a hydrocarbon producing reservoir, the following conditions must exist:

◆ There must be a body of rock having sufficient porosity (ϕ) to contain the reservoir fluids and permeability (k) to permit their movement.

◆ The rocks must contain hydrocarbons in commercial quantities, oil or gas saturation S_o or S_g.

7.1.1 Porosity

Most oil and gas deposits are found in sandstones and coarse-grained limestones, not in underground lake or pools. A piece of sandstone or limestone is very much like a hard sponge, full of holes, but not compressible. These holes, or pores, can contain water or oil or gas, and the rock will be saturated with one of the three. The holes are much tinier than sponge holes(Figure 7.1), but they are still holes, and they are called porosity. The oil and gas become trapped in these holes, stays there, for millions of years, until petroleum geologists come to find it and extract it(Figure 7.2, Figure 7.3).

Porosity is conventionally symbolized by the Greek letter phi (ϕ). Pores are of three morphological types: catenary, cul-de-sac, and closed. Catenary pores are those that communicate with others by more than one throat passage(Figure 7.4). Cul-de-sac, or dead-

7.1 油藏岩石性质

为了使地下油气藏产出碳氢化合物，储层岩石需要具备以下条件：

◆ 储层岩石需要有足够的孔隙度（ϕ）储存油气藏流体以及足够的渗透率(k)允许流体流动；

◆ 储层必须有足够可商业开采的碳氢化合物，即足够的含油饱和度S_o和含气饱和度S_g。

7.1.1 孔隙度

绝大多数的油气储存在砂岩或中粗粒灰岩中发现，而不是在地下湖或池中。一块砂岩或灰岩很像一块硬海绵，充满孔洞但是不可压缩。这些孔、洞能储存水、油、气，而储层中岩石由此三类中的一种流体饱和。这些孔洞比海绵中的孔眼还小（图7.1），但仍然是孔隙，叫孔隙度。油气被困于这些孔隙中达上百万年，直到石油地质学家找到并开发它们（图7.2、图7.3）。

孔隙度通常由希腊字母（ϕ）表示。孔隙有三种微观形态：悬链状孔隙、死端孔隙和

end pores have only one throat connecting with another pore. Closed pores have no communication with other pores(Figure 7.5).

Catenary and cul-de-sac pores constitute effective porosity, in that hydrocarbons can emerge from them. In catenary pores hydrocarbons can be flushed out by a natural or artificial water drive. Cul-de-sac pores are unaffected by flushing, but may yield some oil or

死孔隙。悬链状孔隙是那些连通多个喉道的孔隙（图7.4）。死端孔隙只有一个喉道连通其他孔隙。死孔隙指没有连通的孔隙（图7.5）。

悬链状孔隙、死端孔隙通常为有效孔隙，其中的碳氢化合物能从中流出。悬链状孔隙

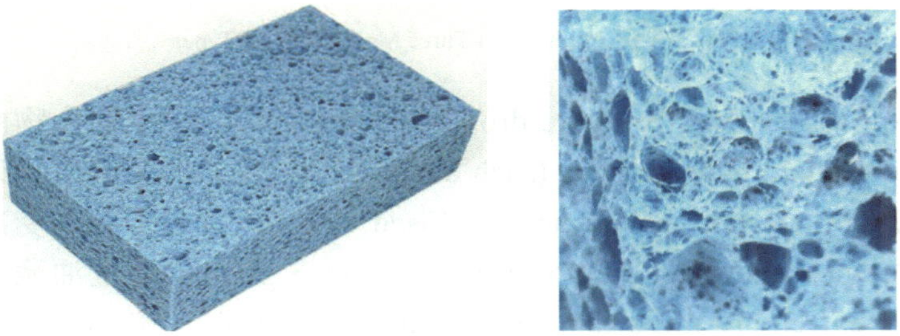

Figure 7.1 Sponge Holes (Source: http://product.suning.com/)

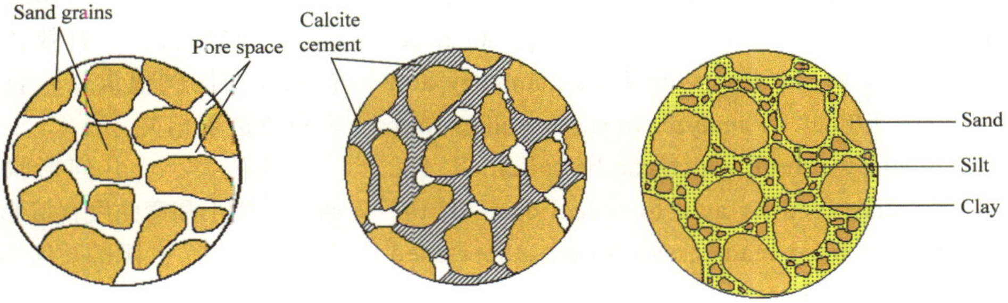

Figure 7.2 Pores (Source: http://homepage.usask.ca)

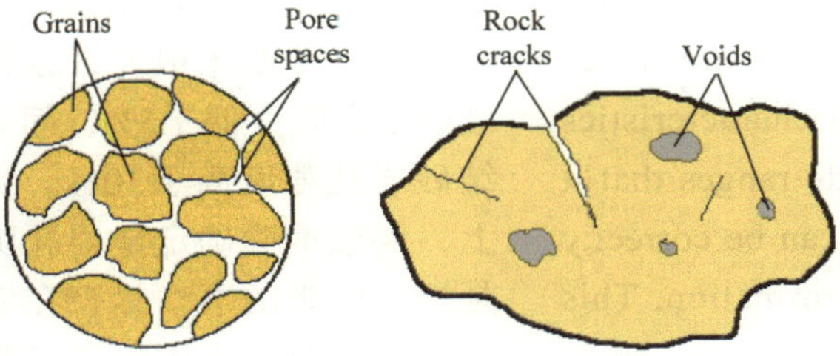

Figure 7.3 Porosity Voids
(Source: http://cmex.ihmc.us/)

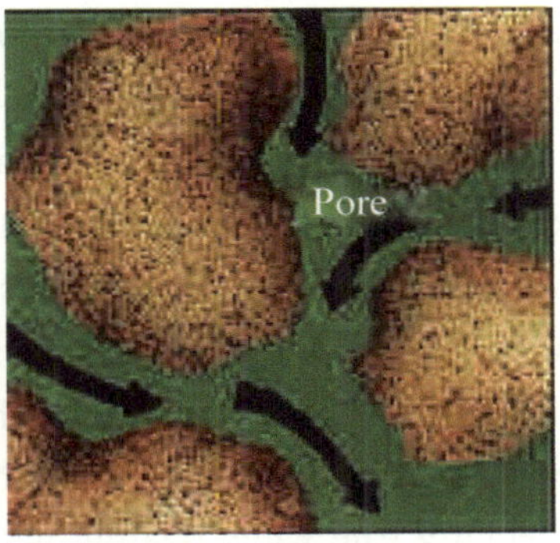

Figure 7.4 Oil Moving Through Pore Space in Sandstone
(Source: http://www.geomore.com/oil-and-gas-traps/)

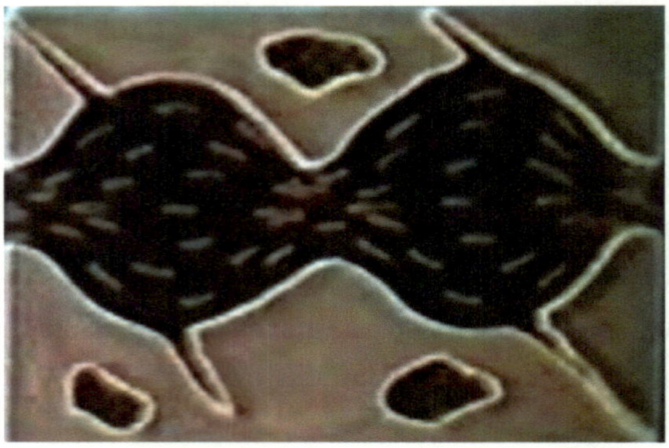

Figure 7.5 Pores are of Three Morphological Types

gas by expansion as reservoir pressure drops. Closed pores are unable to yield hydrocarbons (such oil or gas having invaded an open pore subsequently closed by compaction or cementation). The ratio of effective to total porosity is extremely important, being directly related to the permeability of a rock.

The size and geometry of the pores and the diameter and tortuosity of the connecting throat passages all affect the productivity of the reservoir. Pore geometry and continuity are difficult to analyze. In some studies rock pores are filled by a fluid, which can be solidified. The rock is then dissolved by acid to reveal casts of the pores. Someone used lead in sandstone others used epoxy resin on carbonate.

Porosity can vary from less than 5% in tightly cemented sandstones to more than 30% for unconsolidated sands. In reality, accurate determination of formation porosity is an extremely difficult matter. While it is true that laboratory technicians who specialize in this work can make accurate determinations on cores taken from a reservoir section, the physical characteristics of most reservoirs vary over such wide ranges that it is difficult to arrive at any figure that can be correctly called "average porosity" for that formation. This is especially true when dealing with carbonates, since the porosities and permeabilities are dependent on the nature of the rock developed during deposition and the

中的碳氢化合物能通过自喷或人工水驱流出。死端孔隙中的碳氢化合物不受冲洗的影响，但是能通过油藏压力下降油气膨胀产出部分油气。死孔隙不能产出碳氢化合物（这类油气最终由于压实和胶结作用进入周围开放孔隙）。有效孔隙占总孔隙的比例很重要，直接与岩石渗透率相关。

孔隙大小、几何形状、连通喉道大小和弯曲度直接影响油藏生产能力。孔隙的几何形状及连续性很难分析。在一些研究中，将储层岩石浸入某种液体中充满整个孔隙，固化成模。再用酸溶解岩石取得孔隙模型。有人将砂岩浸入铅液，有人将碳酸盐岩浸入环氧树脂液。

孔隙度变化很大，致密胶结砂岩孔隙度低于5%，而非固结砂岩孔隙度高于30%。实际上，要想准确确定地层孔隙度是极其困难的事。尽管在实验室通过专业分析师能准确确定岩心的孔隙度，但岩心来自于油藏的某一段，而油藏的物理特

diagenetic/solution processes(Figure 7.6).

The pore spaces, or voids, within a rock are generally filled with connate water, but contain oil or gas within a field.The porosity of a formation is its capacity to contain fluids. Porosity is either expressed as the void ratio, which is the ratio of voids to solid rock, or, more frequently as a percentage(Figure 7.7, Figure 7.8):

$$Porosity(\%)=\frac{Volume\ of\ voids}{Total\ Volume\ of\ Rock}\times 100\%$$

Porosity is the volume of the non-solid portion of the rock filled with fluids, divided by the total volume of the rock.There are three types of porosity(Figure 7.9):

◆ Intergranular porosity where the porosity is caused by the space in between the grains.

◆ Vugular porosity where the porosity is caused by holes that form in the rocks.

◆ Fractional porosity where the porosity is created by fractures or large cracks in the reservoir rocks.

In other classification, we can say there are two

性变化很大，这一点与那一处不同，因此很难用某一平均孔隙度代替整个油藏地层。特别是碳酸盐岩，其孔隙度和渗透率由岩石沉积和变质成岩过程决定（图7.6）。

岩石孔隙空间通常由共生水所充填，对于油藏而言孔隙空间还包含油或气。地层孔隙度反映地层岩石储存流体的能力。孔隙度一般由孔隙占岩石固体的百分比例表示（图7.7、图7.8）：

孔隙度（%）=孔隙体积/岩石总体积×100%

孔隙度是岩石中充满地层流体的非固化部分除以岩石的总体积。有三种孔隙类型（图7.9）：

◆ 粒间孔隙，其孔隙空间为砾石间空隙；

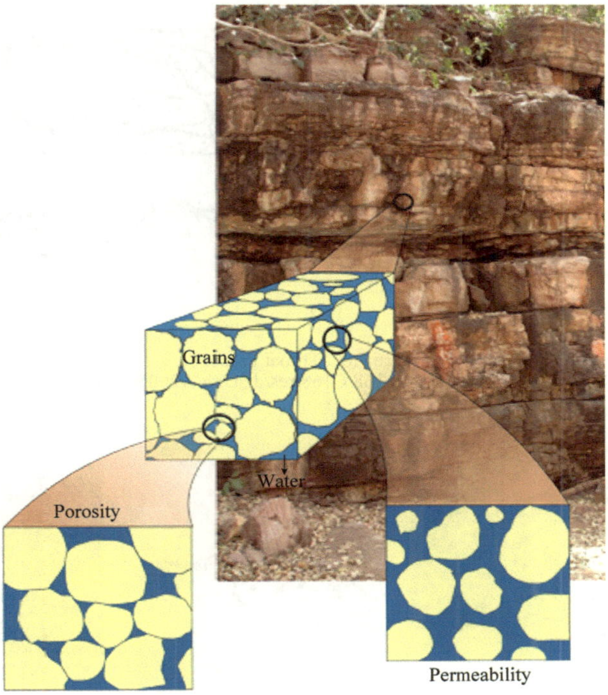

Figure 7.6 Porosity & Permeability
(Source: http://www.acwadam.org/images/hydropedia/big/porosity_permeability_big.jpg)

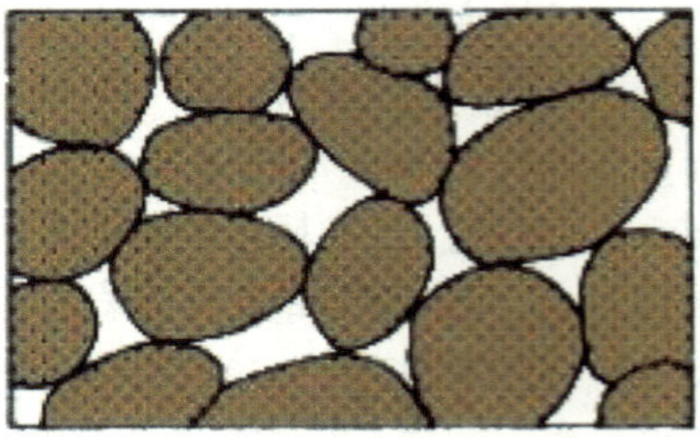

(a) Poorly Graded Uniform Granular Material Which Particles Have Their Own Porosity System

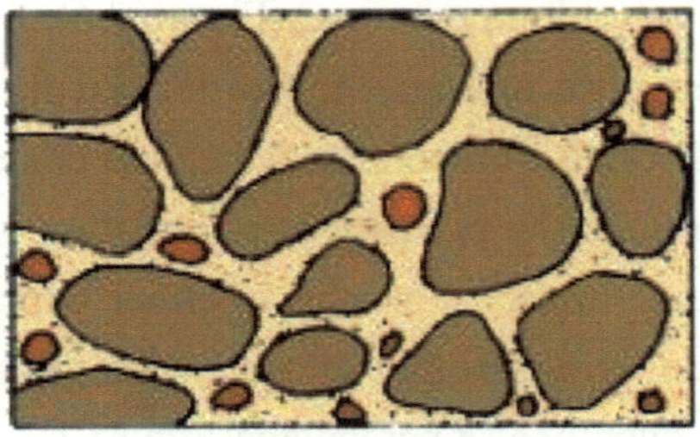

(b) Well Graded Granular Material with small Particles Filling the Big Pores

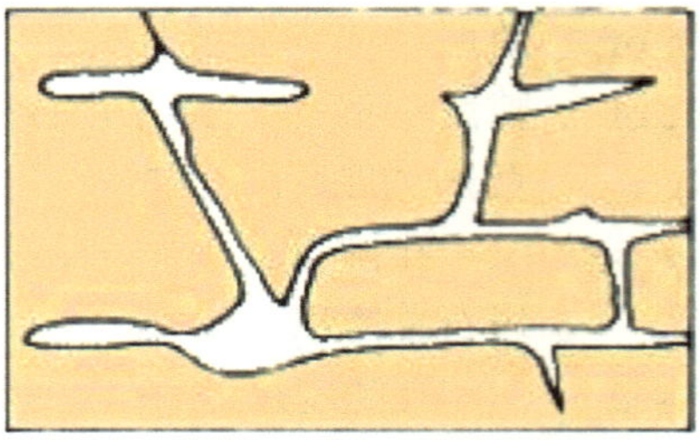

(c) Partially Closed Discontinuity Systems Appeared in an Intact Porous Rock due to Active Water

(d) Open Discontinuity Systems Appeared in an Intact Rock due to Mechanic Fracturing

Figure 7.7 Influence of Structure and Texture of the Porous Medium on Porosity
(Source:http://echo2.epfl.ch/VICAIRE/mod_3/chapt_1/main.htm)

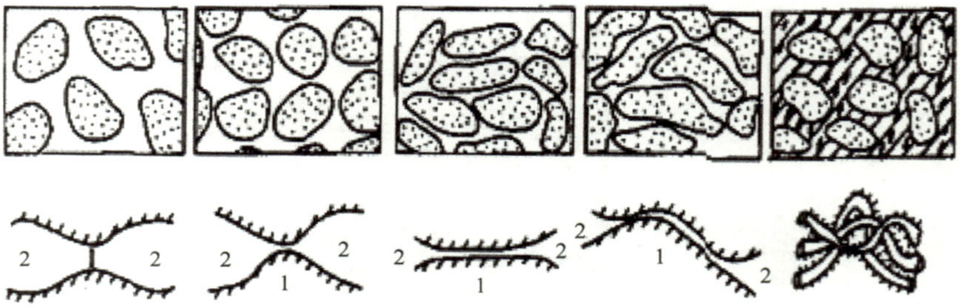

(a) (b) (c) (d) (e)

Figure 7.8 Throat Passage
(Source: http://210.41.240.168/C168/res/dzjc/zh/2/200507242348.htm)

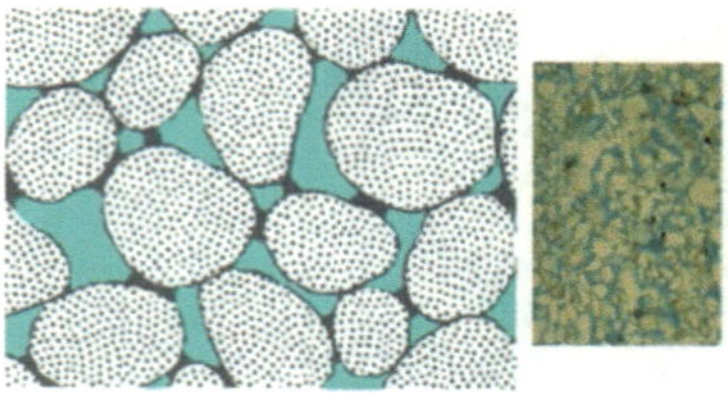

(a) Intergranular Porosity

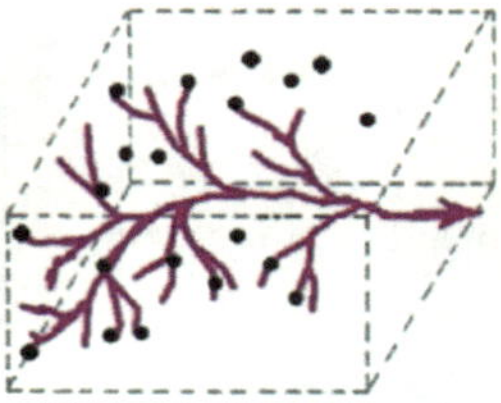

(b) Vugular Porosity

(c) Fractural Porosity

Figure 7.9 Porosity Types
(Source: Overview of the oil industry, Richard Lau)

types of porosity, Primary porosity and Secondary porosity:Primary porosity is the porosity developed by the original sedimentation process by which the rock was created. Secondary porosity is created by processes other than primary cementation and compaction of the sediments(Figure7.10, Figure7.11).

◆ 溶洞孔隙，其孔隙空间为岩石的溶洞空隙；

◆ 裂缝孔隙，其孔隙空间为油藏岩石裂缝和大的断裂带。

还有一种划分孔隙的方法将孔隙分为两种，原生孔隙和次生孔隙。原生孔隙指孔隙由原始沉积过程中的成岩作用形成。次生孔隙指原始固结和压实作用以后形成的孔隙（图7.10、图7.11）。

Pores
- *primary proespores*
  - *interparticle pores*
  - *intraparticle pores*
- *secondary pores*
  - *vuggyandmoldic pores*
  - *fenestral pores*
  - *intercrystalline pores*
  - *fracture pores*

Figure7.10 Pores

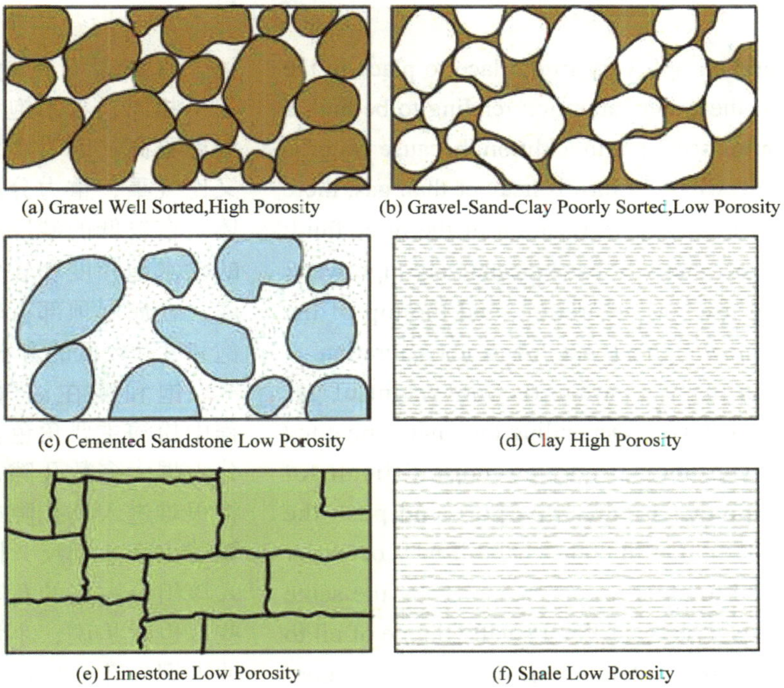

Figure 7.11 Different Rock Types Have Different Levels of Porosity
(Source: http://www.kgs.ku.edu/Publications/Bulletins/ED10/04_occur.html)

7.1.2 Saturation

Another rock characteristic affecting the movement of liquid is saturation, which express how much of the pore space is occupied by gas, oil and water. To say that a portion of the reservoir is saturated with gas means that most of the pores in this part of the reservoir are filled with gas(Figure7.12).

7.1.2 饱和度

另一个影响储层流体流动的参数是饱和度，饱和度指孔隙空间中哪些部分被油、气、水占据。油气藏被气饱和指这部分油气藏中绝大多数孔隙被气所占据（图7.12）。

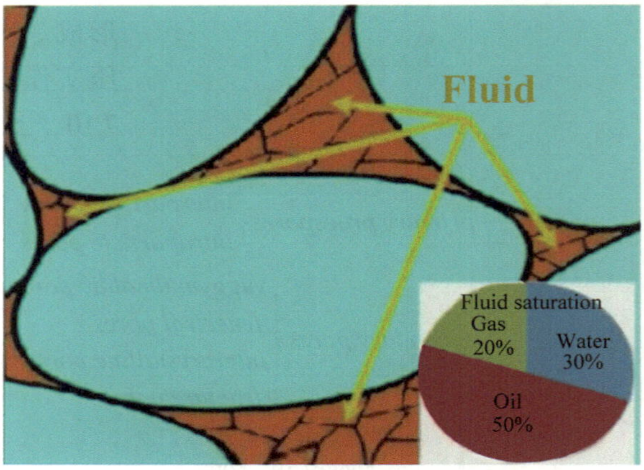

Figure 7.12 Fluid Saturation
(Source: Overview of the Oil Industry, Richard Lau)

The saturation will vary from place to place in the reservoir rock, the water saturation tending to be higher in the less porous sections. In addition, because water is heavier than oil, which in turn is heavier than gas, there will tend to be a gravity segregation of the three fluid; the lower part of sand will have a relatively high water saturation and low gas saturation, and the top of the structure will show high gas and low water saturation.

Saturation plays a part in the movement of gas and liquid in a reservoir. Depending on what compound occupied a given pore, it may be more difficult for different compound to enter that pore and push the first compound out. On the other hand it may be easier for a compound to enter a pore occupied by the same compound. Thus, it is easier for a small volume of oil to enter a pore already containing oil than to enter a pore containing water or gas.

油藏岩石的饱和度各点各不相同，含水饱和度在非孔洞处更高。因为水比油重，油又比气重，重力分离这三类流体。在较低的砂层有相对较高的含水饱和度和较低含气饱和度，在储层顶部有较高的含气饱和度和较低的含水饱和度。

饱和度在油气水流动过程中扮演主要角色。当某一组分占据大多数孔隙，其他组分将很难进入其孔隙或把该组分驱替出来。相反，同一组分进入该组分占据的孔隙更容易一些。也就是说，小量的油进入油占据的孔隙比进入水或气占据的孔隙更容易。

Total fluid saturation = $S_w + S_o + S_g$

100% =30% + 50% + 20%

These saturation percentages: S_w, S_o, S_g are great important to reservoir engineer. Because they give a measurement for the economic value of the reservoir. Formation with high S_o or high S_g may indicate a profitable reservoir.

流体总饱和度 = 含水饱和度+含油饱和度+含气饱和度

100% =30%+50%+20%

含水饱和度S_w、含油饱和度S_o、含气饱和度S_g这三个百分数参数对油藏工程师极为重要。因为他们用此来计算油藏的经济价值。储层的高含油饱和度S_o，含气饱和度S_g指示高回报的油藏。

7.1.3 Permeability

The third essential requirement for a reservoir rock is permeability. Permeability is the ability of a fluid to flow through rock(Figure 7.13). This is contrasted with porosity, which is the space between particles of a rock.

Porosity alone is not enough; the pore must be connected. Permeability refers to the ease with which fluids flow through any substance, it' s the ability of fluids to pass through a porous material. It is not sufficient to have oil or gas in a formation; the hydrocarbons must be able to flow from the reservoir into the well bore in order to be recovered at the surface. *Absolute permeability* is a physical characteristic of the rock. Permeability of a rock for oil, gas, or water is a

7.1.3 渗透率

第三个油藏岩石重要的参数是渗透率，渗透率指流体流经岩石孔隙的容易程度（图7.13）。它与孔隙度不同，表示岩石颗粒间的空间。

Figure 7.13 Permeability
(Source: Overview of the Oil Industry, Richard Lau)

光有孔隙度是不够的，孔隙必须连通起来。渗透率是指流体在任意孔隙介质中的流动难易程度。如果储层没有足够的油或气，碳氢化合物将不能从油藏流入井筒到地面。绝对渗透率是岩石本身的物理特性。岩石对油、气、水的渗透

function of the absolute permeability and the viscosity of the fluid.

H. Darcy, who studied the flow rates of the spring at Dijon in France, carried out the original work on permeability in 1856(Figure 7.14). Muskat and Botset further developed his work. They formulated Darcy's law as follows(Figure 7.15):

$$Q=\frac{K(p_1-p_2)A}{\mu L}$$

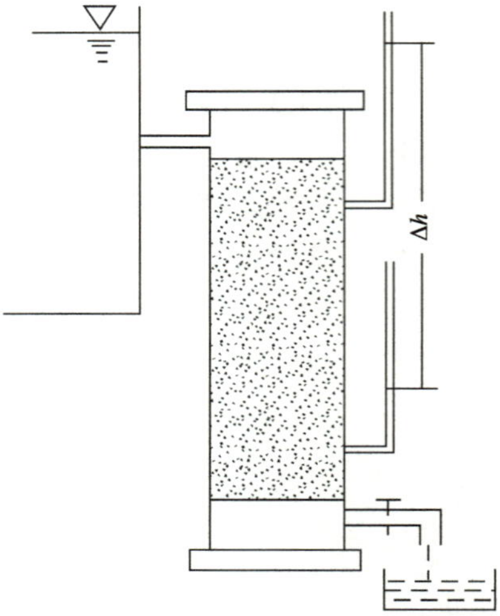

Figure 7.14 The Flow Diagram of Darcy's Experiment
(Source: The Fluid Mechanics in Porous Medium of Reservoirs)

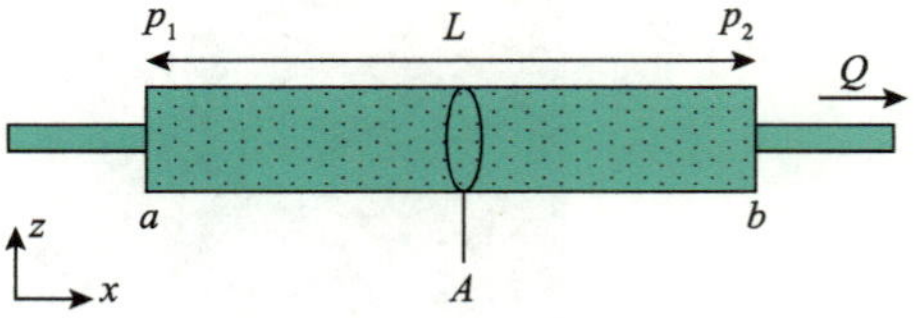

Figure 7.15 Diagram Showing Definitions and Directions for Darcy's Law
(Source: http://en.wikipedia.org/wiki/Darcy's_law)

So $K=\frac{Q\mu L}{(p_1-p_2)A}$

Where: Q——rate of flow;

K——permeability;

p_1–p_2——pressure drop across the sample;

A——cross-sectional area of the sample;

L——length of the sample;

μ——viscosity of the fluid.

Relative permeability is the ratio of the effective permeability for a particular fluid at a given saturation to a base permeability. The relative permeability ranges from 0.0 to 1.0. Thus for oil, gas and water:

率是绝对渗透率和流体黏度的函数。

亨利•达西在法国第戎研究过喷泉流速，于1856年作出渗透率的最初描述（图7.14），Muskat和Botset 在达西的研究基础上做了进一步的研究，他们形成的达西定律如下（图7.15）：

$$Q=\frac{K(p_1-p_2)A}{\mu L}$$

因此 $K=\frac{Q\mu L}{(p_1-p_2)A}$

式中 Q——流速；

K——渗透率；

p_1–p_2——流经岩样两端的压力；

A——岩样的横切面积；

L——岩样长度；

μ——流体黏度。

相对渗透率指基于某一流体在一定饱和度岩样下有效渗

$$K_{ro}=\frac{K_o}{K} \qquad K_{rg}=\frac{K_g}{K} \qquad K_{rw}=\frac{K_w}{K}$$

Where: K——absolute permeability (or permeability at irreducible wetting phase saturation);

K_r——relative permeability;

K_g——effective permeability for gas;

K_w——effective permeability for water;

K_o——effective permeability for oil.

We measure rock permeability in the laboratory using rock cone samples(Figure 7.16).

Figure 7.17, Figure 7.18 illustrates the total relative permeability relationship for different saturations. It is necessary to consider the relative saturations, how the oil and gas are distributed, and the presence or absence of water. Most reservoirs are water-wet, having a film of water separating the pore boundaries from the oil. Few reservoirs are oil-wet, totally lacking a film of connate water at pore boundaries. A mixed, or intermediate, wettability condition may also occur.

透率对基底渗透率（绝对渗透率）的比值。相对渗透率的范围为0～1.0。有油相相对渗透率、气相相对渗透率和水相相对渗透率：

$$K_{ro}=\frac{K_o}{K} \qquad K_{rg}=\frac{K_g}{K}$$

$$K_{rw}=\frac{K_w}{K}$$

式中K——绝对渗透率；

K_r——相对渗透率；

K_g——气相相对渗透率；

K_w——水相相对渗透率；

K_o——油相相对渗透率。

我们用岩石圆柱样本在实验室测得渗透率（图7.16）。

图7.17、图7.18展示总的相对渗透率与不同饱和度的关系。提到渗透率时必须要考虑流体相对饱和度，油气如何分布和水的存在。绝大多数油藏

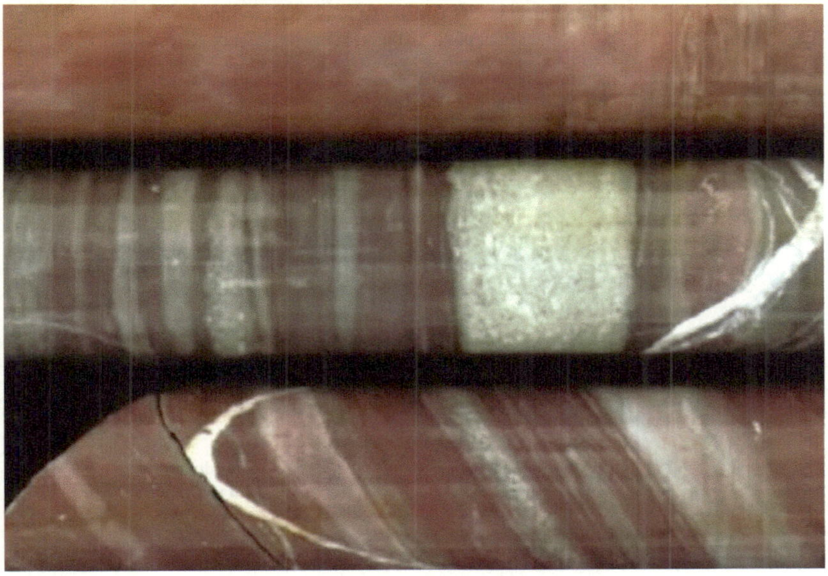

Figure 7.16 Rock Core Samples
(Source: Overview of the Oil Industry, Richard Lau)

The relationship between porosity and permeability for a given formation is not necessarily a close or direct one. However, high porosity is often associate with high permeability.

为水湿油藏，孔隙边缘有一层水膜隔离油藏。少数油藏为油湿油藏，在孔隙边缘缺乏共生水膜。也有水湿和油湿混合或中间油藏出现。

对于某一特定地层的孔隙度大小和渗透率值没有明显直接关系，但是高孔隙度区一般也是高渗透率区。

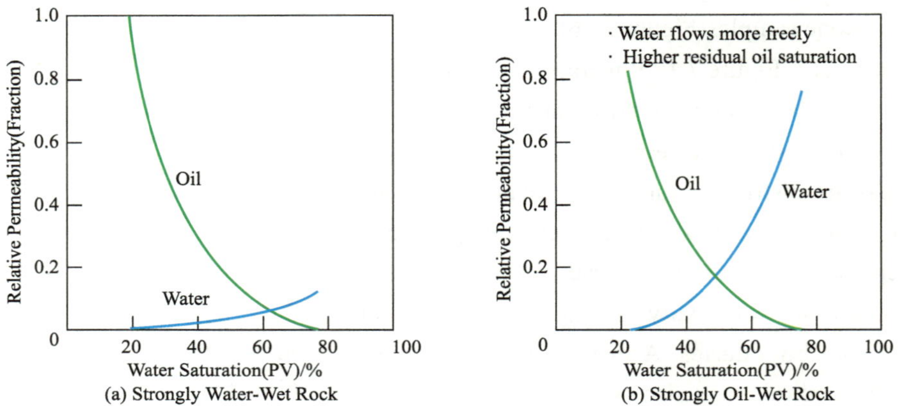

Figure 7.17 Relative Permeability Curves in Water Wet and Oil Wet Reservoirs
(Source: http://www.spec2000.net/09-relperm.htm)

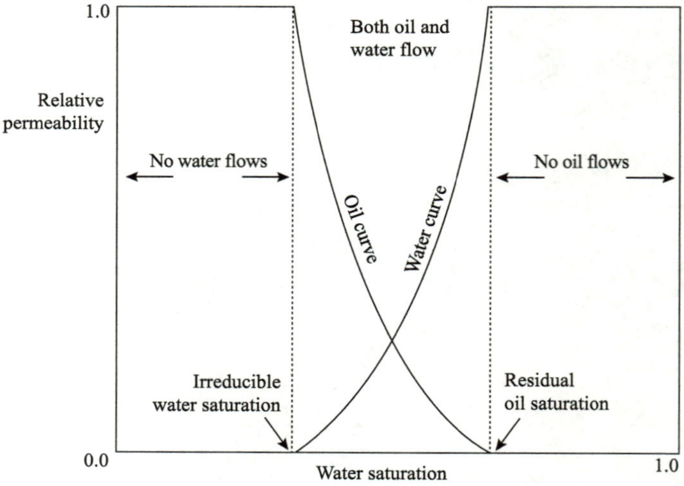

Figure 7.18 Typical Relative Permeability Curves for Oil and Water
(Source: http://www.spec2000.net/15-permbasics.htm)

7.2 Reservoir Pressure

The fluids in the pores of the reservoir rock are under a certain degree of pressure, generally called "*reservoir pressure*" or "*formation pressure*".

The relationship between reservoir fluid pressure and depth may be used to define the interface between fluids (e.g. gas–oil or oil–water interface) or to confirm the observations made directly by wireline logs(Figure 7.19). This is helpful in determining the volumes of fluids in place, and in distinguishing between areas of a field which are in different pressure regimes or contain different fluid contacts. If different pressure regimes are encountered within a field, this is indicative of areas which are isolated from each other either by sealing faults or by lack of reservoir continuity. In either case, the development of the field will have to reflect this lack of communication, often calling for dedicated wells in each separate fault block. This is important to understand during development planning, as later realisation is likely to lead to a sub-optimal development (either loss of recovery or increase in cost).

Normal pressure regimes follow a hydrostatic fluid gradient from surface, and are approximately linear. A normal reservoir pressure at the oil-water contact approximates very closely the hydrostatic pressure gradient varies somewhat, depending upon the amount of

7.2 油藏压力

油藏压力或地层压力指油藏岩石孔隙中流体的压力。

油藏压力与深度的关系用于确定流体界面（例如气油界面或油水界面）或确定测井数据（图7.19），从而帮助确定地下油气体积，识别不同压力体系下的不同油气藏或不同的流体界面。假如在同一油田中有不同压力体系，表面上这些油气藏是相互独立的，它们可能被断层封隔从而缺少连通。在后续开发过程中需为各封隔断块独立钻井。将这些考虑进开发方案中很重要，从而作出最优开发方案（或次优开发方案）。

正常压力指从地面到地下静水压力梯度为线性。油藏在油水界面处的正常压力约等于静水头压力。异常压力指高于静水头压力或低于静水头压力，代表与正常压力的不连续（图7.20）。

在异常压力油藏中，储层岩石碎屑可能部分由流体压力

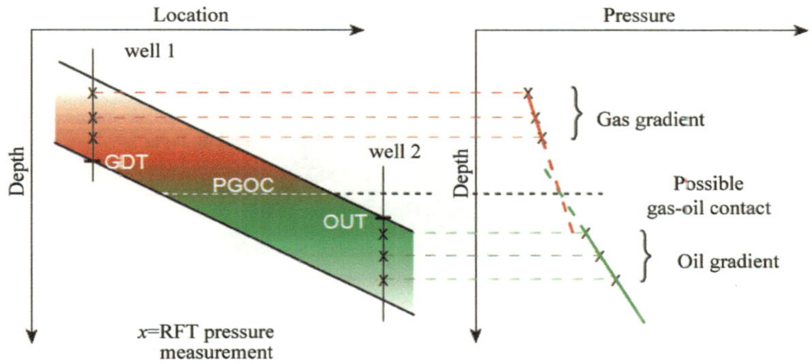

Figure 7.19 The Gradient Intercept Technique
(Source: Hydrocarbon Exploration and Production)

dissolved salts in the formation water. Abnormal pressure regimes include overpressured and underpressured fluid pressures, and represent a discontinuity in the normal pressure gradient(Figure 7.20).

In abnormally pressured reservoirs, clastics can be partially supported by the fluid pressure and partially by the grain-to-grain contact. When this abnormal pressure is reduced through production, compaction of the reservoir bed will occur. If left unchecked, this subsurface compaction can cause serious problems, for example it can cause casing to collapse and be the cause of subsidence, which can be expressed all the way to the surface. Drilling through abnormal pressure regimes requires special care.

承载，部分由岩石颗粒承载。随着开发生产异常压力减小，油藏基岩可能被进一步压实。如果放任不管，地下岩石压实可能会导致严重问题，例如套管破裂和地层沉降，然后波及危害到地面。因此，钻遇异常压力段需要特别小心。

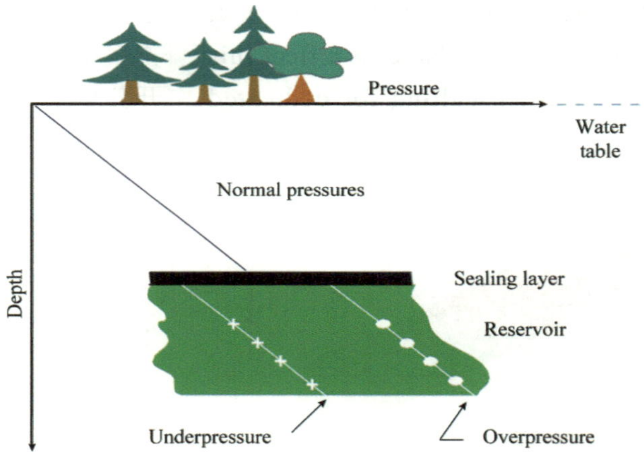

Figure 7.20 Normal and Abnormal Pressure Regimes
(Source: Hydrocarbon Exploration and Production)

7.3 Reservoir Fluids

Petroleum reservoirs usually contain three main fluids: crude oil, natural gas and water, with minor constituents being acid gases (carbon dioxide and hydrogen sulfide) and hydrocarbon solids (bitumen, asphalt, etc.). These components will vary considerably in combination and proportion within each reservoir. In addition, each component' s presence, quality, phase and quantity will vary from reservoir to reservoir(Figure7.21).

7.3 油藏流体

石油通常由三种流体组成：原油、天然气和水，还有部分微组分在酸气（二氧化碳和硫化氢）和碳氢化合物固体（沥青等）中。各个油藏中这些组分和其比例不同。另外，每种组分的质量、相态和数量在各个油藏中也不相同（图7.21）。

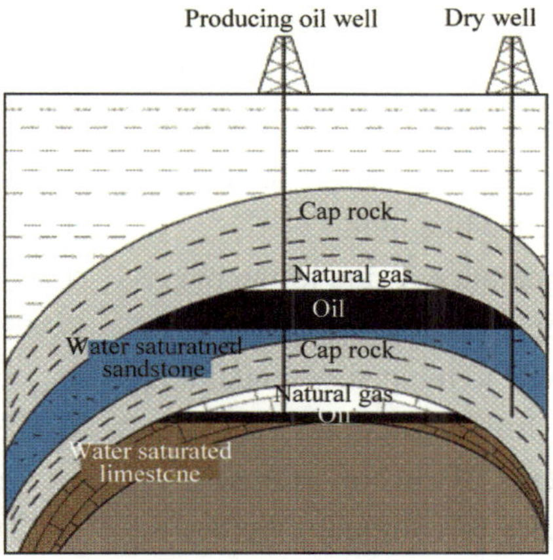

Figure7.21 Typical Anticline Oil and Gas Reservoir
(Source: http://www.ems.psu.edu/)

7.3.1 Hydrocarbon Chemistry

The fluids contained within petroleum accumulations are mixtures of organic compounds, which are mostly hydrocarbons (molecules composed of hydrogen and carbon atoms), but may also include sulphur, nitrogen, oxygen and metal compounds. This section will concentrate on the hydrocarbons, but will also explain the significance of the other compounds in the processing of the fluids(Figure7.22).

Petroleum fluids vary significantly in appearance, from gases, through clear liquids with the appearance of lighter fuel, to thick black, almost solid liquids. In terms of weight percent of crude oil, for example the carbon element represents 84%~87%, the hydrogen element

7.3.1 碳氢化合物的化学组成

油藏中流体是有机物组分的混合物，主要以碳氢化合物（氢原子和碳原子组成的分子）为主。但也包含硫、氮、氧和金属化合物。这章集中介绍碳氢化合物，也会介绍流体中的其他组分（图7.22）。

石油流体在外表形态上变化很大，从气体到液体再到固体各不相同。有透明的轻质燃油，又黑又黏石油沥青。就原油而言，碳元素占到84%~87%，氢元素占到11%~14%，其他元素一

- *Petroleum*
  - *Crude Oil*
    - *Hydrocarbons*
      - *Paraffins* (C_nH_{2n+2})
      - *Naphthenes* (C_nH_{2n})
      - *Aromatics* ($C_6H_5+C_nH_{2n+2}$)
    - *Heterocompounds*
      - *Oxygen Compounds*
      - *Sulfur Compounds*
      - *Nitrogen Compounds*
      - *Others*
  - *Natural Gas*
    - *Dry Gas* *Methane*(CH_4)
    - *Wet Gas*
      - *Ethane* (C_2H_6)
      - *Propane* (C_3H_8)
      - *Butane* (C_4H_{10})

Figure 7.22 Petroleum Hydrocarbon

11%~14% and the other elements typically less than 1%. Despite this fairly narrow range of weight percent of the carbon and hydrogen elements, crude oil can vary from a light-brown liquid with a viscosity close to that of water, to a very high viscosity tar-like fluid.

The diversity of the appearance is due to the many ways in which the carbon atoms are able to bond to each other, from single carbon atoms to molecules containing hundreds of carbon atoms linked together in linear chains, to cyclic arrangements of carbon atoms. It is the ability of carbon molecules to combine together in long chains (catenate) which makes organic (i.e. carbon containing) compounds far more numerous than those of other elements, and the basis of living matter.

The various arrangements of carbon atoms can be categorised into ‘series’ which describe a common molecular structure (Table 7.1). The series are based on four main categories which refer to:

般低于1%。碳、氢元素所占比例的微小变化可以使原油从亮棕色液体变为高黏焦油。

石油外在的不同源于碳原子彼此固结方式的不同，从一个碳原子到成百上千个碳原子以线性连接，到以环形连接在一起。这是一个碳分子以长链结合成有机组分形成生物的基础。

碳原子按照不同的排列方式可以划分为以下4个系列（表7.1）：

◆ 碳分子排列的不同：

（1）直链；

（2）环链。

◆碳分子之间键的不同：

（1）饱和（单键）连接；

(2) 不饱和 (多键) 连接。

Table 7.1　Hydrocarbon Constituents of Petroleum

Hydrocarbons	Examples
Paraffin	$CH_3(CH_2)_6CH_3$ octane　　$(CH_3)_3CCH_2CH(CH_3)_2$ 2,2,4-trimethylpentane(isooctane)
Naphthenes	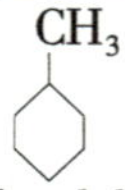 methylcyclohexane　　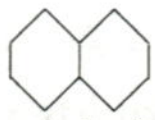 decahydronaphthalene(decalin)
Aromatics	 benzene　　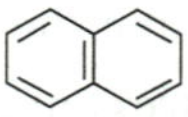 naphthalen　　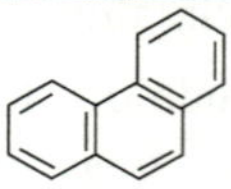 phenanthrene
Non-Hydrocarbons	**Examples**
Sulfur Compounds	CH_3CH_2SH ethanethiol(ethylmercaptan)　　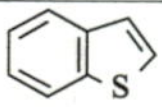 benzothiophene
Oxygen Compounds	 phenol　　 furan
Nitrogen Compounds	 pyridine　　 pyrrole　　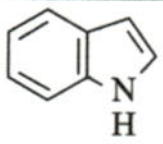 indole
Metals	Fe, Ni, V, Cu

(Source: http://www.wiredchemist.com/environmental/instructional)

◆ the arrangement of the carbon molecules:

(1) open chain (which may be straight chain or branched);

(2) cyclic (or ring).

◆ the bonds between the carbon molecules:

(1) saturated (or single) bond;

(2) unsaturated (or multiple) bond.

7.3.1.1 The Alkanes

The largest series is that of the alkanes or paraffins, which are open chain molecules with saturated bonds, and have the general formula C_nH_{2n+2}.

Figure 7.23 shows the way in which the molecules are visualised, their chemical symbol and the names of the first three members of the series. The carbon atom has four bonds that can join with either one or more carbon atoms (a unique property) or with atoms of other elements, such as hydrogen. Hydrogen has only one bond, and can therefore join with only one other atom.

7.3.1.1 烷烃

碳氢化合物中最大的一个系列是烷烃，直链分子以饱和键相连，分子式通式为C_nH_{2n+2}。

图7.23显示烷烃中头3个的分子形式。碳原子有4个键连接着一个或多个碳原子或别的原子（如氢原子）。氢原子只有一个键，因此每次只能连接一个别的原子。

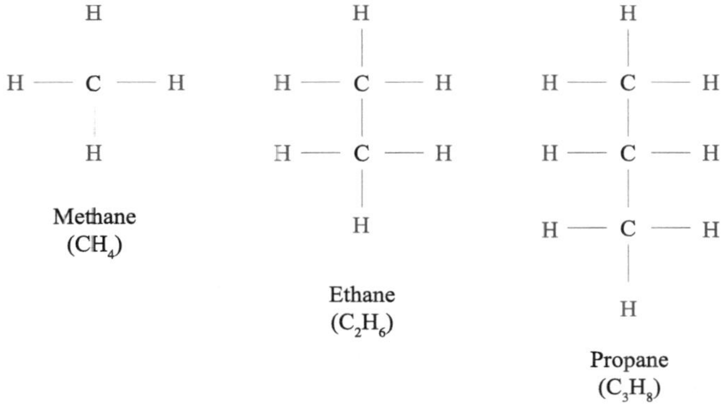

Figure 7.23 Examples from the Alkane (Paraffin) Series
(Source: Hydrocarbon Exploration and Production)

Under standard conditions of temperature and pressure (STP), the first four members of the alkane series (methane, ethane, propane and butane) are gases. As the length of the carbon chain increases, the density of the compound increases: C_5H_{12} (pentane) to $C_{17}H_{36}$ are liquids, and from $C_{18}H_{36}$, the compounds exist as wax-like solids at STP.

The most common prefixes are written below using

在标准温度和压力下，烷烃中的头4个都是气体（甲烷、乙烷、丙烷和丁烷）。随着碳链的增长，复合物的密度增加，C_5H_{12} (戊烷) 到 $C_{17}H_{36}$是液体，$C_{18}H_{36}$以上是蜡状固体。

以碳分子前缀表示烷烃系列，如下：

the alkane series as an example, and the prefixes are italicized:

C_1 *methane*;
C_2 *ethane*;
C_3 *propane*;
C_4 *butane*;
C_5 *pentane*;
C_6 *hexane*.

Beyond propane, it is possible to arrange the carbon atoms in branched chains whilst maintaining the same number of hydrogen atoms. These alternative arrangements are called isomers, and display slightly different physical properties (e.g. boiling point, density, critical temperature and pressure). Some examples are shown in Figure 7.24.

C_1 甲烷;
C_2 乙烷;
C_3 丙烷;
C_4 丁烷;
C_5 戊烷;
C_6 己烷。

从丙烷开始，碳原子上分支的排列可能不同。这种不同叫异构，其物理性质（如沸点、密度、临界温度和压力）也有些许不同（图7.24）。

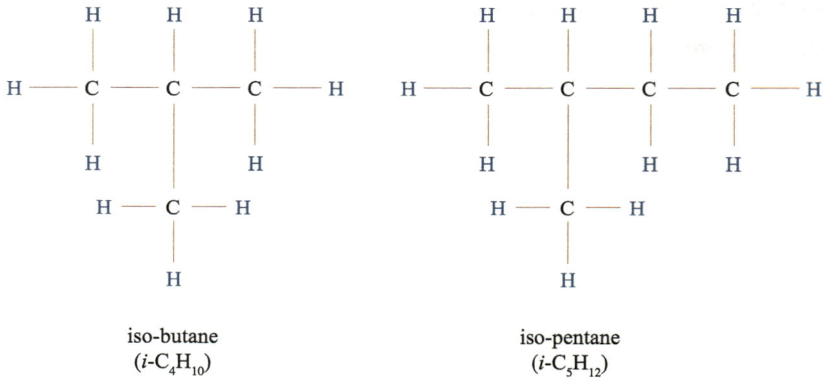

Figure 7.24 Isomers of the Paraffin Series
(Source: Hydrocarbon Exploration and Production)

Alkanes from CH_4 to $C_{40}H_{82}$ typically appear in crude oil, and represent up to 20% of the oil by volume. The alkanes are largely chemically inert (hence the name paraffins, meaning little affinity), owing to the fact that the carbon bonds are fully saturated and therefore cannot be broken to form new bonds with other atoms. This probably explains why they remain unchanged over long periods of geological time, despite their exposure to elevated temperatures and pressures.

烷烃从CH_4 到$C_{40}H_{82}$为典型的原油，代表着20%的石油产量。烷烃有很大的化学惰性（因此叫烷烃，意思是没有亲和力），因为碳原子全部被饱和，不能打断形成新的连接。这就能解释为什么在很长的地质年代中不管温度、压力的改变，烷烃能保持不变。

7.3.1.2 The Olefins

Open chain hydrocarbons which are undersaturated, that is having at least one carbon–carbon double bond, are part of the olefin series, and have the ending '-ene'. Those with one carbon–carbon double bond are called mono-olefins or alkenes, for example ethylene $CH_2=CH_2$.

The double bond is not stronger than the single bond; on the contrary, it is more vulnerable, making unsaturated compounds more chemically reactive than the saturates.

In the longer carbon chains, two carbon–carbon double bonds may exist. Such molecules are called diolefins (or dienes), such as butadiene $CH_2=CH-CH=CH_2$.

7.3.1.2 烯烃

未饱和的开链烃类至少有一个碳碳双键，是烯烃系列的一部分，以“-ene”结尾。一个碳碳双键为单烯烃，以乙烯$CH_2=CH_2$为例。

双键没有单键牢固，相反它们更不稳定，比饱和烃更活跃而容易形成不饱和化合物。

在长碳链中，双碳双键可能存在，叫二烯，如丁二烯。

7.3.1.3 Acetylenes

Acetylenes are another series of unsaturated hydrocarbons which include compounds containing a carbon–carbon triple bond, for example acetylene itself:

$$CH \equiv CH$$

Olefins are uncommon in crude oils due to the high chemical activity of these compounds which causes them to become saturated with hydrogen. Similarly, acetylene is virtually absent from crude oil, which tends to contain a large proportion of the saturated hydrocarbons, such as the alkanes.

Whilst the long-chain hydrocarbons (above 18 carbon atoms) may exist in solution at reservoir temperature and pressure, they can solidify at the lower temperatures and pressures experienced in surface facilities, or even in the tubing. The fraction of the longer chain hydrocarbons in the crude oil is therefore of particular interest to process engineers, who will typically require a detailed laboratory analysis of the crude oil composition, extending to the measurement of the fraction of molecules as long as C_{30}.

7.3.1.3 乙炔

乙炔是另一种不饱和碳氢化合物，它包含碳碳三键，如乙炔：

$$CH \equiv CH$$

烯烃由于其高活性在原油中并不常见，它们容易被氢原子饱和。同样，乙炔在原油中也不存在，它们容易被氢原子饱和变成烷烃。

长碳分子（18个碳原子以上）可能在油藏温度和压力下存在，但在温度和压力下降后的地面或油管中变成固体。因此，这部分长碳链原油引起工程师的极大关注，他们需要进行详细的原油成分实验分析及相关的测试，一直到C_{30}。

7.3.1.4 Ring or Cyclic Structures

The naphthenes (C_nH_{2n}), or cycloalkanes, are ring or cyclic saturated structures, such as cyclohexane

7.3.1.4 环烃

环烃或环烷烃为环状或圆环状饱和结构，如环己烷（C_6H_{12}），然而其他尺寸的

(C_6H_{12}), though rings of other sizes are also possible. An important series of cyclic structures is the arenes (or aromatics, so called because of their commonly fragrant odours), which contain carbon–carbon double bonds and are based on the benzene molecule (Figure 7.25).

环状也是有可能的。具有环状结构的重要碳氢化合物是芳烃（该系列普遍有香味），芳烃包含碳碳双键和苯类型的结构（图7.25）。

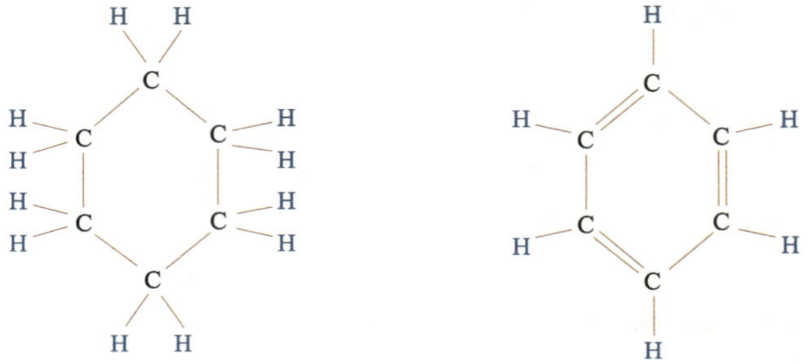

Cyclo-Hexane(C_6H_{12})
(Naphthene Series)

Benzene(C_6H_6)
(Aromatic Series)

Figure 7.25 Ring or Cyclic Structures
(Source: Hydrocarbon Exploration and Production)

Although benzene contains three carbon–carbon double bonds, it has a unique arrangement of its electrons (the extra pairs of electrons are part of the overall ring structure rather than being attached to a particular pair of carbon atoms) which allows benzene to be relatively unreactive. Benzene is, however, known to be a cancerinducing compound.

尽管苯包含3个碳碳双键，但由于其电子排列的独特结构（额外的电子对是整个环状结构的一部分，而不是连接到一对特殊的碳原子上），使苯相对不活跃。苯，众所周知，它是一种致癌化合物。

Some of the common aromatics found in crude oil are the simple derivatives of benzene in which one or more alkyl groups (CH_3) are attached to the basic benzene molecule as a side chain which takes the place of a hydrogen atom. These arenes are either liquids or solids under standard conditions (Figure 7.26).

在原油中发现的其他芳烃是苯的简单衍生物，烷基代替氢原子连接到苯分子上作为一个侧链。在标准状况下，芳烃为液体或固体（图7.26）。

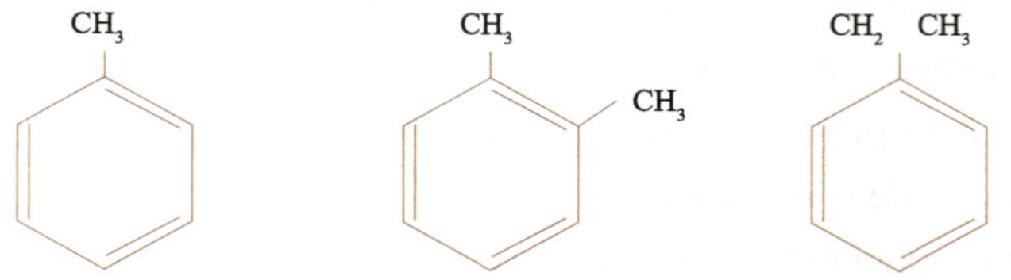

Toluene or
Methylbenzene
$C_6H_{10}(CH_3)$

O-Oxylene or
1,2-Dimethylbenzene
$C_6H_{10}(CH_3)_2$

Ethylbenzene
$C_6H_{10}(CH_2CH_3)$

Figure 7.26 Derivatives of Benzene
(Source: Hydrocarbon Exploration and Production)

7.3.1.5 Non-hydrocarbon Components of Petroleum Fluids

The non-hydrocarbon components of crude oil may be small in volume percent, typically less than 1%, but their influence on the product quality and the processing requirements can be considerable. It is therefore important to identify the presence of these components as early as possible, and certainly before the field development planning stage, to enable the appropriate choice of processing facilities and materials of construction to be made.

Sulphur and its products are the most common impurity in crude oil, ranging from 0.2% to over 6% in some Mexican and Middle Eastern crudes, with an average of 0.65% by weight. Corrosive sulphur compounds include free *sulphur*, hydrogen sulphide (H_2S, which is also highly toxic) and mercaptans of low molecular weight (e.g. ethyl mercaptan, C_2H_2SH). Mercaptans are formed during the distillation of crude oil, and require special alloys in plant equipment to avoid severe corrosion. The noncorrosive sulphur compounds are the sulphides [e.g. diethyl sulphide $(C_2H_5)_2S$], which are not directly corrosive, but require careful temperature control during processing to avoid decomposition to the corrosive products. Sulphur compounds have a characteristic bad smell, and both corrosive and non-corrosive forms are generally undesirable in crude oils. Corrosion due to H_2S is known as 'sour' corrosion.

Some natural gases contain a high H_2S content; above 30% in some Canadian producing wells, where the sulphur is recovered from the product stream and is sold commercially.

Nitrogen content in crude oil is typically less than 0.1% by weight, but can be as high as 2%. The nitrogen compounds in crude oil are complex, and remain largely unidentified. Gaseous nitrogen reduces the calorific value and hence sales price of the hydrocarbon gas.

Oxygen compounds are present in some crude

7.3.1.5 非烃组分

原油中的非碳组分所占比例很小，一般不到1%，但是它们却影响产品的品质和处理成本。因此，尽快地识别这些成分，并于开发方案阶段考虑进去，为选择合适的处理设备和基建做准备。

硫及其相关产物是原油中的最常见的杂质，在某些墨西哥和中东原油中占到0.2%～6%，平均占总质量的0.65%。腐蚀性的硫化物包括硫、硫化氢（有剧毒）和低分子硫醇（如乙硫醇）。硫醇会在原油蒸馏中形成，因此为避免严重腐蚀需要特殊设备。非腐蚀性硫的化合物叫硫化物（如二乙基硫化物），硫化物虽不产生直接腐蚀，但要求处理过程中要小心控制温度以免降解成腐蚀性产品。硫的相关产物都带有难闻的气味，原油中具有腐蚀性或非腐蚀性的硫产物都不是想要获得的。硫化氢的腐蚀性一般叫酸腐。

一些天然气井中包含很高的硫化氢成分，在加拿大的一些气井中，硫化氢含量高达30%，因此硫被采出然后商业出售。

氮含量在原油中一般低于0.1%，但也可能高达2%。氮的化合物在原油中很复杂，还有大部分未识别。氮气减少了燃烧值，因此降低了天然气的价格。

氧化物在某些原油中存在，在蒸馏中会分解成环烷

oils, and decompose to form naphthenic acids upon distillation. These may be highly corrosive.

Carbon dioxide (CO_2) is a very common contaminant in hydrocarbon fluids, especially in gases and gas condensate, and is a source of 'sweet' corrosion problems. CO_2 in the gaseous phase dissolves in any water present to form carbonic acid (H_2CO_3) which is highly corrosive. Its reaction with iron creates iron carbonate ($FeCO_3$):

$$Fe + H_2CO_3 \rightarrow FeCO_3 + H_2$$

The corrosion rate of steel in carbonic acid is faster than in hydrochloric acid.

Other compounds which may be found in crude oil are metals such as vanadium, nickel, copper, zinc and iron, but these are usually of little consequence.

Natural gas may contain helium, hydrogen and mercury, though the latter is rarely a significant contaminant in small quantities.

酸，也是高腐蚀性的。

二氧化碳是碳氢化合物中的主要污染物，特别在气体和凝析气中，它是“甜气”腐蚀的来源。二氧化碳气相溶于水变成碳酸，具有高腐蚀性，能和铁反应生成碳酸铁：

$$Fe + H_2CO_3 \rightarrow FeCO_3 + H_2$$

碳酸对铁的腐蚀速率比盐酸还快。

在原油中还有其他金属成分，如钒、铌、铜、锌和铁，但这些一般含量很少。

天然气中可能含有氦气、氢气和水银，水银量少对天然气的品质有很大污染。

7.3.2 Reservoir Fluids Properties

7.3.2.1 General Hydrocarbon Phase Behaviour

As the conditions of pressure and temperature vary, the phases in which hydrocarbons exist and the composition of the phases may change. It is necessary to understand the initial condition of fluids to be able to calculate surface volumes represented by subsurface hydrocarbons. It is also necessary to be able to predict phase changes as the temperature and pressure vary both in the reservoir and as the fluids pass through the surface facilities, so that the appropriate subsurface and surface development plans can be made.

Figure 7.27 shows the phase boundaries between the component in the solid, liquid and gas (vapour) states. Starting with the liquid (water) at point *A*, as the temperature is increased the boiling point is approached until the boiling point curve is reached, at which point the water boils and turns to steam (gas). Starting from the situation of the gaseous phase at point *B*, if the temperature is reduced the dew point curve is approached, and when

7.3.2 油藏流体性质

7.3.2.1 碳氢化合物的相态变化

当压力温度变化时，油气藏中的碳氢化合物的相态和组分也随之变化。因此，理解初始状态下的流体状态能计算出地下碳氢化合物的表面体积。从而预测在碳氢化合物从油气藏内开采到地面过程中随温度、压力变化的流体相态变化，谋划出更适宜的地下和地面的开采方案。

图7.27显示组分为固体、液体、气体状态时的相态边界。从液相（水）*A*点开始，当温度上升到沸点线，在沸点液态水沸腾成蒸汽。从气相点*B*开始，假如温度降低到露点，气相组分转换成液相。对于单组

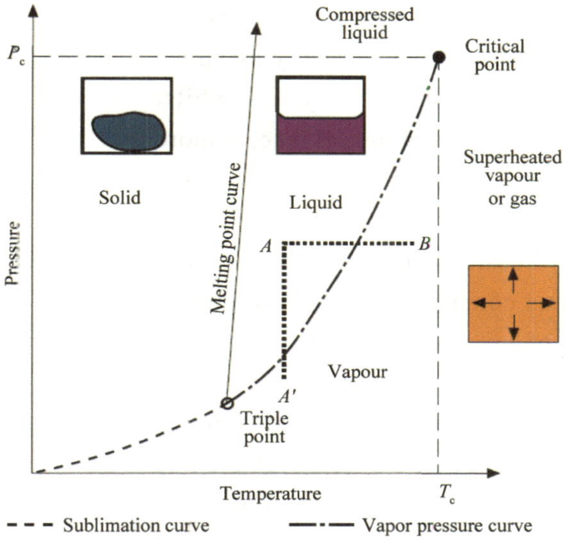

Figure 7.27 Pressure-Temperature Phase Diagram
(Source: Hydrocarbon Exploration and Production)

the dew point is reached, the component changes from the gaseous phase to the liquid phase. For a single component, the boiling point curve and the dew point curve are coincident, and are known as the vapour pressure curve. Of course the phase boundary between the liquid and solid phases is the melting point curve.

At the triple point all three phases can co-exist, and this point is a unique property of pure substances. At the critical point, defined by the critical temperature (T_c) and pressure (P_c), it becomes impossible to distinguish between the gaseous and liquid phases; the highly compressed gas has the same density and appearance as a high temperature liquid. The effect of the increased pressure and attractive forces between molecules is to move molecules together and increase the density (as when a gas becomes a liquid), but the increasing temperature increases the kinetic energy of the molecules and tends to drive them apart, thus reducing the density (as when a liquid becomes a gas). At the critical point, the phases become indistinguishable, and beyond the critical point just one state exists, and is usually referred to as a supercritical fluid.

分物质，沸点线和露点线重合，即蒸汽压力线。而液相和固相之间的边界线是融化点线。

在三点汇集处是三相共存点，这点是纯物质的独特性质。在临界点上，定义临界温度(T_c)和临界压力(P_c)，在临界点上不可能区分液相还是气相，高度压缩的气体和高温下的液体在密度和外表上是一样的。增加压力以增加分子间的吸引力和密度（气体变成液体），但温度增加也增加了分子间的能动性驱使分子分开，从而减小密度（液体变成气体）。在临界点上气液相变得模糊，在临界点上一般只有单相存在，通常为超临界液相。

油气藏生产烃类的过程通常认为是等温衰竭过程，即系统压力减小而保持常温。因

In the production of hydrocarbon reservoirs, the process of isothermal depletion is normally assumed, that is reducing the pressure of the system whilst maintaining a constant temperature. Hence, a more realistic movement on the pressure–temperature plot is from point *A* to *A*'.

此，在图7.27上一个较为现实的压力-温度变化为*A* 到*A*'。

Now using a hydrocarbon component, say ethane, as an example, let us consider the other parameter, volume, using a plot of pressure vs. specific volume (i.e. Volume per unit mass of the component, the inverse of the density). The process to be described could be performed physically by placing the liquid sample into a closed cell (*PVT* cell), and then reducing the pressure of the sample by withdrawing the piston of the cell and increasing the volume contained by the sample.

以碳氢化合物组分乙烷为例，考虑压力随体积变化作图（即单位质量该组分的体积变化，密度的倒数）。这过程可以用液相样品在封闭活塞（*PVT*活塞）里进行物理展示，向外拉活塞可以减小样品压力并增大样品的容积。

Starting at condition *A* with the ethane in the liquid phase, and assuming isothermal depletion, then as the pressure is reduced, the specific volume increases as the molecules move further apart. The relationship between pressure and volume is governed by the compressibility of the liquid ethane.

乙烷样品在*A*点液相开始，假定为等温衰减，随活塞内压力减小，体积增大，分子间距增大。压力与体积之间的关系由液态乙烷的压缩性决定。

Once the bubble point is reached (at point *B*), the first bubble of ethane vapour is released. From point *B* to *C*, liquid and gas co-exist in the cell, and the pressure is maintained constant as more of the liquid changes to the gaseous state. The system exhibits infinite compressibility until the last drop of liquid is left in the cell (point *C*), which is the dew point. Below the dew point pressure only gas remains in the cell, and as pressure is reduced below the dew point, the volume increase is determined by the compressibility of the gas. The gas compressibility is much greater than the liquid compressibility, and hence the change of volume for a given reduction in pressure (the gradient of the curve on the pressure–volume plot) is much lower than for the liquid. Eventually the point *A*' is reached (Figure 7.28).

一旦到达泡点（*B*点），第一批乙烷蒸汽释放出。从*B*点到*C*点，液相和气相在活塞内共存，压力保持常数，更多的液相转换为气相。认为系统具有无限压缩性，直到活塞内残存最后一滴液体（*C*点），也就是露点。在露点压力下活塞内只有气相。露点压力下随压力减小，气相体积的减少由气体的压缩性决定。气体的压缩性比液体的压缩性大得多，因此给定压力减少下（压力-体积图上梯度）气相比液相更低。最终达到*A*'点（图7.28）。

If the experiment is now reversed, starting from *A*' and increasing the pressure, the first drop of ethane

该实验是可反转的，从*A*'开始增加压力，在*C*点出现第一滴

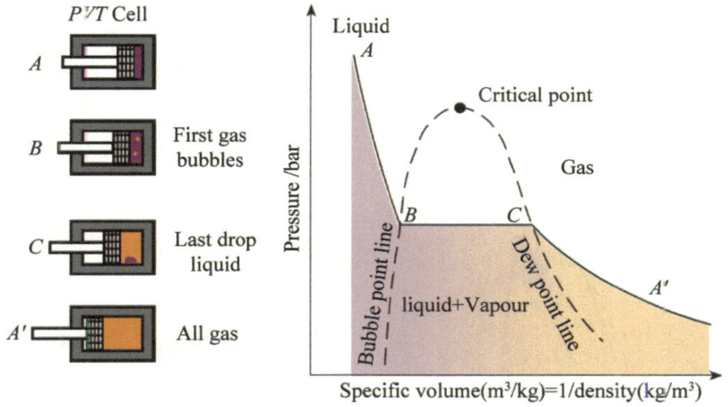

Figure 7.28 Pressure VS. Specific Volume
(Source: Hydrocarbon Exploration and Production)

liquid would appear at point *C*, the dew point of the gas. Remember that throughout this process, isothermal conditions are maintained.

The experiment could be repeated at a number of different temperatures and initial pressures to determine the shape of the two-phase envelope defined by the bubble point line and the dew point line. These two lines meet at the critical point, where it is no longer possible to distinguish between a compressed gas and a liquid.

So far we have considered only a single component. However, reservoir fluids contain a mixture of hundreds of components, which adds to the complexity of the phase behaviour. Now consider the impact of adding one component to the ethane, say n-heptane (C_7H_{16}). We are now discussing a binary (two-component) mixture, and will concentrate on the pressure–temperature phase diagram.

Figure 7.29 shows that each component has its own vapour pressure curve and critical point when we consider the components in isolation. The n-heptane vapour pressure curve is shifted down and to the right on the diagram, indicating that it requires higher temperatures and lower pressure to move n-heptane from the liquid to the gaseous phase. This is generally true for longer chain hydrocarbon components.

When the two components are mixed together (say in a mixture of 10% ethane, 90% n-heptane), the

乙烷液相，即气相的露点。记住整个过程为等温过程。

此实验可以在不同温度和初始压力下重复，只是两相包络线形状由泡点线和露点线决定。泡点线和露点线相交于临界点，在临界点不在区分液相和气相。

以上考虑的为单组分。而油藏流体为成百种组分的混合物，相态变化更为复杂，如考虑增加一种组分正庚烷到乙烷中，在双组分混合物中压力-温度变化如下。

图7.29显示每种组分都有各自的蒸汽压力线和临界点，正庚烷的蒸汽压力线在右下方，表明正庚烷需要较高的温度和较低的压力从液相转换为气相，一般长链烃类都在相图的右下方。

当两种组分混合在一起（如10%的乙烷和90%的正庚

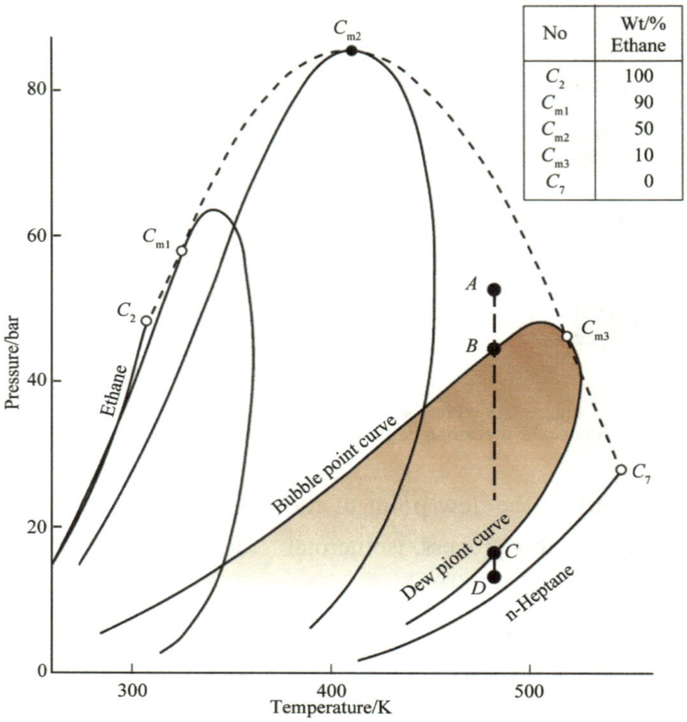

Figure 7.29 Pressure-Temperature Phase Diagram; Mixture of Ethane and N-heptane (Source: Hydrocarbon Exploration and Production)

bubble point curve and the dew point curve no longer coincide, and a two-phase envelope appears. Within this two-phase region, a mixture of liquid and gas exists, with both components being present in each phase in proportions dictated by the exact temperature and pressure, that is the composition of the liquid and gaseous phases within the two-phase envelope is not constant. The mixture has its own critical point C_{m3}.

Using this mixture as an example, consider starting at pressure *A* and isothermally reducing the pressure to point *D* on the diagram. At point *A* the mixture exists entirely in the liquid phase. When the pressure drops to point *B*, the first bubble of gas is evolved, and this will be a bubble of the lighter component, ethane. As the pressure continues to drop, the gaseous phase will acquire more of the heavier component and hence the liquid volume decreases. At point *C*, the last drop of liquid remaining will be composed of the heavier

烷），泡点线和露点线将不再相交而形成两相包络线。在包络线里面为两相区，液相和气相共存，每一相按比例在相应的温度和压力存在，即在两相包络线里的液相和气相组分并不是常数。混合物有其自己的临界点C_{m3}。

用此混合物作例子，考虑从压力*A*点开始等温减压到*D*点。在*A*点混合物完全为液相，当压力降落到*B*点，第一批气泡产生，出现轻组分乙烷的气泡。当压力继续下降，气相将增加较重的组分，因此液相体积减少。到*C*点，最后一滴液体为较重组分，它超过露点后蒸发以致露点后完全为气

component, which itself will vaporise as the dew point is crossed, so that below the dew point the mixture exists entirely in the gaseous phase. Outside the two-phase envelope the composition is fixed, but varies with pressure inside the two-phase envelope.

相。两相包络线外沿由组分确定，在两相包络线内随压力而变化。

Moving back to the overall picture, it can be seen that as the fraction of ethane in the mixture changes, the position of the two-phase region and the critical point change, moving to the left as the fraction of the lighter component (ethane) increases.

纵观全图，乙烷组分在混合物中改变，如轻组分（乙烷）增加使两相区和临界点位置向左移。

The example of a binary mixture is used to demonstrate the increased complexity of the phase diagram through the introduction of a second component in the system. Typical reservoir fluids contain hundreds of components, which makes the laboratory measurement or mathematical prediction of the phase behaviour more complex still. However, the principles established above will be useful in understanding the differences in phase behaviour for the main types of hydrocarbon identified.

上述双组分混合物例子表明系统内增加第二组分后增加了相图复杂性。一般油藏流体有成百种组分，实验测量和数学预测相态行为更复杂。因此，建立法则以主要烃类相态理解不同相态差异更为有用。

7.3.2.2 Gas’s Properties

A gas can be defined as a homogenous fluid of low density and low viscosity, which has neither independent shape nor volume. It expands to complete fill the vessel in which it is contained. The properties of gas differ from liquids mainly because the molecules in gas are much further apart than liquids. The ideal gas law states:

$$pV=nRT$$

The equation has limited practical value since no known gas behaves as an ideal gas. However, this equation does describe the behavior of most real gases at low pressures and it serves as a starting point to develop equations of states for real gases at elevated pressures. Furthermore, the behaviors of most real gases do not deviate drastically from the behavior predicted by this equation. By inserting a correction factor (Z) into the ideal gas equation, real gas behavior can be accurately predicted.

7.3.2.2 气相性质

气体可以被定义为无独立形状和体积的低密度和低黏度均一流体。气体充满任何容纳它的容器。气体形状与液体不同，因为气体分子比液体分子间距更大。理想气体状态方程如下：

$$pV=nRT$$

以上方程限制了其实用价值，因为真实气体与理想气体不一样。因此，上述方程只能描述在低压状态下大多数真实气体的状态，随压力升高真实气体状态将严重偏离理想气体状态方程，通过插入偏差因子在理想气体状态方程能较准确地预测真实气体状态。

$$pV=Z\,nRT$$

The correction factor is called the gas compressibility factor; the deviation from ideal gas behavior.

◆ Density/Gravity

Gas gravity is the molar mass (molecular weight) of the natural gas divided by the molar mass of air (28.94kg/kmol). It ranges from 0.55 for dry sweet gas to approximately 1.5 for wet sour gas. Petroleum gases typically have a gravity of 0.65.

◆ Compressibility Factor

The gas compressibility factor is a measure of its deviation from ideal gas behavior. The gas compressibility factor is the ratio of the volume actually occupied by a gas at a given pressure and temperature to the volume the gas would occupy at the same pressure and temperature if it behaved like an ideal gas.

$$Z=\frac{V_{\text{actual}}}{V_{\text{ideal}}}$$

The gas compressibility factor is usually between 0.8 and 1.2, but it can be as low as 0.3 and as high as 2.0. It is used in the calculation of gas pseudo-pressures, and in converting gas volumes and rates from standard conditions to reservoir conditions (and vice-versa). The gas compressibility factor directly affects the gas compressibility.

◆ Viscosity

Gas viscosity is a measure of the resistance to flow exerted by the gas and is given in units of centipoises (cp). Higher values indicate more resistance to flow. For gas, the viscosity increases with increasing temperature and pressure. As pressure decreases, gas viscosity decreases. The molecules are simply further apart at lower pressure and move past each other more easily.

Experimental determination of gas viscosity is difficult. Usually it is not measured but is obtained from correlations, which include corrections for H_2S, CO_2 and N_2. Gas viscosity is used in numerous equations, most

$$pV=Z\,nRT$$

偏差因子也叫气体压缩因子，用以描述真实气体与理想气体之间的偏差状态。

◆ 密度/重度

气体重度指天然气被空气除后的分子质量，从0.55的干甜气到1.5左右的湿酸气。石油气相一般重度为0.65。

◆ 压缩系数

气体压缩系数描述真实气体与理想气体之间的偏差。气体压缩系数指真实气体在一定压力和温度下与理想气体的体积之比。

$$Z=\frac{V_{\text{actual}}}{V_{\text{ideal}}}$$

气体压缩系数一般为0.8~1.2，但也可能低于0.3和高于2.0。压缩系数用于计算气体拟压力、从标准状态转换油藏状态气体体积（反之亦然）。气体压缩系数直接影响气体压缩性。

◆ 黏度

气相黏度用以测量气相流动时的阻力，单位为厘泊（cp）。值越高流动阻力越大。气相黏度随温度压力的增加而增加。当压力下降时，气相黏度下降。分子在低压下间距更大，移动更容易。

实验测气相黏度很困难。通常气相黏度不是通过实验测定，而是通过相关计算而得，包括对H_2S、CO_2和N_2气体的校正。气相黏度在多个方程中用到，特别在拟压力和拟时间定义方程中。一

notably in definitions of pseudo-pressure and pseudo-time. Typically, gas viscosity is in the range of 0.015 to 0.03 or 15 to 30 micro-Pa-s.

般气相黏度从0.015~0.03mPa·s或15 ~ 30mPa·s。

7.3.2.3 Oil's Properties

◆ Density/Gravity

The density of crude oil is usually measured in "specific gravity". Specific gravity is the ratio of one substance to an equal volume of pure water, at a standard temperature [generally 60° F (15℃)]. The *API* gravity scale, used in the petroleum industry, is an arbitrary one, and does not have a straight-line relationship with specific gravity or any other physical property, such as viscosity. High value of *API* gravity correspond to low specific gravities, and low values of *API* gravity correspond to high specific gravities.

Oil density at surface conditions is commonly quoted in *API*:

$$API = \frac{141.5}{\gamma_o} - 131.5$$

where γ_o is the specific gravity of oil (relative to water =1, measured at STP).

◆ Volume

Oil within the reservoir contains dissolved gas, and the volume of oil in the solution depends on the "gas-oil ratio" (*GOR*) and reservoir pressure. The *GOR* is the number of cubic feet of gas per barrel of oil. The amount of gas that may be dissolved in oil depends on pressure, increasing the pressure will increase the volume in solution, and this will continue until the saturation pressure is reached. This pressure is called "the bubble point", and when the pressure (P_b) falls below this point, gas will come out of solution.

◆ Shrinkage Factor

The factor that must be applied to a barrel of reservoir oil to convert it to an equivalent amount of "stock-tank" oil is known as the shrinkage factor (i.e. the fractional volume of oil in the stock-tank obtainable from one barrel of oil and its dissolved gas in the reservoir). Shrinkage factors range between 0.63 to 0.88.

7.3.2.3 油相性质

◆ **密度/重度**

原油的密度一般指重度，用标准状况下［一般为60°F（15℃）］与同样体积的纯水重力比值表示。*API* 重度是石油行业采用的一个标准，它并不与某一重力值或物理值（如黏度）成线性关系。*API* 重度高，表明低重力，*API* 重度低，表明高重力。

原油密度在地表用*API* 重度表示：

$$API = \frac{141.5}{\gamma_o} - 131.5$$

式中，γ_o是原油相对于水的比重：

◆ **体积**

原油在油藏中溶解大量气体，油的体积由气油比和油藏压力决定。气油比指每立方米气与每桶油的比值。溶解于油中的气与压力有关，增加压力将增加油中气的溶解量，压力越大，气溶解得越多，直到达到饱和压力。这时的压力称为泡点，压力低于泡点，溶解气将从原油中逸出。

◆ **回缩系数**

而另一个参数用以描述油藏油转换成同等数量地面储罐油的体积变化，叫回缩系数（指一桶地下原油及其溶解气到地面可得多少体积储罐油）。回缩系数的变化范围为0.63 ~ 0.88。

◆ Formation Volume Factor (*FVF*, *β*)

Beta is the factor for the volume change undergone by the reservoir fluids when they are produced. In the case of oil, it is the ratio of the space occupied by a barrel of oil at reservoir conditions to the space occupied by a stock tank barrel (STB) of the oil. Most oils shrink when their solution gas dissipates at the surface and because of the cooler surface conditions, so there are greater than 1,The factor that is applied to a barrel of stock-tank oil to convert it to an equivalent amount of reservoir oil is known as the formation-volume factor (i.e. the volume of reservoir oil necessary to yield one barrel of stock-tank oil). formation-volume factors range between 1.14 to 1.60(Figure 7.30).

$$\beta_o = \frac{V_{res}}{V_{surface}}$$

Oil formation volume factor is influenced by two main factors. The dominant factor is solution gas. As pressure increases, the amount of solution gas that the oil can dissolve increases such that the oil swells, and so the formation volume factor exceeds 1.0. Once there is no remaining free gas available to dissolve in the oil, further increases in pressure result in decline in formation volume factor due to the second influencing factor——the compressibility of oil. As shown in the Figure 7.31 below, oil formation volume factor is

◆ 地层体积系数

*β*是油藏流体产出后的体积变化系数。以油为言，地层体积系数是一桶油在油藏条件下的体积与地面标准条件下体积之比。绝大多数原油体积缩小因为其中溶解的气在地面逸散掉，因此地层体积系数比1大。而地面储罐油转换为同等数量地下油藏油的体积变化用地层系数表示（指地下多少体积油藏油产出1桶地面储罐油），地层系数变化范围为1.14~1.60（图7.30）。

$$\beta_o = \frac{V_{res}}{V_{surface}}$$

油地层体积系数主要受两个因素影响。第一是溶解气。随压力上升，油中溶解的气量增加，因此油体积膨胀，地层体积系数超过1.0。一旦没有自由气溶解进油中，继续增加压力只会导致地层系数受第二因素影响——油的压缩性。如图7.31所示，油的地层体积系数在泡点压力下由溶解气膨胀

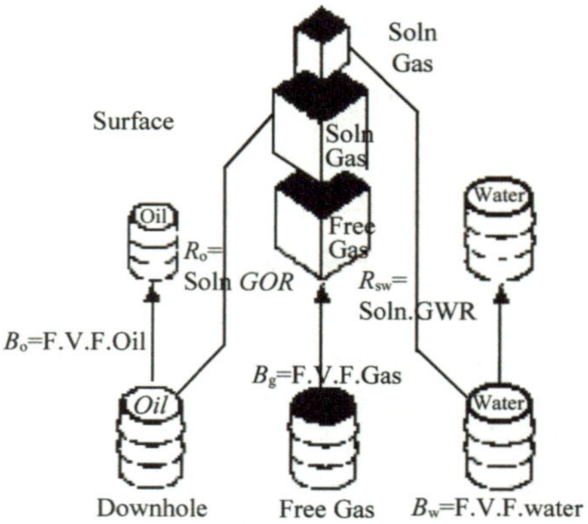

Figure 7.30 Formation Volume Change
(Source: www. fekete.com)

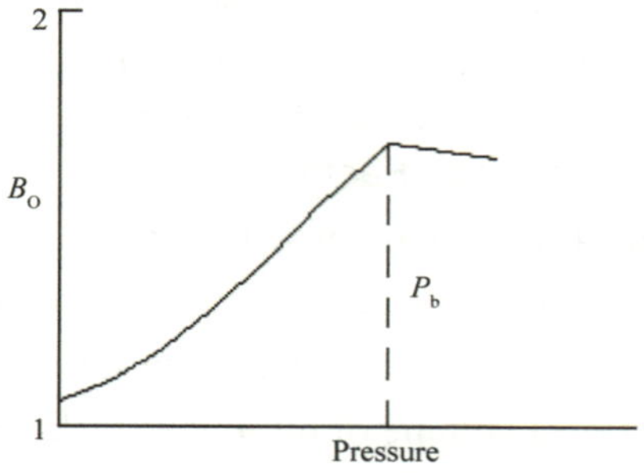

Figure 7.31 Oil Formation Volume Factor
(Source: www. fekete.com)

dominated by swelling below the bubble point pressure (due to dissolved gas), and by compressibility above the bubble point pressure (since all available gas in now dissolved).

决定，在泡点压力之上由压缩性决定（所有气体已经溶解进油里）。

◆ Viscosity

◆ 黏度

The viscosity of crude oil is generally dependent on the amount of dissolved gas and the temperature (i.e. the more gas in solution and the higher the temperature, the lower the viscosity). As more and more gas gets dissolved in crude oil, its viscosity gets lower, the API gravity gets higher, and its specific gravity gets lower. The viscosity of oil will be at its minimum at its saturation pressure.

原油的黏度与溶解气量和温度有关（也就是说溶解气越多，温度越高，黏度越低）。溶解气量越多，原油黏度越低。在饱和压力下黏度达到最小值。

◆ Solution Gas

◆ 溶解气

Solution gas is dissolved in the oil under the initial reservoir pressure and temperature. Under these conditions, the gas is in the liquid state, not the gaseous state.

溶解气指在原始油藏压力和温度下溶解在油中的气，在地下条件下，溶解气是液态而不是气态。

◆ Critical Saturation

◆ 临界饱和度

The minimum saturation of oil, water, or gas in the reservoir that cause the fluid to be a continuous medium, and therefore producible.

油、气、水在油藏中成为连续介质，可流动产出最小的饱和度。

◆ Condensate

◆ 凝析油

Condensate is light hydrocarbon liquid formed by condensation of petroleum compounds that were in the gaseous phase under initial reservoir conditions. It is highly volatile and from clear to a light yellow in color.

凝析油是轻质烃类液体，由初始油藏条件下气相石油成分凝析而成。凝析油具有高挥发性，颜色透明或带浅黄色。

◆ Gas-Oil Ratio (*GOR*)

◆ 气油比

The producing *GOR* (R) is the standard cubic feet of gas produced per stock tank barrel of oil produced (SCF/STB).

生产气油比指标准状态下产出天然气与油之比R。

The solution *GOR* (R_s) is the standard cubic feet of gas in solution in as stock tank barrel of oil at initial reservoir conditions. R_s may or may not be the same R.

溶解气油比R_s指在原始油藏条件下溶解气与油之比，溶解气油比可以与生产油气比相同或不同。

◆ Associated and Non-associated Gas

◆ 伴生气和非伴生气

Associated gas includes solution gas and gas cap gas. They both occur only in association with oil.

伴生气指溶解气和气顶气。它们伴随油生产。

Non-associated gas is from solely gas reservoirs

非伴生气指单纯气藏气，

where no "black" (non-condensate) oil is present.

其中没有黑油产出。

◆ Refractive Index

◆ 反射系数

The refractive index (*RI*) is dependent on the density of a substance. It is essentially a measure of how a beam of light is bent when passing through the substance. Hand-held refractometers are available which can determine the *API* gravity (density) of oils. These refractometers require only a small amount of oil and are relatively accurate. The refractive indices for oil range from 1.39 to 1.49 (the lower the *API* gravity, the higher the *RI*).

反射系数由物质的密度决定，它用来表征光线穿过该物质的容易程度。手持反射测量仪可以确定油的密度。手持反射测量仪只需很少的油样结果却相对准确。原油的反射系数为1.39～1.49。

◆ Recovery Factor

◆ 采收率

The recovery factor is the percent of the reservoir's original oil in place (OOIP) or original gas in place (OGIP) that will be recovered.

采收率为油藏中原始油和气采出的百分比。

◆ Fluorescence

◆ 荧光性

The fluorescence property of oil is commonly used in mud logging to determine the *API* gravity of oil. Fluorescence color is observed under an ultra-violet light, with the higher gravity oils having a lighter color and the lower gravity oils having a darker color.

原油的荧光性一般用于泥浆测井中确定油的重度。在紫外线下，高重度的油有较浅的荧光色，较轻的油有较深的荧光色。

◆ Pour Point

◆ 凝固点

The pour point of an oil is the temperature at which it will no longer flow. The pour point of oils range from −70° F to +90° F (−56.7℃ to +32℃). If an oil's pour point is above the surface temperature (for example, in winter months), the paraffins in the oil will precipitate and the oil will not flow until it is heated.

原油的凝固点指原油不再流动的温度点。原油的凝固点范围为−70~90° F (−56.7~32℃)。假如原油凝固点高于地表温度，则原油中的石蜡成分析出，原油不加热将会停止流动。

◆ Flash Point

◆ 闪点

This is the temperature at which vapors rising off the surface of a heated oil will ignite with a flash, when a flame is passed over the surface.

闪点指加热的原油蒸汽遇上点火，会发生短促的闪光。

7.3.3 Natural Gas System

7.3.3 气藏系统

Natural gas is usually classified as "associated", when it occurs with oil or "non-associated" when it occurs alone. In an associated reservoirs, when the oil becomes saturated, the excess gas will form a "gas cap" above the oil and be produced as "free gas".

当天然气和原油一起产出时，我们称为"伴生气"；而只产天然气不产油时，我们称为"非伴生气"。在伴生气油藏中，当油藏处于饱和状态

When the oil is under-saturated, then the bubble point (the temperature and pressure when the gas comes out of solution) must be reached before gas is produced. With high enough temperatures and pressures, some of the alkanes will become a liquid and become indistinguishable from oil.

时，多余的天然气会在油层上形成气顶，能被以自由气的方式采出。而当油藏处于未饱和状态时，油藏压力必须降到泡点（气体从液体中析出的温度和压力）才能产气。在足够的高温、高压下，部分烷烃会液化从原油中析出。

Natural gas is composed primarily of 4 hydrocarbon molecules(Figure 7.32). These range from 1 to 4 carbons in length and methane (1 carbon), ethane (2 carbons), propane (3 carbons), and butane (4 carbons). The composition of natural gas from all of world varies considerably, but methane is by far the most common hydrocarbon.

天然气主要由4种碳分子组成（图7.32）。从1个碳到4个碳分别为甲烷、乙烷、丙烷和丁烷。全世界的天然气组成各有不同，但甲烷是目前为止最主要的碳氢化合物。

Many natural gas fields are almost pure methane. The remainder of the low end, propane, butane, pentane

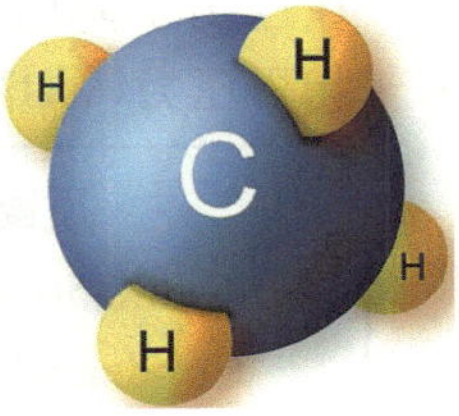

Figure 7.32 Natural Gas Molecule
(Source: http://ecofriend.com/wp-content/uploads/2012/07/natural-gas-molecule_69.jpg)

and hexane will either be gases or liquids, depending on the temperatures and pressures (since the temperature of a reservoir is more or less constant, pressure changes usually dictate whether the fluid is a gas or liquid). And propane and butane burn so hot that they are often refined from natural gas, compressed to liquefy them, and sold separately as liquefied petroleum gas (LPG).

许多气田产几乎纯的甲烷气。剩下的乙烷、丙烷、丁烷和戊烷可以为气相或液相，这取决于油藏压力和温度（由于油藏温度或多或少维持恒定，压力变化通常指示储层流体为气体或液体）。丁烷和戊烷燃烧值很高，通常从天然气中提出，压缩液化成LNG单独出售。

Natural gas is usually sold to be burned. Because of this, natural gases are often priced or a bonus is given for the heat content of the natural gas. A unit used to measure heat quantity in a fuel is the British Thermal Unit (BTU), which is approximately equivalent to the heat given off by burning one wooden match.

天然气一般作为燃料出售。所以，天然气的价格取决于天然气的燃烧值。一个用以描述燃料热值单位是英国热值单位BTU，它指大约燃烧多少根火柴可以达到同样的热值。

The natural gas of commerce generally has the following composition:

Methane	72.3%

商业天然气一般有以下构成：

甲烷	72.3%
乙烷	14.4%

Ethane	14.4%
Carbon Dioxide	0.5%
Nitrogen	12.8%

and ranges from 900 to 1200 BTU per cubic foot.

Classification of gas include "dry gas" (less than 0.1 gallons of natural liquid per 1,000 cubic feet) or "wet gas" (0.3 or more natural gas liquid per 1,000 cubic feet).

The hydrocarbons in the reservoir are gaseous at initial conditions. During initial gas production, large quantities of condensate drop out in the surface facilities. Later on, as reservoir pressure declines, the condensate begins to drop out in the formation, "Retrograde" refers to the fact that these gases liquefy when pressure is reduces, while normal fluid behavior is for gas to liquefy when pressure is increase.

Retrograde Condensate Gas
Middle 50' s < API < 70
Condensate $\mu \approx 0.25$ cp
2000 to 6000< R_s initial <15000

All hydrocarbon in a wet gas reservoir are in a gaseous state. When the gas is produced, the associated cooling causes condensate to drop out in the surface facilities.

Wet Gas
Condensate $API > 60$
Condensate $\mu \approx 0.25$ cp
2000 to 6000< R_s initial <15000

The gas is composed primarily of methane with only small amounts of ethane, propane and butane. These reservoirs produce no condensate.

Dry Gas

Figure 7.33 helps in explaining how the phase diagrams of the main types of reservoir fluid are used

二氧化碳	0.5%
氮气	12.8%

每立方英尺天然气热值 为900～1200BTU。

天然气可以分为“干气”和“湿气”。干气指每1000ft^3 (1ft^3= 2.831685×10^{-2}m^2)天然气其天然液相小于0.1gal(1gal= 3.7854dm^3)；湿气指每1000ft^{3}天然气其天然液相多于0.3gal。

凝析气在油藏原始条件下是气相，刚开始产气，在地面设施中凝析出大量的油滴。后来随着油藏压力的下降，油滴凝析在储层中，“反”指气相在压力下降时液化，而通常气相再压力增加液化。

反凝析气
Middle 50's < API < 70
Condensate $\mu \approx 0.25$ cp
2000～6000< R_s initial <15000

在湿气藏中所有碳氢化合物为气相，当产气后由于温度压力下降，在地面设备中凝析成液相。

湿 气
Condensate $API > 60$
Condensate $\mu \approx 0.25$ cp
2000～6000< R_s initial <15000

气藏中主要为甲烷，只有少量乙烷、丙烷和丁烷。干气藏没有凝析油。

干 气

图7.33帮助解释油藏主要

to predict fluid behaviour during production and how this influences field development planning. It should be noted that there are no values on the axes, since the scales will vary for each fluid type. Figure 7.33 shows the relative positions of the phase envelopes for each fluid type.

The four vertical lines on the diagram show the isothermal depletion loci for the main types of hydrocarbon:

流体相态变化，可预测生产过程中流体相态变化和油气藏开发方案对其影响。值得注意的是没有坐标轴心，因为随流体类型不同、比例不同，图7.33显示对每项流体的相包络线相对位置。

图7.33中4条垂直线显示主

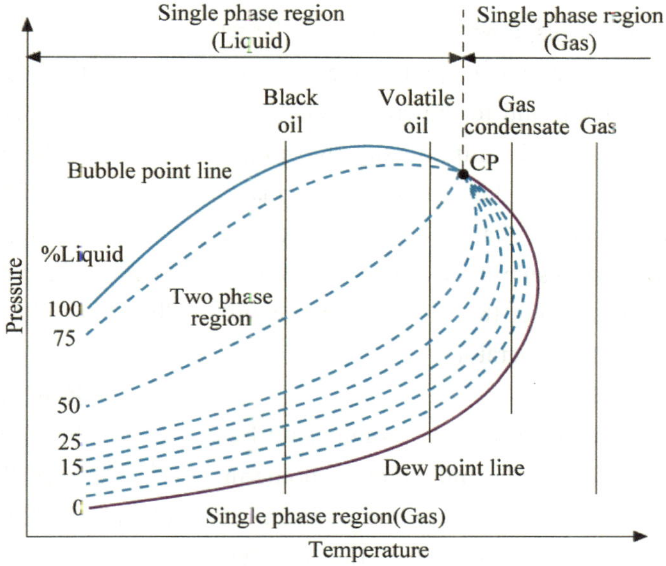

Figure 7.33 Pressure-Temperature Phase Envelopes for Main Hydrocarbon Types -Showing Initial Conditions Relative to the Phase Envelope Only
(Source: Hydrocarbon Exploration and Production)

gas (incorporating dry gas and wet gas), gas condensate, volatile oil and black oil. The starting point, or initial conditions of temperature and pressure, relative to the two-phase envelope are different for each fluid type.

◆ Dry Gas

The initial condition for the dry gas is outside the two-phase envelope, and is to the right of the critical point, confirming that the fluid initially exists as a single-phase gas. As the reservoir is produced, the pressure drops under isothermal conditions, as indicated by the vertical line. Since the initial temperature is higher than

要烃类流体等温衰减轨迹，气相（合并干气和湿气）、凝析气相、挥发性油相和黑油相。开始点或初始点的温度和压力不同及各相流体类型不同两相包络线不同。

◆ 干气

干气的初始条件在两相包络线外面，临界点右边，流体初始为单相气态。当气藏生产后，等温压降，显示为垂直线。因为初始温度高于两相包

the maximum temperature of the two-phase envelope, the reservoir conditions of temperature and pressure never fall inside the two-phase region, indicating that the composition and phase of the fluid in the reservoir remains constant.

In addition, the separator temperature and pressure of the surface facilities are typically outside the two-phase envelope, so no liquids form during separation.

◆ Wet Gas

Compared to a dry gas, a wet gas contains a larger fraction of the C_2~C_6 components, and hence its phase envelope is moved down and to the right. Whilst the reservoir conditions remain outside the two-phase envelope, so that the reservoir fluid composition remains constant and the gaseous phase is maintained, the separator conditions are inside the two-phase envelope. As the dew point is crossed, the heavier components condense as liquids in the separator.

◆ Gas Condensate

The initial temperature of a gas condensate lies between the critical temperature and the cricondotherm. The fluid therefore exists at initial conditions in the reservoir as a gas, but on pressure depletion the dew point line is reached, at which point liquids condense in the reservoir. As can be seen from Figure 7.33, the volume percent of liquids is low, typically insufficient for the saturation of the liquid in the pore space to reach the critical saturation beyond which the liquid phase becomes mobile. These liquids therefore remain trapped in the reservoir as an immobile phase. Since these liquids are valuable products, there is an incentive to avoid this condensation in the reservoir by maintaining the reservoir pressure above the dew point. This is the reason for considering recycling of gas in these types of reservoir (Figure 7.34).

络线最高温度，气藏的温度和压力绝不会掉入两相区，气藏流体成分和相态保持不变。

而且，地面分离器的温度和压力一般也在两相包络线外，所以在整个分离过程中也没有液体产生。

◆ 湿气

相比干气，湿气包含更多的C_2 ~ C_6成分，从而其相包络线向下和向右移。气藏条件在两相包络线外，因此气藏流体组分为常数和气相，而分离器内条件在两相包络线内。当降到露点时，较重组分在分离器里凝析成液体。

◆ 凝析气

凝析气藏初始温度在临界温度和临界凝结温度之间，因此流体在地下初始条件为气相，但是随压力衰竭到露点线气藏开始凝析出液相。从图7.33可以看出，液相比例很低，一般在孔隙中的液相含量不到可以流动的液相含量。这些凝析油在气藏中保持可不流动的束缚状态。因为这部分凝析油具有高价值，因此有必要避免气藏发生凝析现象，一般考虑注循环气进气藏保持气藏压力在露点压力之上（图7.34）。

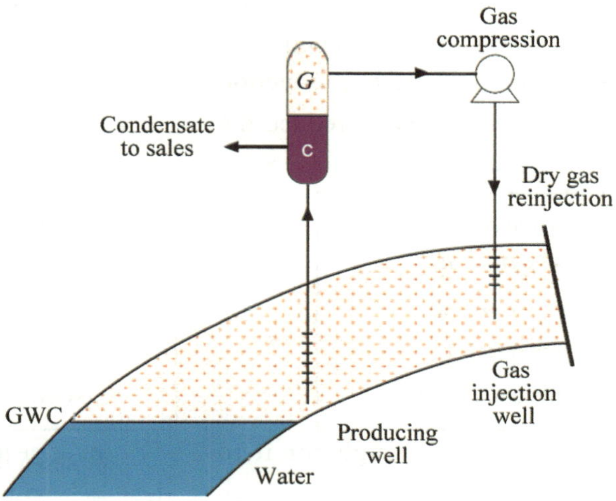

Figure 7.34 Gas Recycling Process
(Source: Hydrocarbon Exploration and Production)

7.3.4 Oil System

Since each reservoir contains a unique blend of many different hydrocarbon compounds, no two reservoirs behave identically when produced. Instead, they cover a wide spectrum of variation.

The following classification of hydrocarbon systems is an attempt to reduce this complexity to workable levels. It is based on the reservoir's relative mix of lighter and heavier hydrocarbons and starts with the heaviest mix.

Because it has few small molecules, bitumen's viscosity is so high that for practical purposes it is immovable in a reservoir. Only if the reservoir is shallow enough for a mining/retorting operation can bitumen be commercial. It is therefore only of interest to be the mining industry, not to the oil and gas industry.

Bitumen
$4 < API < 10$
$1000000 > \mu > 5000$
$\beta \approx 1$
$R_s = 0$

7.3.4 原油流体系统

每个油藏包含独特碳氢化合物组分，没有两个油藏生产情况相同。相反，它们的变化范围很广。

以下定义碳氢化合物系统试图减少工作中的混淆。基于油藏相对重的组分到相对轻的组分。

因为沥青没什么小分子，沥青的黏度非常高，以至于沥青在油藏中是不可移动的。只有油藏足够浅可以挖掘/干馏沥青才有商业价值。因此，只有矿产业对沥青感兴趣，而油气行业却对沥青没有兴趣。

沥青
$4 < API < 10$
$1000000 > \mu > 5000$
$\beta \approx 1$
$R_s = 0$

With more lighter ends than bitumen and a small amount of gas in solution, heavy oil flows sluggishly through a reservoir. To attain commercial production rates, however, it' s usually necessary to reduce the crude' s viscosity by injecting steam.

Tar or Heavy Oil
$10 < API < 20$
$5000 > \mu > 100$
$1.0 < \beta < 1.1$
Negligible $< R_s < 50$

比沥青含的小分子多，还有少量溶解气，重油在油藏中移动缓慢。为获得商业产量，一般有必要注入热蒸汽以减少原油黏度。

焦油或重油
$10 < API < 20$
$5000 > \mu > 100$
$1.0 < \beta < 1.1$
Negligible $< R_s < 50$

This oil' s viscosity is low enough for it to move through the formation readily, so low-shrinkage reservoirs are normally commercially producible. When production starts, however, the limited quantity of gas in solution blows off quickly and causes reservoir pressure to decline rapidly. As a result, the primary recovery factor is generally low and secondary measures are indicated.

Low-Shrinkage Oils ("Black Oils")
$20 < API <$ Low 30
$100 > \mu > 2$ to 3
$1.1 < \beta < 1.5$
$50 < R_s < 500$

黑油黏度够低能在储层中移动，因此低收缩油藏一般具有商业可采价值。开始生产后，有限的溶解气迅速逸散导致油藏压力急剧下降。结果初始采收率通常很低，需要二次采油方法。

低收缩原油（"黑油"）
$20 < API <$ Low 30
$100 > \mu > 2 \sim 3$
$1.1 < \beta < 1.5$
$50 < R_s < 500$

Because of its low viscosity, volatile oil produces at high rates and its high R_s results in good recovery factors. It also contains a maximum of higher-valued gasoline-range compounds. It is therefore the most desirable fluid system to encounter in a reservoir.

High-Shrinkage Oils ("Volatile Oils")
Low 30 $< API <$ Low 50
$2 \sim 3 > \mu > 0.25$
$1.5 < \beta < 2.5$ to 3.5
$500 < R_s < 2000$ to 6000

因为高收缩油的低黏度，挥发油产量高，溶解气油比高，采收率高。其中，包含最大量的高价值的汽油组分。因此，挥发油是此种油气藏最想得到的流体。

高收缩原油（"挥发油"）
Low 30 $< API <$ Low 50
$2 \sim 3 > \mu > 0.25$
$1.5 < \beta < 2.5 \sim 3.5$
$500 < R_s < 2000 \sim 6000$

◆ Volatile oil and black oil

For both volatile oil and black oil, the initial reservoir temperature is below the critical point, and the fluid is therefore a liquid in the reservoir. As the pressure drops, the bubble point is eventually reached, and the first bubble of gas is released from the liquid. The composition of this gas will be made up of the more volatile components of the mixture. Both volatile oils and black oils will liberate gas in the separators, whose conditions of pressure and temperature are well inside the two-phase envelope.

A volatile oil contains a relatively large fraction of lighter and intermediate components which vaporise easily. With a small drop in pressure below the bubble point, the relative amount of liquid to gas in the two-phase mixture drops rapidly, as shown in the phase diagram (Figure 7.35) by the wide spacing of the iso-

◆ 挥发性油藏和黑油藏

挥发性油藏和黑油藏初始温度都在临界温度之下，因此在地下为液相。随压力下降，最终达到泡点压力，第一批气体从液相中释放。气相组分将由混合物中的挥发性成分组成。地面分离器温度和压力在两相包络线之外，挥发性油和黑油在分离器里都会释放出气体。

挥发性油藏包含相对较多相对更容易挥发的轻质和中间质组分。在泡点压力下，压力降落一小步，在两相混合物中液体相对量下降迅速，如图7.35所示，在等温–体积线上的

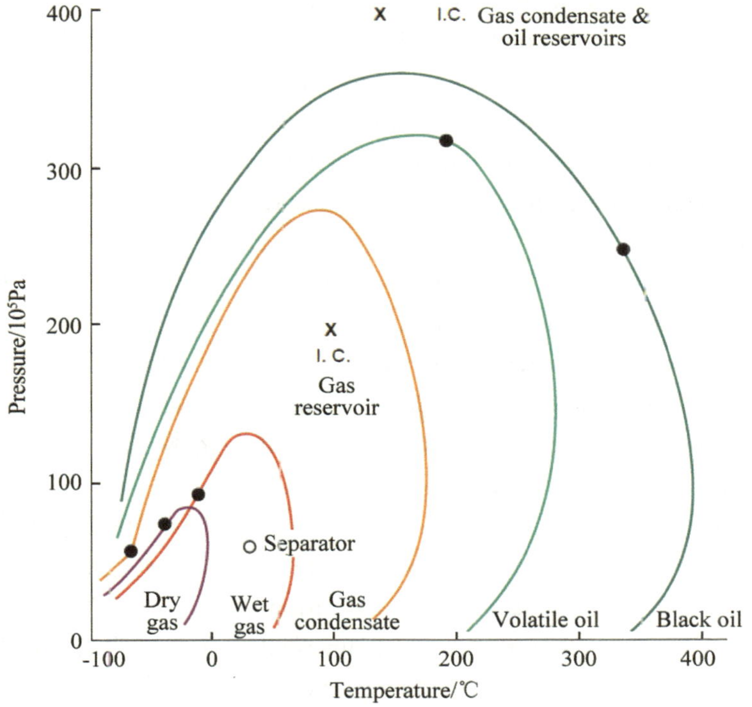

Figure 7.35 Relative Positions of Phase Envelopes
(Source: Hydrocarbon Exploration and Production)

vol lines. At reservoir pressures below the bubble point, gas is released in the reservoir, and is known as solution gas, since above the bubble point this gas was contained in solution. Some of this liberated gas will flow towards the producing wells, whilst some will remain in the reservoir and migrate towards the crest of the structure to form a secondary gas cap.

宽间距。当油藏压力低于泡点压力，油藏中有气体释放，这叫溶解气，因此高于泡点压力油中都有溶解气。一些释放后的气流入生产井，而另一些保留在油藏中，迁移到构造顶部形成二次气顶。

Black oils are a common category of reservoir fluids, and are similar to volatile oils in behaviour, except that they contain a lower fraction of volatile components and therefore require a much larger pressure drop below the bubble point before significant volumes of gas are released from solution.

黑油藏是油藏流体常见的分支，其流体行为同挥发性流体相似。黑油只包含少量挥发性组分，因此要求在泡点下大压力降才有大量溶解气释放。

Volatile oils are known as high shrinkage oils because they liberate relatively large amounts of gas either in the reservoir or the separators, leaving relatively smaller amounts of stabilised oil compared to black oils (also called low shrinkage oils).

挥发性油也叫高回缩油，因为释放相对大量的气体在油藏或在分类器内，相比黑油残留相对少量稳定油，黑油也叫低回缩油。

When the pressure of a volatile oil or black oil reservoir is above the bubble point, we refer to the oil as *under saturated*. When the pressure is at the bubble point we refer to it as saturated oil, since if any more gas were added to the system it could not be dissolved in the oil. The bubble point is therefore the saturation pressure for the reservoir fluid.

当挥发性油藏和黑油藏压力在泡点压力之上，称油为未饱和油。当挥发性油藏和黑油藏压力在泡点，称为饱和油，无法有更多的气可溶解在油中。因此，泡点为油藏流体的饱和压力点。

An oil reservoir which exists at initial conditions with an overlying gas cap must by definition be at the bubble point pressure at the interface between the gas and the oil, the gas-oil contact (GOC). Gas existing in an initial gas cap is called free gas, whilst the gas in solution in the oil is called dissolved or solution gas.

油藏初始条件下带气顶可以被定义为气相和油相界面的泡点压力，气-油界面（GOC）。初始气顶中的气叫自由气，而溶解在油中的气叫溶解气。

7.3.5 Reservoir Water System

7.3.5 油藏水系统

Water is the most common fluid encountered in the subsurface. In a most uncharacteristic definition, a "dry hole" is one which contains no commercial hydrocarbons, but contains water. As the fluids separate out within a reservoir, water having the highest specific gravity, will form the base of the fluid column. The

水是地下遇到的最常见流体。在石油用语中，"干井"不是指没有水的井，而是指没有商业价值油的井。水因为有最大的比重，所以处在油藏底部。位于油/水界面之下的是

water below the oil/water contact is known as "bottom-water" or "edge-water".

"底水"或"边水"。

Most hydrocarbons reservoirs form in aquifers. These aquifers being formed either meteoric water (water fallen as rain and percolating downwards) or connate water (the sea water in which the sediments were deposited) filling the pore spaces within the rocks.

油藏中大多含有水层。这些水可能是"共生水"（如雨水落下后渗透下来的水）也可能是"原生水"（在沉积环境时一起沉积的水），它们填满岩石的孔隙。

There will be interstitial water filling the pore spaces within the formation that is absorbed on the mineral surface of the rock or held in capillary openings by capillary pressure. The amount of this interstitial water will range from 10 to 50 percent, or more, of the pore space, the remainder being filled by hydrocarbons(Figure 7.36).

这些空隙中的水被岩石矿物表面吸收或因毛管压力吸附在毛细管口上，充填在地层孔隙中。这些空隙水占据10%~50%，或更多的孔隙空间，余下部分才被碳氢化合物充填（图7.36）。

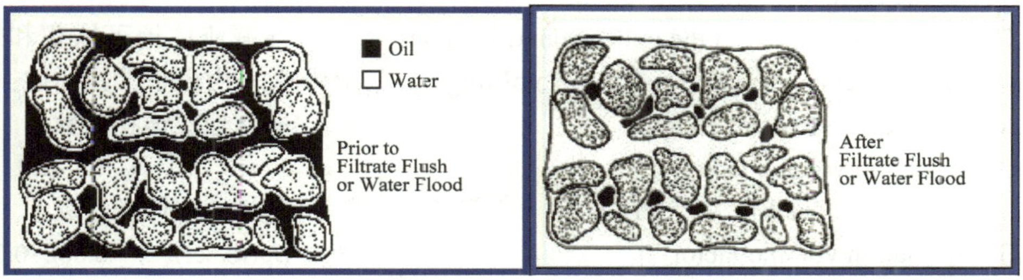

(a) Black is Hydrocarbon, White is Irreducible Water
(b) Same Reservoir after depletion with Residual Oil and Higher Water Saturation

Figure 7.36 Reservoir at Initial Saturation Conditions
(Source: http://www.spec2000.net/01-quickmath.htm)

7. 4 Oil Reservoirs— Primary Drives

7.4 油藏天然能量驱动机理

Reservoir fluids (oil, water, gas) and the rock matrix are contained under high temperatures and pressures; they are compressed relative to their densities at standard temperature and pressure. Any reduction in pressure on the fluids or rock will result in an increase in the volume, according to the definition of compressibility.

油藏流体（油、水、气）和岩石基质在地下处于高温、高压下。在标准状况下，它们被相对压缩压实。因此，任何压力的下降都会造成岩石和流体体积的释放。

Gas has a much higher compressibility than oil or water, and therefore expands by a relatively large amount for a given pressure drop. As underground fluids are withdrawn (i.e. production occurs), any free gas present expands readily to replace the voidage, with only a small drop in reservoir pressure. If only

气体相对于油或水有更高的压缩性，因此在相同压降下有更大的体积释放。当地下流体采出，压力降落时，自由气会膨胀占领空出来的空间。假如油藏中只有油和水两相，需

oil and water were present in the reservoir system, a much greater reduction in reservoir pressure would be experienced for the same amount of production.

要更多的压降才能采出相同的流体。

The expansion of the reservoir fluids, which is a function of their volume and compressibility, act as a source of drive energy which can act to support primary production from the reservoir. Primary production means using the natural energy stored in the reservoir as a drive mechanism for production. Secondary recovery would imply adding some energy to the reservoir by injecting fluids such as water or gas, to help to support the reservoir pressure as production takes place.

油藏流体的膨胀是它们压缩性和体积的函数，作为第一驱动能量支撑油藏一次开采。一次开采是利用油藏储存的天然能量作为驱动力进行生产。二次开采指注入流体，注水或注气以增加油藏能量进行生产。

Figure 7.37 shows how the expansion of fluids occurs in the reservoir to replace the volume of fluids produced to the surface during production.

图7.37显示开采后油藏流体膨胀占领之前产出到地面流体的空间。

The relationship between the underground volumes (measured in reservoir barrels) and the volumes at surface conditions were denoted by:

油藏流体地下体积和地面体积关系由以下参数描述：

Oil formation volume factor

B_o (rb/stb) 1.1~2.0

Gas formation volume factor

B_g (rb/scf) 0.002~0.0005

Water formation volume factor

B_w (rb/stb) 1.0~1.1

原油地层体积系数

B_o (rb/stb) 1.1~2.0

气体地层体积系数

B_g (rb/scf) 0.002~0.0005

水地层体积系数

B_w (rb/stb) 1.0~1.1

One additional contribution to drive energy is by

另一种驱动能量是孔隙压实。当由于生产孔隙压力减

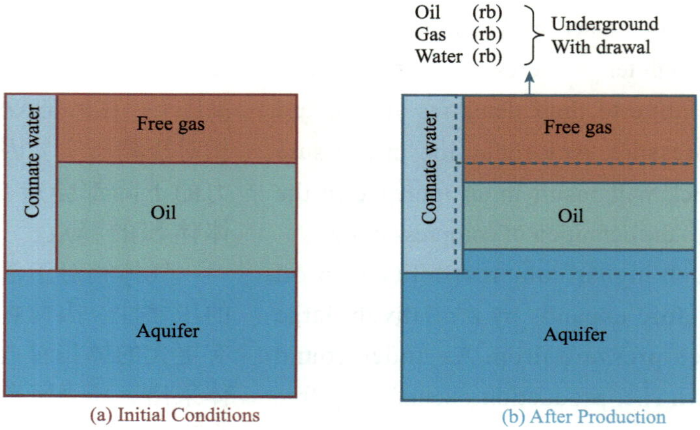

Figure 7.37 Expansion of Fluids to Replace Produced Volumes
(Source: Hydrocarbon Exploration and Production)

pore compaction. As the pore fluid pressure reduces due to production the grain to grain stress increases, which leads to the rock grains crushing closer together, thereby reducing the remaining pore volume, and effectively adding to the drive energy. The effect is usually small (less than 3% of the energy contributed by fluid expansion), but can lead to reservoir compaction and surface subsidence in cases where the pore fluid pressure is dropped considerably and the rock grains are loosely consolidated.

小后，会增加颗粒之间的压缩力。这促使岩石颗粒越来越紧密，减少原有的孔隙空间，有效地驱动油流。但这种压实作用的总作用微乎其微（流体膨胀产生的能量贡献小于3%），并且有时能导致油藏的压缩和地面的塌陷，这是由孔隙流体压力下降而压缩颗粒又相对松散所致。

Reservoir engineers describe the relationship between the volume of fluids produced, the compressibility of the fluids and the reservoir pressure using material balance techniques. This approach treats the reservoir system like a tank, filled with oil, water, gas and reservoir rock in the appropriate volumes, but without regard to the distribution of the fluids (i.e. the detailed distribution or movement of fluids inside the system). Material balance uses the PVT properties of the fluids described, and accounts for the variations of fluid properties with pressure. The technique is firstly useful in predicting how reservoir pressure will respond to production. Secondly, material balance can be used to reduce uncertainty in volumetrics by measuring reservoir pressure and cumulative production during the producing phase of the field life.

油藏工程师用物质平衡法描述流体生产后孔隙空间在流体压缩和油藏压力变化后的变化。该方法把油藏系统看作一个充满油气水的密封罐，而忽略其中流体的具体分布（即系统中详细的流体分布和运移）。物质平衡法利用PVT性质描述流体，解释流体随压力改变的变化。该方法首先用于预测油藏压力改变后的生产情况。其次，物质平衡用于模拟油藏压力和累积产量在油田生产过程中的变化。

To be commercially productive, a petroleum reservoir must have some form of natural drive mechanism to force the hydrocarbons to the surface through the wellbore. There are several natural drive mechanisms(Table 7.2).

为实现商业开采，油藏必须有某种形式的天然驱动以迫使碳氢化合物流入井筒中。以下是几种天然驱动机理（表7.2）。

Table 7.2 Drive Mechanisms

Drive mechanism	Fluid initial condition
Solution gas drive (or depletion drive)	Undersaturated oil (no gas cap)
Water drive (via injection or with a large underlying aquifer)	Saturated or undersaturated oil
Gas cap drive	Saturated oil with a gas cap
Gravity drainage drive	Saturated or undersaturated oil
Compaction drive	Saturated or undersaturated oil
Combination drive	Saturated or undersaturated oil

7.4.1 Solution Gas Drive

Solution gas drive occurs in a reservoir which contains no initial gas cap or underlying active aquifer to support the pressure and therefore oil is produced by the driving force due to the expansion of oil and connate water, plus any compaction drive. The contribution to drive energy from compaction and connate water is small, so the oil compressibility initially dominates the drive energy. Because the oil compressibility itself is low, pressure drops rapidly as production takes place, until the pressure reaches the bubble point.

Once the bubble point is reached, solution gas starts to become liberated from the oil, and since the liberated gas has a high compressibility, the rate of decline of pressure per unit of production slows down.

The characteristic production profile for a reservoir developed by solution gas drive is shown in Figure 7.38.

As for all production profiles, there are three distinct phases, defined by looking at the oil production rate (for an oil field). After the first production date, there is a build-up period, during which the development wells are being drilled and brought on stream, and its shape is dependent on the drilling schedule. Once the plateau is reached, the facilities are filled and any extra production potential from the wells is choked back.

7.4.1 溶解气驱

溶解气驱发生在油气藏中没有原始气顶和活跃水层，用以保持压力的情况下，油流由油和共生水的膨胀驱动和压实驱动。来自压实和共生水膨胀的驱动力很小，主要靠油流的膨胀作用。因为，油本身的压缩膨胀能力较小，因此生产后压力会迅速下降，直到降到泡点。

一旦降到泡点压力，溶解气开始从油中逸出，逸出的气有很大的压缩膨胀性，因此由生产导致的压力下降将减弱。

图7.38显示溶解气驱开发油藏过程的生产特征：

在整个生产过程中，根据油产量分为3个阶段。开始生产时，产量递增阶段，随开发井不断增加井产量不断增加，产量的增加取决于钻井的速度。之后到达平台期，地面建设完善，产量保持恒定，多余的产能被抑制。一旦井产能不能维

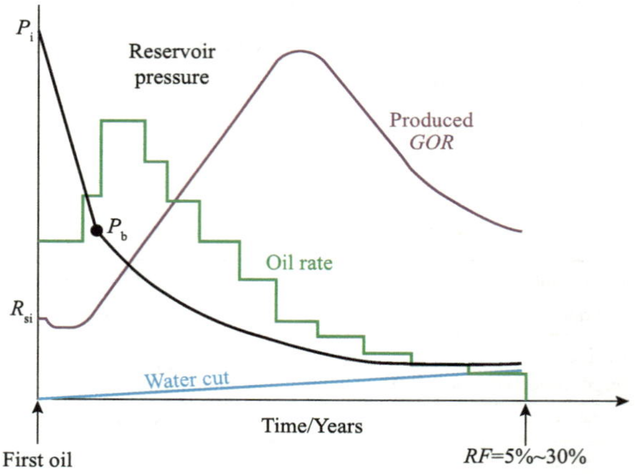

Figure 7.38 Production Profile for Solution Gas Drive Reservoir
(Source: Hydrocarbon Exploration and Production)

Once the well potential can no longer sustain the plateau oil rate, the decline period begins and continues until the abandonment rate is reached. Abandonment, better referred to as decommissioning, occurs when the cost of production is greater than the revenues from the production.

In the solution gas drive case, once production starts the reservoir pressure drops very quickly, especially above the bubble point, since the compressibility of the system is low. Consequently, the producing wells rapidly lose the potential to flow to surface, and not only is the plateau period short, but the decline is rapid.

The producing *GOR* starts at the initial solution *GOR* (R_{si}), decreases until the critical gas saturation is reached, and then increases rapidly as the liberated gas is produced into the wells. The producing *GOR* may decline in later years as the remaining volume of gas in the reservoir diminishes.

Commonly the water cut remains small in solution gas drive reservoirs, assuming that there is little pressure support provided by the underlying aquifer. Water cut is also referred to as BS&W (base sediment and water), and is defined as

$$\text{Water cut (BS\&WT)} = \frac{\text{Water production(stb)}}{\text{Oil plus water production(stb)}} \times 100(\%)$$

The typical RF from a reservoir developed by solution gas drive is in the range 5%~30%, depending largely on the absolute reservoir pressure, the solution *GOR* of the crude, the abandonment conditions.

This rather low RF may be boosted by implementing secondary recovery techniques, particularly water injection, or gas injection, with the aim of maintaining reservoir pressure and prolonging both plateau and decline periods. The decision to implement these techniques (only one of which would be selected) is both technical and economic. Technical considerations

持平台期产量，下降阶段不可避免，直到废弃。废弃指停产，当生产投入比产出收益更高，需要停产。

溶解气驱时，开始生产后油藏压力下降非常快，特别是在泡点压力上由于液体压缩性很小，油藏压力下降特别快。因此，生产井消耗大量产能促进油流入地面，因此平台期和下降期都非常短。

生产气油比*GOR*以初始气油比*GOR* (R_{si})开始不断上升，直到达到临界气饱和度，而后生产气油比迅速下降，因为大量气体逃逸进井。而后很多年生产气油比*GOR*下降，油藏中气体体积减少。

在溶解气驱油藏中，含水量一般较少，认为储层下水层无支撑压力。含水量也指BS&W (基层沉积水量)，定义为：

$$\text{含水量（BS\&WT）} = \frac{\text{产水量(stb)}}{\text{产液量(油+水量)(stb)}} \times 100(\%)$$

溶解气驱开发油藏一般采收率为5%~30%，很大部分取决于油藏绝对压力、生产气油比*GOR*和废弃油藏条件。

对于相对低采收率，可以通过二次开采方式提高采收率，二次开采特别是注水或注气，以保持油藏压力为目标，能延长开采平台期和下降期。以油藏的技术条件和经济条件选择具体的二次开采方式。技术条件指有没有外在气源和可

would be the external supply of gas, and the feasibility of injecting the fluids into the reservoir(Figure 7.39).

供注入的流体进入油藏（图7.39）。

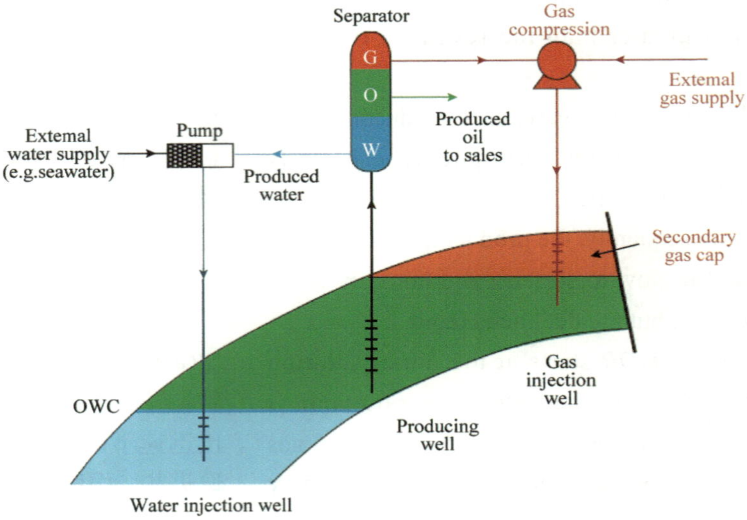

Figure 7.39 Secondary Recovery: Gas or Water Injection Schemes
(Source: Hydrocarbon Exploration and Production)

7.4.2 Water Drive

Although water is virtually incompressible, large amounts of potential energy can be stored in large volumes of water in extensive regional reservoirs. Expansion of this water to replace flowing oil will produce a less powerful, but longer enduring drive than a gas drive.

Water production has a lower GOR, since only the gas dissolved in oil is produced. As the oil/water transition moves upwards, increasing amounts of water will be produced with the oil. Water drives can produce up to 70% of the reservoir fluids before enhanced recovery methods need to be used(Figure 7.40).

Natural water drive occurs when the underlying aquifer is both large (typically greater than ten times the oil volume) and the water is able to flow into the oil column that is it has a communication path and

7.4.2 水压驱动

尽管水实际上是不可压缩的，但大量能量储存在巨大体积的水中，水体的膨胀驱替油流进入井筒，虽然没有气驱有力却比气驱更持久。

水驱产油比溶解气驱产油有更低的气油比。因为只有溶解在油中的气产出。油水向上移动时增加了油井的产水。不使用增产措施，水驱就能产出高达70%的油藏流体（图7.40）。

当水层体积巨大尤其是油体积的10倍以上时，又有良好的通道和足够的渗透率，水能容易地流入油层中就会发生天然水驱。一旦油流生产产生压

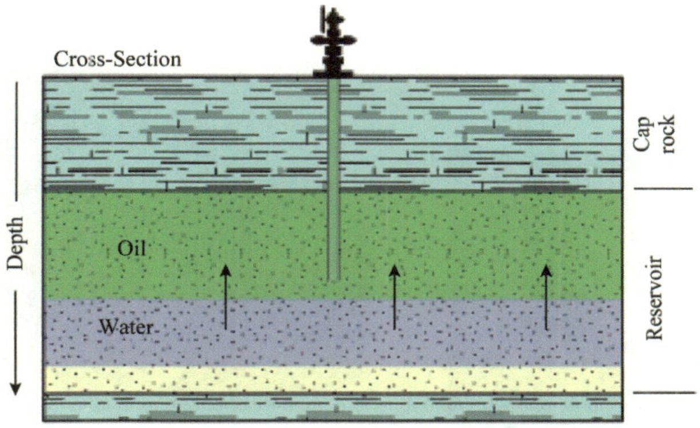

Figure 7.40 Water Drive
(Source: http://www.gg.uwyo.edu/media/energy/fossilfuels/oil/production/)

sufficient permeability. If these conditions are satisfied, then once production from the oil column creates a pressure drop the aquifer responds by expanding, and water moves into the oil column to replace the voidage created by production. Since the water compressibility is low, the volume of water must be large to make this process effective, hence the need for the large connected aquifer. In this context, ‘large’ would be 10 to 100 × the volume of oil in place.

The prediction of the size and permeability of the aquifer is usually difficult, since there is typically little data collected in the water column; exploration and appraisal wells are usually targeted at locating oil. Hence the prediction of aquifer response often remains a major uncertainty during reservoir development planning. In order to see the reaction of an aquifer, it is necessary to produce from the oil column, and measure the response in terms of reservoir pressure and fluid contact movement; use is made of the material balance technique to determine the contribution to pressure support made by the aquifer. Typically 5% of the STOIIP must be produced to measure the response; this may take a number of years(Figure 7.41).

The aquifer response (or impact of the water injection wells) may maintain the reservoir pressure close to the initial pressure, providing a long plateau

力降并传到水层，水通过油流通道填补因生产产生的孔隙空隙。因为水的压缩性很小，因此必须有很大的水体量才能达到水驱效果，所以连接水层必然是大水层。这里指的大水层指原油储量的10~100倍的水层量。

预测水层体积和渗透率通常很困难，因为勘探井和评价井通常都是以确定油层位置为目标，关于水层一般只能收集到很少的数据。因此，在油藏开发计划阶段预测水层反应常保持很大的不确定性。为了明确水层反应，有必要在油层生产过程中，测量油藏压力和油水接触面的运动情况，用物质平衡法确定水层对压力的影响。一般而言，生产5%的原油地质储量STOIIP后才能观测到水层的影响，需要多年的时间（图7.41）。

水层反应或注水井效应会保持生产过程中的油藏压力在初始压力附近，并有一个很长的平

Figure 7.41 Injection Water Drive
(Source: http://www.amerexco.com/recovery.html)

period and slow decline of oil production (Figure 7.42). The producing *GOR* may remain approximately at the solution *GOR* if the reservoir pressure is maintained above the bubble point. The outstanding feature of the production profile is the large increase in water cut over the life of the field, which is usually the main reason for abandonment. Water cut may exceed 90% in the final part of the field life. This is important to the process engineers, who need to design a facility capable of handling large water throughputs in later field life. As water cut increases, so oil production typically declines;

台期和很慢的油产量下降期（图7.42）。当油藏压力保持在泡点压力之上时，生产气油比*GOR*保持在溶解气油比*GOR*附近。水驱生产有一个突出特征是水侵量在整个油田生命过程中一直保持大的增长，这也是终止开发，结束油田开采的主要原因。在油田寿命最后水侵量可能超过90%。这对基建工程师很重要，他需要设计出一整套在油田

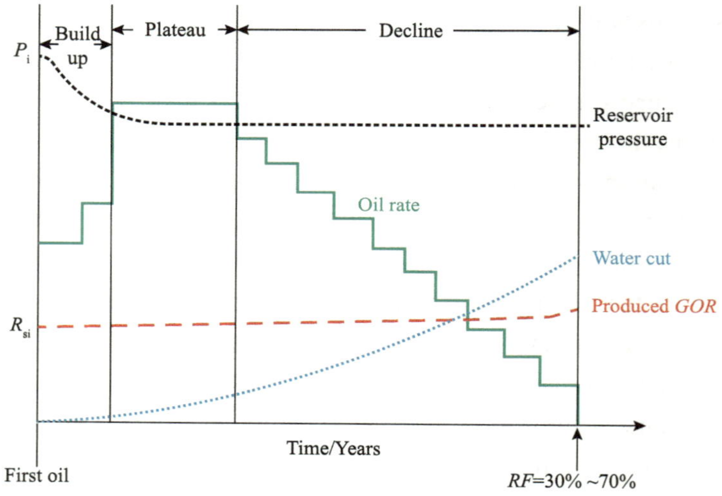

Figure 7.42 Characteristic Production Profile: Water Drive
(Source:Hydrocarbon Exploration and Production)

a constant gross liquids (oil plus water) production may be maintained.

The *RF* is in the range 30%~70%, depending on the strength of the natural aquifer, or the efficiency with which the injected water sweeps the oil. The high RF is an incentive for water injection into reservoirs which lack natural water drive.

7.4.3 Gas Cap Drive

When a reservoir is totally enclosed by impermeable beds, it has a constant volume and a constant pressure. Gas in the reservoir may be as a separate "gas cap" or in solution, depending on the in-situ pressures. Once a well has been drilled into the reservoir, the volume and pressure will decrease. Gas expansion and gas coming out of the solution will maintain the natural drive of the reservoir(Figure 7.43).

A production continues, the column of gas increase, measured at the surface by the GOR which measures volumes of gas after separation. Large amounts of potential energy can be stored in compressed gas, and production can continue for long periods with only minor change in pressure and production rates.

The reservoir is a closed system, however, and after a while the gas volume will decrease, resulting in decreased flow at the surface. This reduction in pressure can leave up to 50% of the reservoir fluids in

后期能处理巨大产水量的地面设备。当水侵量增加时，产油量一般下降，总体产液量（产油量加产水量）保持不变。

水驱的采收率在30%~70%，由天然水层的能量或注入水扫油效率决定。水驱的高采收率促使在缺乏天然水驱的油藏中采用注水开发。

7.4.3 气顶驱

当油藏由不可渗透层围起来，油藏体积和压力保持不变。油藏中气体相当于一个独立的气顶或以溶解气的形式存在，这取决于当地的压力。一旦钻入该油藏，油藏体积和压力下降，气体膨胀并从溶解状态分离出来从而天然驱动油藏（图7.43）。

随着生产继续，气体体积增加，地面测得的分离后气体体积的气油比增加。大量潜在的能量存储在压缩气体中，长时间生产中压力和产量下降很慢。

由于油藏是一个封闭系统，一段时间以后，气体的体积减少导致地面流体减少。在提高采收率措施采用之前，油

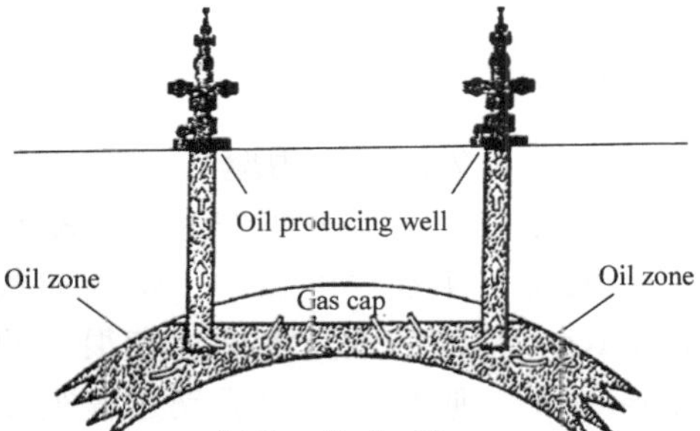

(a) Cross Section View

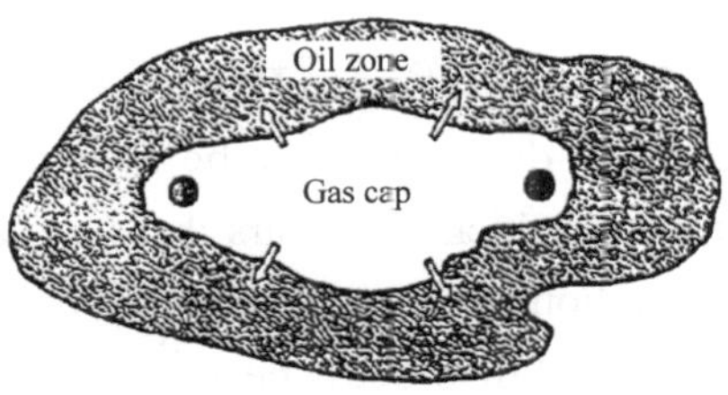

(b) Map View

Figure 7.43 Gas Cap Drive
(Source: http://www.assignmenthelp.net/assignment_help/gas-cap-drive)

the formation before enhanced recovery methods are employed.

The initial condition required for gas cap drive is an initial gas cap. The high compressibility of the gas provides drive energy for production, and the larger the gas cap, the more energy is available.

Compared to the solution gas drive case, the typical production profile for gas cap drive shows a much slower decline in reservoir pressure, due to the energy provided by the highly compressible gas cap, resulting in a more prolonged plateau and a slower decline (Figure 7.44). The producing *GOR* increases as the expanding gas cap approaches the producing wells, and gas is coned or cusped into the producers. Again, it is assumed that there is negligible aquifer movement, and water cut remains low (in the order of 10% at the end of field life). Typical RFs for gas cap drive are in the range 20%~60%, influenced by the field dip and the gas cap size.

藏流体压力降低可达50%。

原始状况下的气顶驱为原始气顶。气体的高压缩性为生产提高驱动能量，气顶越大，驱替能量越多。

与溶解气驱相比，典型的气顶驱生产表现出更慢的压力降落，因为驱动能量由高压缩性的气体提供。生产平稳期更长，下降更缓慢（图7.44）。生产气油比随气顶膨胀而增加，气锥现象出现。假定有可忽视的水层，含水率保持较低的水平（在油田后期大约为10%）。典型的气顶驱采收率为20%~60%，受油田和气顶大小的影响。

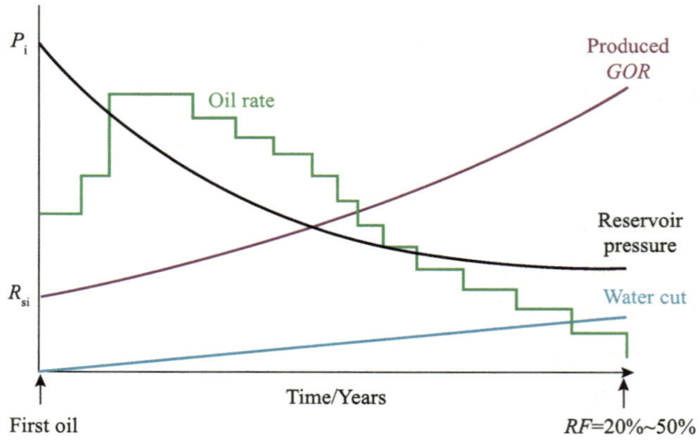

Figure 7.44 Characteristic Production Profile: Gas Cap Drive
(Source: Hydrocarbon Exploration and Production)

7.4.4 Gravity Drainage Drive

Gravity drainage may be a primary producing mechanism in thick reservoirs that have a good vertical communication or in steeply dipping reservoirs. Gravity drainage is a slow process because gas must migrate up structure or to the top of the formation to fill the space formerly occupied by oil. Gas migration is fast relative

7.4.4 重力驱

重力驱作为第一驱动力发生在厚油藏中，这类油藏有很好的垂直渗透性和很大的油藏倾角。重力驱动过程很慢，因为气体向上运移到构造中或到达地层顶部填满原油占据的空

to oil drainage so those oil rates are controlled by the rate of oil drainage.

间是个非常缓慢的过程。气体相对油流运移较快，因此主要由油泄流速度控制油产量。

7.4.5 Compaction Drive

Compaction drives are found in abnormal pressure reservoirs. Overpressure reservoir compacts or pushes in on itself, pushing out the oil into the producing well.

7.4.5 压实驱

压缩驱动在异常压力油藏中出现，过压油藏压缩和驱动油流进入生产井中。

7.4.6 Combination Drives

It is possible that more than one of these drive mechanisms occur simultaneously; the most common combination being gas cap drive and natural aquifer drive. Some reservoirs have a combination of a gas drive acting downwards and a water drive acting upwards on the reservoir. To maintain production rates and percentage recovery, some enhancement to the natural drives is incorporated by adding either water or gas to the reservoir. The source of water or gas may be from another well, or that previously recovered from the same well and pumped back down to replenish the reservoir(Figure 7.45).

7.4.6 混合驱

在油藏驱动中可能不止一种驱动作用，最普遍的是气顶驱和天然水驱一起作用。在一些油藏中，气顶驱向下作用和水驱向上作用。为保持产量和提高采收率，一些注水或注气增产措施和天然驱动联合作用。水源或气源可能来自别的井，也可能是之前从同一口井中采出又注回油藏中（图7.45）。

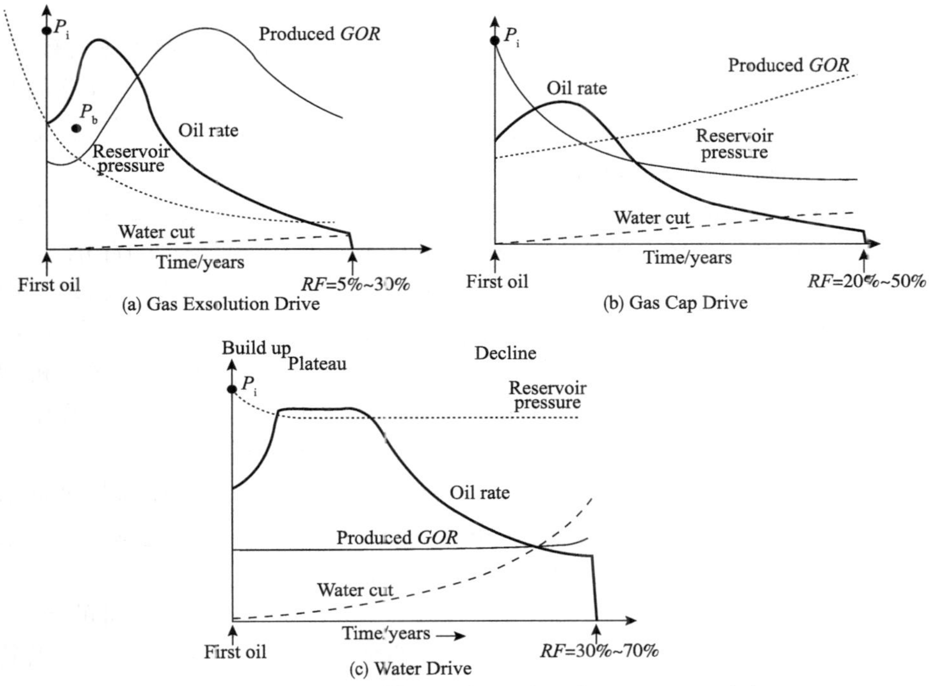

Figure 7.45 Drive Mechanisms (Source:http://pages.geo.wvu.edu/)

Material balance techniques are applied to historic production data to estimate the contribution from each drive mechanism(Figure 7.46).

物质平衡原理用于计算历史产量和评估每种驱动能量的贡献（图7.46）。

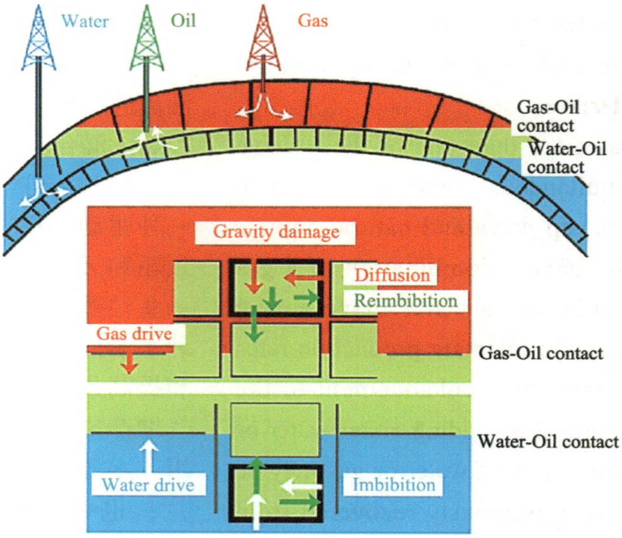

Figure 7.46 Main Drive Mechanisms in Fractured Reservoirs
(Source: http://omicsonline.org/2157-7463/images/2157-7463-2-109-g001.html)

7. 5 Gas Reservoirs

7.5.1 Wet & Dry Gas Reservoirs

The reservoir mechanics of wet and dry gas reservoirs are identical because in both cases there are no liquid hydrocarbons in the formation.

Gas has an extremely low viscosity (less than a hundredth of most oil viscosities), so it flows readily through the formation to the wellbore. Gas wells are therefore drilled on much wider spacing (distance between the wells) than oil wells.

The operation of the gas-expansion drive mechanism could not be simpler. As reservoir pressure drops, the gas expands into the wellbore. This continues until the reservoir pressure is essentially zero. In some cases, compressors are installed to suck out the last few MCF. Recovery factors often reach 95%.

Unlike oil, gas is highly compressible so its surface volume is much greater than its reservoir volume. The

7. 5 气藏

7.5.1 湿气藏和干气藏

湿气藏和干气藏的开发机理是相同的，因其在地层中都呈气态。

天然气的黏度非常低（一般低于原油黏度的百分之一），所以在地层中容易流动。因此，气井的井距较油井大得多。

气藏的膨胀驱动机理很简单。气体在开采过程中不断膨胀直到压力释放完毕。一些情况下会采用压风机来抽出最后剩余的地下气体。气藏采收率通常达到95%。

与原油不同的是，气体的膨胀系数很大，所以其地面体积要比地下体积大很多。根

ideal gas law states that the SCF of gas occupying a given volume of reservoir space is directly proportional to the pressure. For example, a cubic foot of gas volume at 5000 psi in the reservoir would actually contain 340 (5000 ÷ 14.7) standard cubic feet of gas.

Not all gas reservoirs are constant volume. Active water drives are common but are considered a negative because the encroaching water usually bypasses considerable gas. Compaction drives also occur.

据理想气体方程可知，天然气在地下的体积与地层压力直接相关。比方说，在气藏压力为5000psi（1psi=6.895kPa）的情况下，地下每立方英尺可以容纳标准条件下地面上340ft^{3}的天然气。

不是所有气藏都有恒定的体积。在气藏的开采过程中，也经常会发生水驱作用。但这种作用一般认为是不利的，因为会有大量的气被绕过。压实驱动机理同样也是存在的。

7.5.2 Retrograde Condensate Gas Reservoirs

Production from retrograde condensate reservoirs is initially very profitable because of the rich condensate yields of two or three hundred barrels per MMCF. Problems start, however, when the reservoir pressure drops to its dew point. The condensate then starts dropping out in the formation. This is a two-pronged problem. First, the operator loses most of the revenue from condensate sales. Second, and more serious, is that the liquids building up in the formation restrict its permeability to gas. As a result, gas production drops off sharply. Quite suddenly, this reservoir is not nearly as profitable as it first appeared to be.

One remedy is to initiate a gas-cycling project. This entails stripping the natural gas liquids from the produced gas stream, then compressing and reinjecting the residue gas. The dry gas circulates through the formation and re-vaporizes some of the liquids which are then stripped out at the surface. Carried to its conclusion, cycling can result in very high recoveries.

The distressing reality of gas-cycling is, however, that if there is a market for the residue gas—no matter how low-priced—the time-value-of money renders cycling uneconomic. The present value of the reservoir is optimized by selling off the residue and then

7.5.2 反凝析气藏

反凝析气藏初始阶段的利润很高，每百万立方英尺天然气可产出两三百桶凝析油。然而，当气藏温度降到露点以下时，问题就来了。这时，会在地层中形成凝析油。这就带来了两方面的问题：其一，凝析油开采带来的利润大部分损失掉了；其二，也是更为严重的，由于气藏地层中凝析油的存在，大大减少了气的相对渗透率，导致气产量严重下降。这样一来，该气藏的利润便较之以前的估计大打折扣了。

这一问题的解决办法之一是循环注气。这一过程为，首先去掉产出气的凝析成分。然后将剩余的干气注入地层。这些干气可以带出地层中的凝析油，这部分凝析油在地面分离后，干气再度注入地层，如此循环往复。这种方法可以达到很高的采收率。

循环注气的一个弱点是，

abandoning the reservoir that is loaded up (rendered unproducable) with liquids.

Retrograde condensate reservoirs are therefore candidates for recycling only if discovered in areas remote from market.

如果有对干气的市场需求，哪怕价格很低，资金的时间价值都会使得循环注气这一方式变得不经济。这种情况下，要想让气藏的现值达到最高，最佳方式是卖掉产出的干气，然后放弃被凝析油堵塞的气藏。

因此，循环注气工艺只有在气藏远离市场的情况下，才是经济可行的。

7. 6 Oil and Gas Recovery Methods

In primary recovery—the initial approach to produce oil—natural reservoir pressure or simply mechanical pumps are used to raise oil to the surface. Most oil wells in the world today have to be pumped(Figure 7.47).

7.6 油气采收率

一次采油指刚开始采油，利用天然油藏压力或简单的机械泵将油举升到地面。今天世界上绝大多数油井采用泵采油(图7.47)。

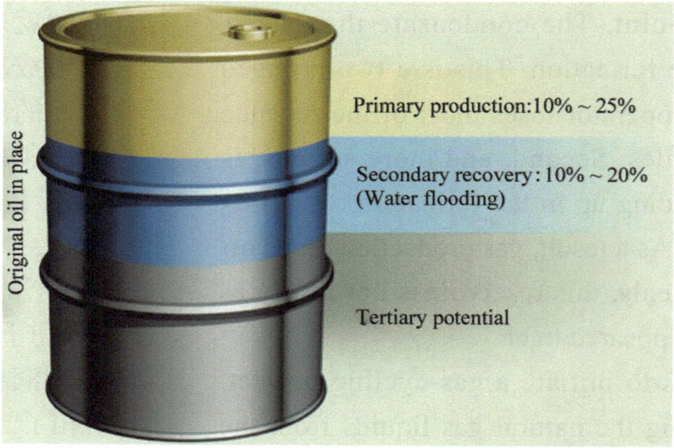

Figure 7.47 Oil Recovery Overview
(Source: http://www.midconenergypartners.com/oil-recovery-overview.php)

A number of methods can improve primary recovery. The most common is infill drilling, which involves drilling more wells into the same pool so the oil does not have to travel to as far through the rock to reach a wellbore. Directional wells are often used for infill drilling. More than one directional well can be drilled from a common platform. Horizontal drilling, which extends the wellbore into a much large portion of

有很多种方法可以提升一次采油的采收率。最普遍的方法是钻加密井，在同一油田上钻更多的井，使原油从油层中不必运移太远就能流入井筒。加密井通常为定向井。从一个平台上不仅钻一口定向井。水平井，指延伸井筒进入更多的

the oil-bearing formation, has been employed since the late 1980s to improve production and enhance recovery.

Further oil production can be obtained by injecting water (waterflooding) or natural gas to maintain reservoir pressure and push oil out of the rock. This is called secondary recovery. Typical recoveries are 25%~45% OOIP after secondary recovery (average 32%).

More advanced methods are referred to as tertiary recovery. The most common tertiary recovery method for light and medium crude oil is miscible flooding. In this procedure, natural gas liquids (ethane, propane and butane) are injected into special injection wells. When dissolved, these liquids reduce surface tension and viscosity to help release the oil from the reservoir rock. Carbon dioxide has also been used for miscible floods.

Even with all these techniques, the average recovery in the light oil fields is a little more than 30 percent of the original oil in place (OOIP). The remaining resource represents billions of cubic meters of oil that has been discovered in the world but cannot be produced economically with existing technology.

油层。水平井钻井开始于1980年代，提升了产量和采收率。

进一步采油可以通过注水（注水驱）和注天然气保持油藏压力从油层中挤出原油，这叫二次采油。二次采油一般能采出原油储量的25%~45%，平均采收率为32%。

比二次采油更进步的采油方法是三次采油。对于轻质到中质油最常用的三次采油方法是混相驱。在混相驱中，天然气液体（乙烷、丙烷和丁烷）被注入特殊的注入井中，与地下原油混相，减少了表明张力和黏度，从而促进原油从油藏岩石中流出。二氧化碳也被用于混相驱。

用二次采油和三次采油，能采出原油储量的30%以上，剩下的成百上千亿原油不能被现有的技术经济地开采出来。

8 FIELD APPRAISAL 油气田评价

After completion and before production we need appraise the field to answer the following questions:

◆ How much oil is in place in this area?

◆ How big are the boundaries?

◆ How many wells are we going to drill?

◆ What is the spacing between the wells?

◆ Are they going to be close together or far apart?

these questions need to answer so we can maximize the recovery factor over a reasonable period time(Figure 8.1).

We Need Appraise the Field before Develop-ment and Production

Acommon procedure of filed appraisal is (Figure 8.2):

完井以后生产之前需要评价油田以回答下面问题：

◆ 有多少油气在此?

◆ 油气边界在哪?

◆ 需要钻多少口井?

◆ 井间距多少?

◆ 井与井之间的距离是远一点好还是近一点好?

回答这些问题能帮助确定在合理生产时间下的最大采收率（图8.1）。

在开发和生产之前需要进行油田评价。

油田评价的一般程序如下（图8.2）：

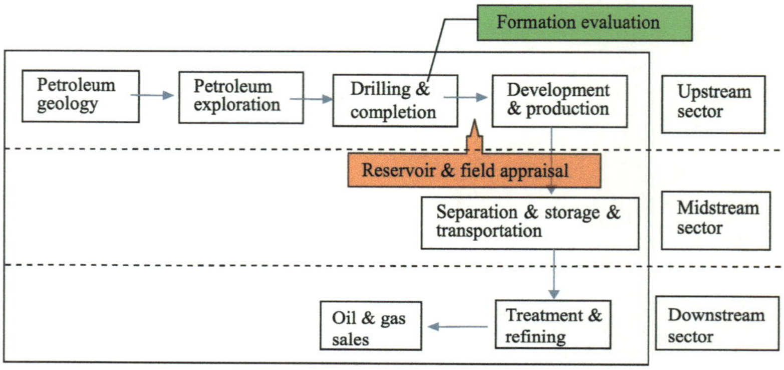

Figure 8.1 Petroleum Industry

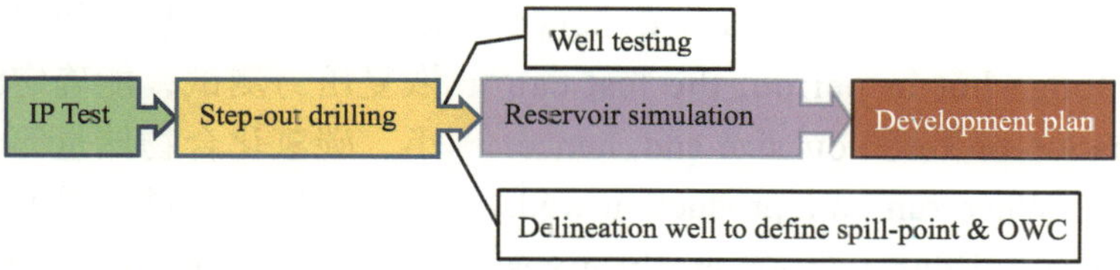

Figure 8.2 Field Appraisal

8.1 Production Testing

Unless there has been a flowing drill stem test on the well, there is considerable uncertainty as to what the well' s production rate will be. After a few days of production to recover the treating fluids and clean up the well, the Initial Potential Test (IP) is taken. The IP is of great interest, so it is carried on the morning drill report.

Well testing is a widely used technique for reservoir assessment. As the term implies, well testing meaning a systematic procedure for a well so as to obtain representative samples of the reservoir fluids and to collect simultaneous measurements of the production rates for individual phase and the corresponding bottom-hole pressures. The parameters available for the data analysis depend upon the type of test performed(Figure 8.3).

8.1 生产测试

除非该井曾做过中途测试，否则无从了解井的生产能力。当井清理干净回收完处理液，初始产能将被测试。初始产能测试时如此重要，以至于它写在钻井日志上。

试井广泛用于油藏评估。如名所示，试井指对一口井的一系列测试，包括收集代表性的油藏流体样品、各相流体的生产测试和井底压力测试等。依据不同的测试类型进行参数分析（图8.3）。

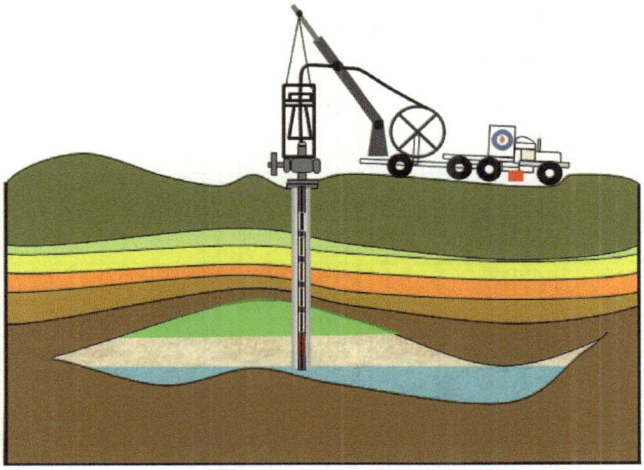

Figure 8.3 Well Testing
(Source: http://www.petro-vision.com/pressure-transient-analysis.html)

The single-well pressure buildup test run on exploration wells can provide information, in general, on flowing well efficiency, formation transmissibility,

在探井中进行的压力恢复测试一般能提供该井的流动效率，地层传导率和油藏非均

and reservoir heterogeneity. If it is specially designed for long flow and long shut-in period, the test can also provide information on reservoir size and, hence, hydrocarbons in place. When run on a production well, a single-well pressure buildup test generally gives information only on the well’s flow efficiency and the transmissibility within the well’s drainage area.

质性方面的信息。如果长关井恢复压力测试，能确定油藏大小、碳氢化合物储量等信息。如果边生产边进行单井压力恢复试井，能确定该井的流动效率，地层传导率和泄油面积方面的信息。

A single-well test will not yield any information on reservoir anisotropy or any directional permeability trends. For these purposes, multiple-well tests, such as interference or pulse tests, must be employed.

单井试井不能得到关于油藏各相异性和任何渗透率方向性趋势的信息。因此多井测试，如干扰试井和脉冲试井必须采用。

◆ Single-Well Test

◆ 单井测试

In a single-well test, rate and pressure measurements are taken at the same well. Pressure and rate measurement collected while the well is on production constitute the drawdown data. Buildup measurements are much less influenced by rate variation (during the flow period) than are the measurements for drawdown. The objectives of a single-well test are different from exploration and production wells(Figure 8.4).

在单井测试中，产量和压力在同一口井中测试。油气井在生产的同时收集产量数据和压力下降数据。压力恢复测试比压力降落测试受产量变化影响小（在生产过程中）。在勘探井和生产井中单井测试的目的不一样。

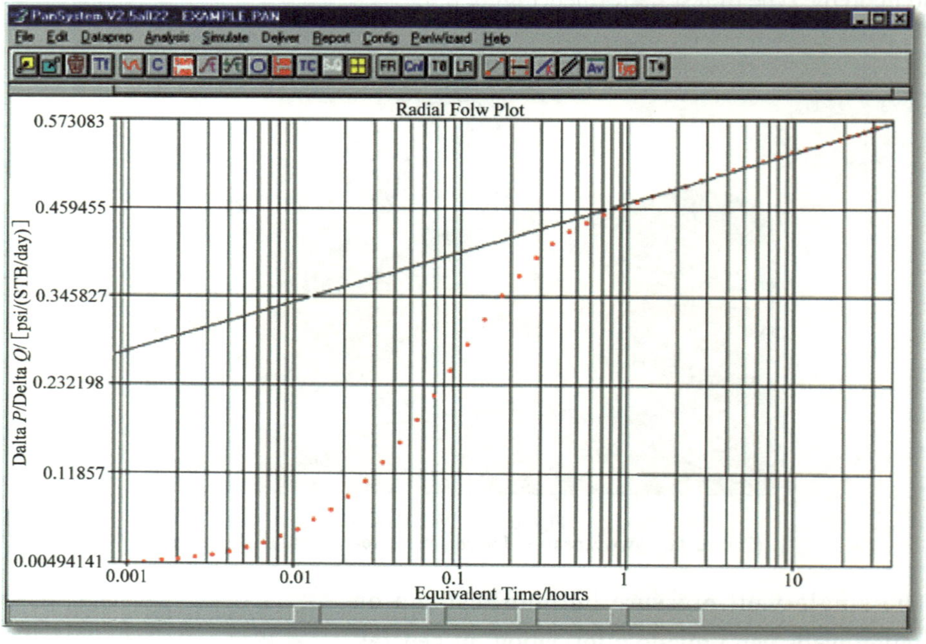

Figure 8.4 Well Testing: Radial Flow
(Source: http://www.ep-weatherford.com/Solutions/Consult/Well_Testing.htm)

8.2 Reservoir Simulation

Reservoir simulation is a technique in which a computer-based mathematical representation of the reservoir is constructed and then used to predict and appraise its dynamic behavior. The reservoir is gridded up into a number of grid blocks. The reservoir rock properties (porosity, saturation and permeability), and the fluid properties (viscosity and the *PVT* properties) are specified for each grid block(Figure 8.5).

8.2 油藏数值模拟

数值模拟是基于计算机上数值构建油藏从而预测其动态行为的技术。油藏被分割成无数小网格，油藏的岩石特征和流体特征赋值在每个小网格中（图8.5）。

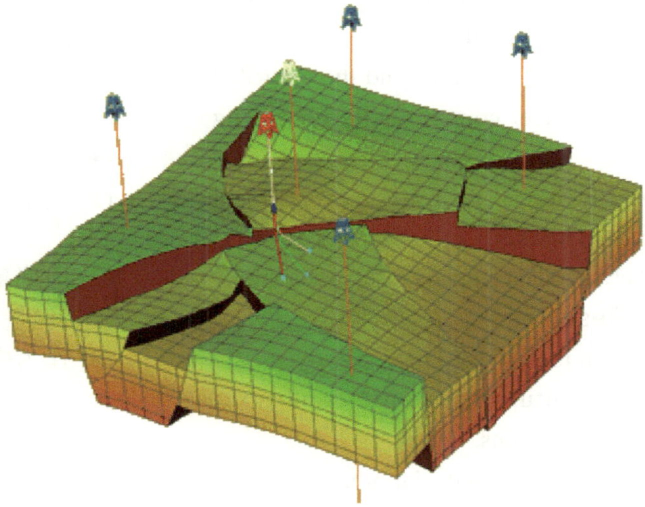

Figure 8.5 Reservoir Simulation
(Source: http://www.coatsengineering.com/sensor_reservoir_simulator.htm)

The number and shape of the grid blocks in the model depend on the objectives of the simulation. A 100 grid block model may be sufficient to confirm rate dependent processes described in the previous section, but a full field simulation to be used to optimise well locations and perforation intervals for a large field may contain up to a million grid blocks. The larger the model, the more time consuming to build, and slower to run on the computer.

网格的数量和形状由模拟的客体决定。100个小网格也许够模拟生产过程中产量的一段，但大油田全油藏的模拟用于确定最优化部井位置和射孔段则需要上百万个网格。模型越大，构建的时间许多，计算机运算越慢。

8.3 Oil and Gas Reserves

Reserves, as defined by the Society of Petroleum Evaluation Engineers (SPE), are estimated volumes of crude oil, condensate, natural gas, natural gas

8.3 油气储量

美国石油评估工程师协会定义储量为当前经济技术及政策条件下可以采出的地下原

liquids, and associated substances anticipated to be commercially recoverable from known accumulations from a given data forward, under existing economic conditions, by established operating practices, and under current government regulations.

油、凝析油或天然气的体积。

8.3.1 Reserve Categories

◆ Proved Reserves

(1) Can be estimated with reasonable certainty.

(2) Generally must be supported by actual production or formation tests. Indirect indications like electric logs are usually not adequate.

(3) Include both drilled locations and the undrilled locations that directly offset commercially producing locations.

(4) Must have operational facilities to process and transport the reserves to market.

(5) Have publicly-quoted company reserve statistics.

◆ Unproved Reserves

(1) Are less certain than proved reserves.

(2) May require future economic conditions different from current conditions.

(3) Are estimated for internal planning or special evaluations, but not routinely compiled.

(4) May be divided into two risk classifications: probable and possible.

8.3.1 储量分类

◆ 证实储量

(1) 测算时有一定的把握；

(2) 有生产或地层测试检验，只有测井曲线是不够的；

(3) 包括生产区的邻近地区；

(4) 应具备地面输油设施；

(5) 向大众公开。

◆ 未证实储量

(1) 比证实储量的可靠度差些；

(2) 经济条件变化后才可能成为证实储量；

(3) 作为内部参考的测算，未公开；

(4) 可以分为概算储量和可能储量。

8.3.2 Reserve Detemination Methods

Early in a field' s life, little is known about the extent, quality, and drive mechanism of the reservoir. Reserve estimates are therefore likely to be highly inaccurate—either on the plus or minus side. As the field is developed, and particularly as data on production performance accumulates, the range of error narrows. It is, however, not until the last barrel is produced and the field abandoned that the reserves are known precisely(Figure 8.6).

Listed below are various methods that are used either singly or in combination to estimate reserves.

8.3.2 储量计算方法

在油田生命周期的初始阶段，油田的范围、品质和驱动机理都不是十分清楚。因此在这一阶段，关于油田储量的估算都很不精确。不过，随着油田开发的不断进行，生产中得到越来越多的资料和数据，估算的误差就会不断得到缩小。储量的精确值则只有在油田开采结束后才能真正知道（图8.6）。

以下是储量计算的一些方法：

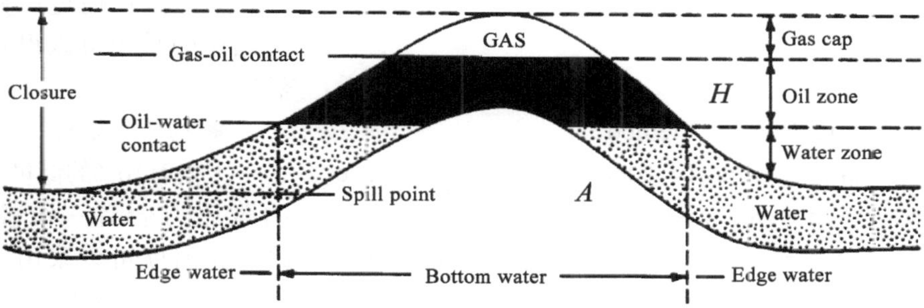

Figure 8.6 Cross-section of A Simple Anticlinal Trap
(Source: Petrolfum Geology)

◆ Volumetric Method

The volumetric method is primarily used early in the field' s life when the rough dimensions of the reservoir are known but performance data is not yet available. It is a relatively crude method, subject to significant error.

Step 1: The original oil in place is calculated by the formula

$$OOIP = 7758 \times A \times H \times \phi \times S_O$$

Where *OOIP*= Original Oil In Place, bbls

7758 = Bbl / acre-foot constant.

A = Reservoir Area, acres.

H = Reservoir Thickness, feet.

ϕ = Porosity, decimal.

S_O = Oil Saturation, decimal.

Step 2: The barrels of recoverable oil is calculated by the formula

$$STB = (OOIP \times RF)/\beta$$

Where *STB* = Stock Tank Barrels

RF = Recovery Factor, reflecting the fraction of *OOIP* similar fields in the vicinity have recovered.

β = Reservoir volume factor, reflecting shrinkage the oil undergoes as it is produced.

Step 3: Associated gas reserves are calculated by multiplying the oil reserves by the solution Gas-Oil-Ratio (*GOR*).

◆ 容积法

这种方法适用于油田开发的初期。在此阶段，油田的范围已知，不过生产数据缺乏。这是一种相对粗糙的方法，有较大的误差。

第一步：采用如下公式计算原始地质储量：

$$OOIP = 7758 \times A \times H \times \phi \times S_O$$

式中 *OOIP*——原油地质储量；

7758——每桶/每英亩英尺，常数；

A——油气藏面积，英亩；

H——油气藏高度，英尺；

ϕ——孔隙度，小数；

S_O——含油饱和度，小数。

第二步：按如下公式计算可采储量：

$$STB = (OOIP \times RF)/\beta$$

式中 *STB*——储罐桶数；

RF——采收率，反映地下原油地质储量被开采了多少的系数；

β——油藏体积系数，反映地下原油生产到地面上的体积收缩。

第三步：通过计算储量与气油比的乘积可以得到伴生天然气储量。

◆ Material Balance Method

After approximately 10% of the reserves are produced——enough to measurably reduce the reservoir pressure——the material balance method yield greater accuracy than the volumetric method.

The reservoir is viewed as a constant-volume tank from which known volumes have been extracted. Having analyzed reservoir samples to gain knowledge of the pressure-temperature-volume behavior of the hydrocarbon system, the remaining reserves can be calculated from the observed reservoir pressure. If water influx is occurring, adjustments can be made to account for it.

◆ 物质平衡法

大概有10%的储量采出后，油藏压力会有一个较大的下降，这时就可以采用比容积法更精确的物质平衡方法了。

在这一方法中，油藏被看作是一个体积恒定的容器，从中已经有一定体积的油气被采出。通过对采出的油气进行*PVT*数据分析，可以根据当前的油藏压力估算出剩余可采储量。

◆ Reservoir Modeling / Simulation

When a great deal of high-quality data is available on a reservoir, a more sophisticated material balance technique requiring computer solutio can be applied. A multiple cell computer model of the reservoir is created and the depletion process is mathematically simulated. Under the right conditions, these simulations can be quite reliable.

◆ 油藏数值模拟法

当取得了有关储量的大量精确数据后，就可以采用一种复杂的计算机辅助的物质平衡计算。这时，在计算机上，油藏被分成许多个小单元，然后进行开发数值模拟。在条件可靠的情况下，这些计算的结果是相当准确的。

◆ Production Decline Curve Method

As a reservoir reaches a more advanced state of depletion, its production rate typically declines along a predictable path. In many cases, the decline is close to exponential; that is, at a constant percent. When the rate is plotted on a logarithmic scale versus time on a Cartesian scale, the decline becomes a relatively straight line which can be extrapolated to yield reserves and a forecast of future production.The extrapolation ends when the line reaches the economic limit——a horizontal line on the graph. This represents the break-even point where revenue equals direct operating expenses and the project is abandoned.

◆ 产量递减曲线法

在油藏进一步衰减后，其产量一般就会按一定规律曲线下降。许多情况下，这一曲线会是指数递减曲线，也就是按照一定的百分比下降。在半对数坐标图上可以看出产量呈直线方式下降，进行外推后，可以得出储量并预测未来产量。对直线的外推在到达经济极限时停止，经济极限是由一条水平线代表的。经济极限代表收益与操作成本的平衡点，达到此点后，项目不再具有经济效益。

This method is effective for one well, a group of wells, or the entire field and is considerably more

该方法适用于单井、井组

accurate than other methods. It therefore dominates reserves estimation. A number of excellent computer programs are availble that display the decline curves, facilitate curve fitting, and calculate the results.

8.3.3 Reserve Revisions

When new information becomes available, companies revise their reserve estimates. For example:

◆ Completion of successful new wells extend the reservoir' s limits; dry holes contract the limits.

◆ Movement in crude oil or gas prices changes the economic limit with higher prices increasing reserves and lower prices decreasing reserves.

◆ Initiation of a waterflood or other enhanced recovery project increases reserves.

◆ New evidence can change reservoir drive assumptions. For example, if it becomes apparent that an active water drive is present, the recovery factor may be increased.

Companies also routinely review the reserve estimates on all their fields once a year for inclusion in the annual report.

及油田，比其他方法都要精确得多。因此，该方法在储量估算上很重要。有许多优秀的软件专门用来绘制递减曲线并进行计算。

8.3.3 储量调整

当有更多的数据信息时，油公司会相应地对储量估算进行调整，如：

◆ 新井或干井使得油田范围扩大或缩小；

◆ 油价的变化使得油田的经济极限发生变化；

◆ 水驱或三次采油方式增加储量；

◆ 新的驱动机制的发现，如活跃水驱，可能使储量估算增加。

油公司会对储量进行年度例行审计并以年报形式上报。

9 PRODUCTION & DEVELOPMENT 开发与生产

When an onshore wildcat well encounters what appears to be commercial hydrocarbons, the well is quickly completed and placed on production.

The development of a newly discovered field and production may be considered as that the oil industry concerned with conveying hydrocarbons from the reservoir to surface. The target is clearly defined: to recover as much oil as possible at as low cost as possible.

In many wells, the natural pressure of the subsurface reservoir is high enough for the oil or gas to flow to the surface. So we called naturally flowing well(Figure 9.1). However, this is not always the case, especially in depleted fields where the pressures

当一口陆上油田的探井钻遇工业性油气流时，该探井会被迅速完井并投入生产。

对于一个新近开发的油田，其生产可以认为是将地下碳氢化合物带到地面。目标明确，开采尽可能多的油气，同时将成本降到最低。

在很多井中，天然的地层压力足够将油气举升到地面，也就是我们所说的自喷井（图9.1）。但是，这不是常有的情况，特别是当油田后期压力下降和在低渗透油藏中都不会自

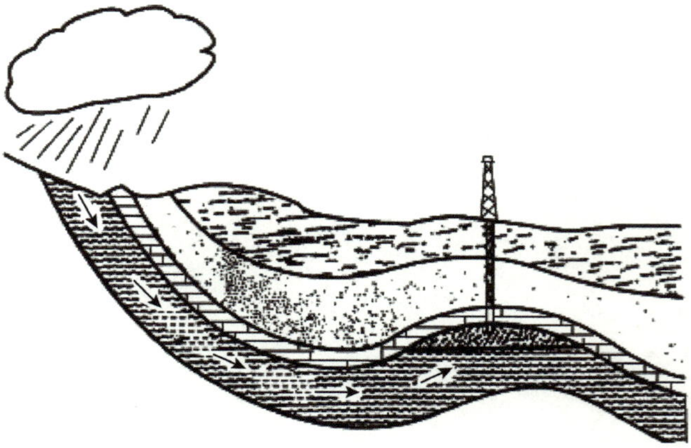

Figure 9.1 Reservoir with Recharge Area
(Source: The Fluid Mechanics in Porous Medium of Reservoirs)

have been lowered by other producing wells, or in low permeability oil reservoirs. Installing a smaller diameter tubing may be enough to help the production, but artificial lift methods may also be needed. Common solutions include downhole pumps, gas lift, or surface pump jacks.

喷。安装小直径的油管可以帮助生产，但是人工举升才是解决问题之道。一般是在地下安装泵、气举或地面深井泵。

9.1 Natural Flow in a Well

Natural flow in a well or naturally flowing well means a well in which the formation pressure is sufficient to produce oil at a commercial rate without requiring a pump. Most reservoirs are initially at pressures high enough to allow an oil or gas well to flow naturally, just like an artesian well(Figure 9.2，Figure 9.3).

9.1 自喷井

自喷井指地层压力足够生产工业性油流而不需安装泵进行举升的井。大多数油藏最初具有高压，能够允许油、气自然流入井中，像自来水井一样（图9.2、图9.3）。

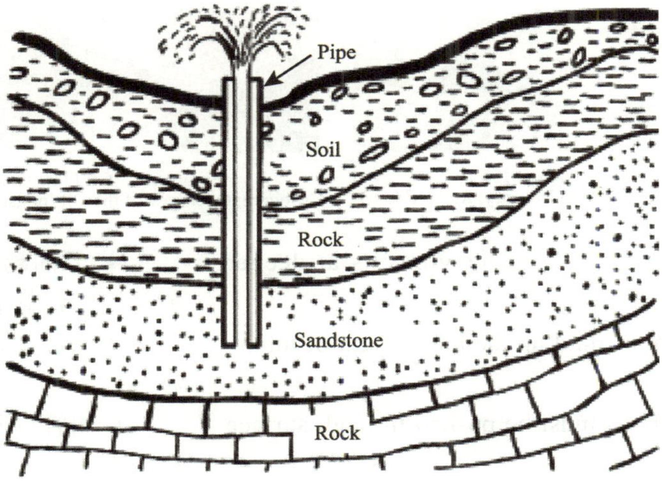

Figure 9.2 Schematic of an Artesian Well
(Source: http://en.wikipedia.org/wiki/Artesian_aquifer)

Figure 9.3 Hydrostatic Column of a Well
(Source: Overview of the Oil Industry, Richard Lau)

Nature lift means the well’s own natural energy must be sufficient to (Figure 9.4):

◆ Push oil and gas to the surface.

◆ Overcome hydrostatic borehole pressure and frictions.

◆ Push hydrocarbons through surface facilities.

There are four main obstacles can cause the nature lift to slow down and failure(Figure 9.5):

◆ The more viscous the fluid,the more energy is needed to overcome friction.

当油层压力大于静水柱压力是就会产生自喷（图9.4）。

◆ 自喷意味着油井自身的天然能量足够推动油气到地面；

◆ 有足够的能量克服静水柱压力和摩擦力；

◆ 有足够的能量推动油气在地面管道设备中流动。

天然能量衰减主要来源于四方面的障碍（图9.5）：

Figure 9.4 Natural Lift
(Source: Overview of the Oil Industry, Richard Lau)

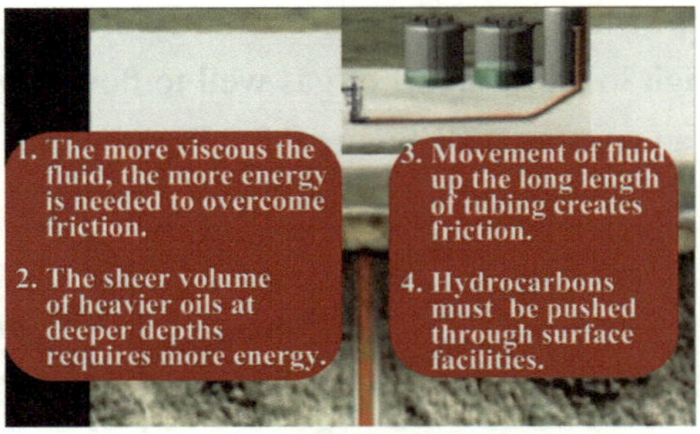

Figure 9.5 Oil Flow Obstacles
(Source: Overview of the Oil Industry, Richard Lau)

◆ The sheer volume of heavier oils at deeper depths requires more energy.

◆ Movement of fluid up the long length of tubing creates friction.

◆ Hydrocarbons must be pushed through surface facilities.

◆ 原油越粘稠需要越多的能量用以克服摩擦力；

◆原油埋藏越深需要能量越多；

在油管内长距离的运移的摩擦力；

◆ 地面管道中流动阻力。

9.2 Artificial Lift

Artificial lift is a process used on oil wells to increase pressure within the reservoir and encourage oil to the surface. When the natural drive energy of the reservoir is not strong enough to push the oil to the surface, artificial lift is employed to recover more production(Figure 9.6).

Although there are several methods to achieve artificial lift, the two main categories of artificial lift include *gas lift sand pumping systems*.

9.2 人工举升

人工举升用于增加油井油藏压力，将油流举升到地面。当天然驱动能量不够驱动油流到地面时，人工举升用于采收更多油气（图9.6）。

尽管人工举升有多种方法，但主要分为两类：气举系统和泵举系统。

- *Petroleum Fluid Lift*
  - *Natural Lift*
  - *Artificial Lift*
    - *Gas Lifts*
    - *Pumping Systems*
      - *Rod Pumps*
      - *Subsurface Pumps*
        - *ESP*
        - *Hydraulic Pump*
        - *Subsurface Jet Pump*
        - *Plunger Lift Pump*
        - *Progressing Cavity Pump*
        - *Others*

Figure 9.6 Petroleum Fluid Lift

9.2.1 Gas Lift

Gas lift injects compressed gas into the well to reestablish pressure, making it produce. Even when a well is flowing without artificial lift, it many times is using a natural form of gas lift(Figure 9.7).

The injected gas reduces the pressure on the bottom of the well by decreasing the viscosity of the fluids in the well. This, in turn, encourages the fluids to flow more easily to the surface. Typically, the gas that is injected is recycled gas produced from the well.

With very few surface units, gas lift is the optimal choice for offshore applications. Occurring downhole, the compressed gas is injected down the casing tubing

9.2.1 气举

气举指压缩气体进入井筒重建压力系统，使油流继续生产。即使没有人工举升，其实井筒中也有天然的气体举升（图9.7）。

注入气体降低了井筒流体的黏度，从而降低井底压力，从而使流体更容易地流向地面。一般而言，注入的气体为该井的生产气体，可循环使用。

气举在地面只有很少设施，是海上油田的优先选择。井下压缩气体注入油管环空，

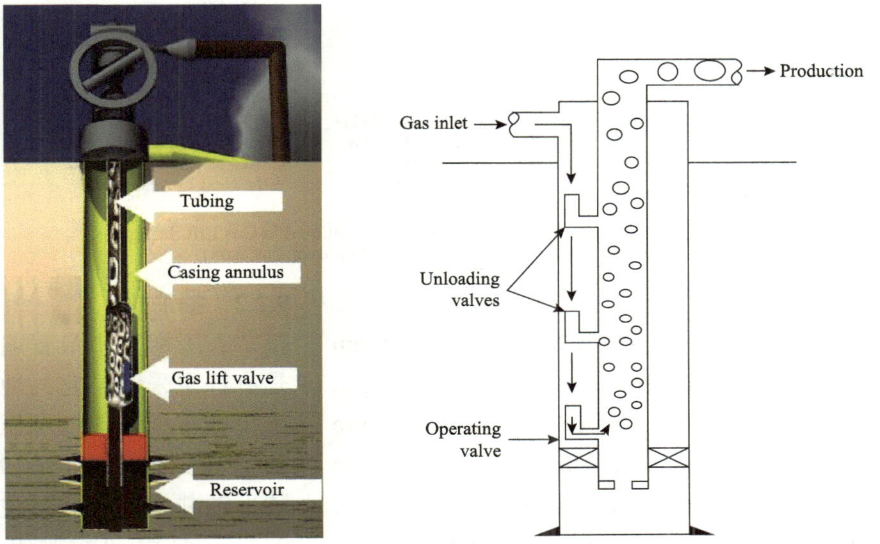

Figure 9.7 Gas Lift
(Source: http://s255.photobucket.com/user/nogososro/media/gaslift-1.jpg.html)

annulus, entering the well at numerous entry points called gas-lift valves. As the gas enters the tubing at these different stages, it forms bubbles, lightens the fluids and lowers the pressure.

依次进入井筒无数个气举阀。当气体进入油管的不同段，形成泡沫流，从而减轻流体，降低压力。

Gas lift is a very flexible method, well adapted to the deviated wells common in offshore operations, and rapidly adjustable to both changes in required production rates and changes in character of the produced fluids. The rise in water content with time, common in many wells, has clearly a significant effect on the pressure system within the wellbore(Figure 9.8).

气举是一种很灵活的人工举升办法，在直井、斜井、陆上、海上油田都可适用，可产量和流体变化时可以迅速调整气举方式。含水量随生产时间增加时，气举能很好地调节井筒中压力系统（图9.8）。

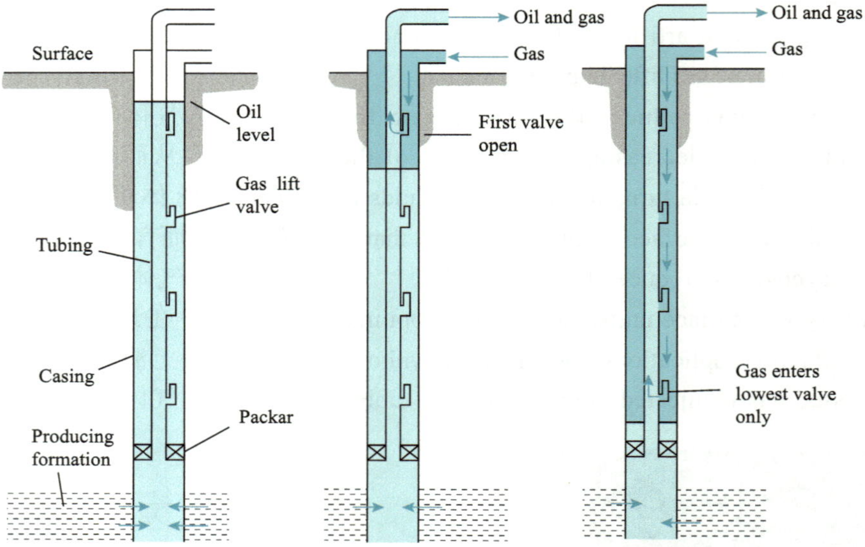

(a) Oil Level above First Valve
(b) First Valve Open and Gas Entering Tubing; Oil Level in Casing/Tubing Annulus Moving Downward
(c) Oil Level Has Moved Downward,and Each Valve Has Closed as Gas Has Entered the Next Lowest Valve

Figure 9.8 Schematic Representation of Operation of a Gas Lift String
(Source: http://www.accessscience.com/overflow.aspx)

9.2.2 Rod Pumps

The most common type of artificial lift pump system applied is beam pumping (rod pumps), which engages equipment on and below the surface to increase pressure and push oil to the surface(Figure 9.9). Consisting of a sucker rod string and a sucker rod pump, beam pumps are the familiar jack pumps seen on onshore oil wells(Figure 9.10).

9.2.2 抽油机

人工举升中最普遍的类型是有杆泵采油，它的地面和地下设施帮助增加油藏压力，将油提升到地面上（图9.9）。陆上油井中，有杆泵主要由抽油杆、泵筒和游梁组成（图9.10）。

Figure 9.9 Beam Pump
(Source: Calsac Corporation)

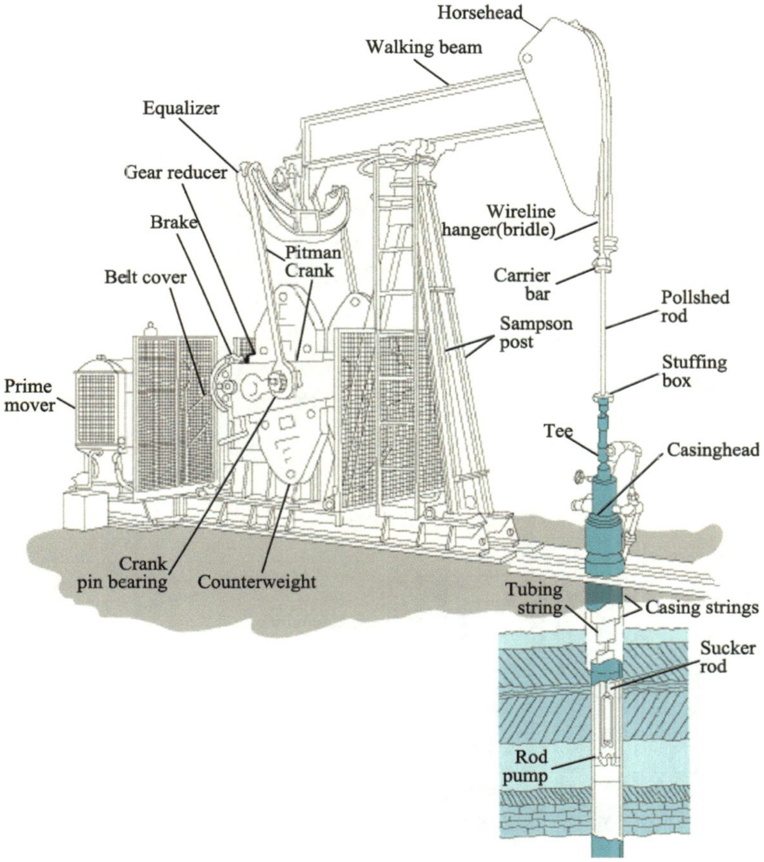

Figure 9.10 Schematic Diagram of Rod Pumping Systems
(Source: http://www.accessscience.com/)

Above the surface, the beam pumping system rocks back and forth. This is connected to a string of rods called the *sucker rods*, which plunges down into the wellbore. The sucker rods are connected to the sucker

地面上，游梁前后起伏运动。游梁连接抽油杆，抽油杆深入井筒中。抽油杆连接抽油泵，泵筒安装在油管底部。当

rod pump, which is installed as a part of the tubing string near the bottom of the well. As the beam pumping system rocks back and forth, this operates the rod string, sucker rod and sucker rod pump, which works similarly to pistons inside a cylinder. The sucker rod pump lifts the oil from the reservoir through the well to the surface(Figure 9.11).

Rod pumps use a down-hole piston operated by

游梁运动时带动抽油杆运动，抽油杆在抽油泵中的上下往返运动就像活塞在气缸中运动。抽油泵将油藏中的油通过井筒举升到地面（图9.11）。

地面上柴油机或电动马达带动曲柄旋转，曲柄连杆将旋转运动改变为游梁上下运动，

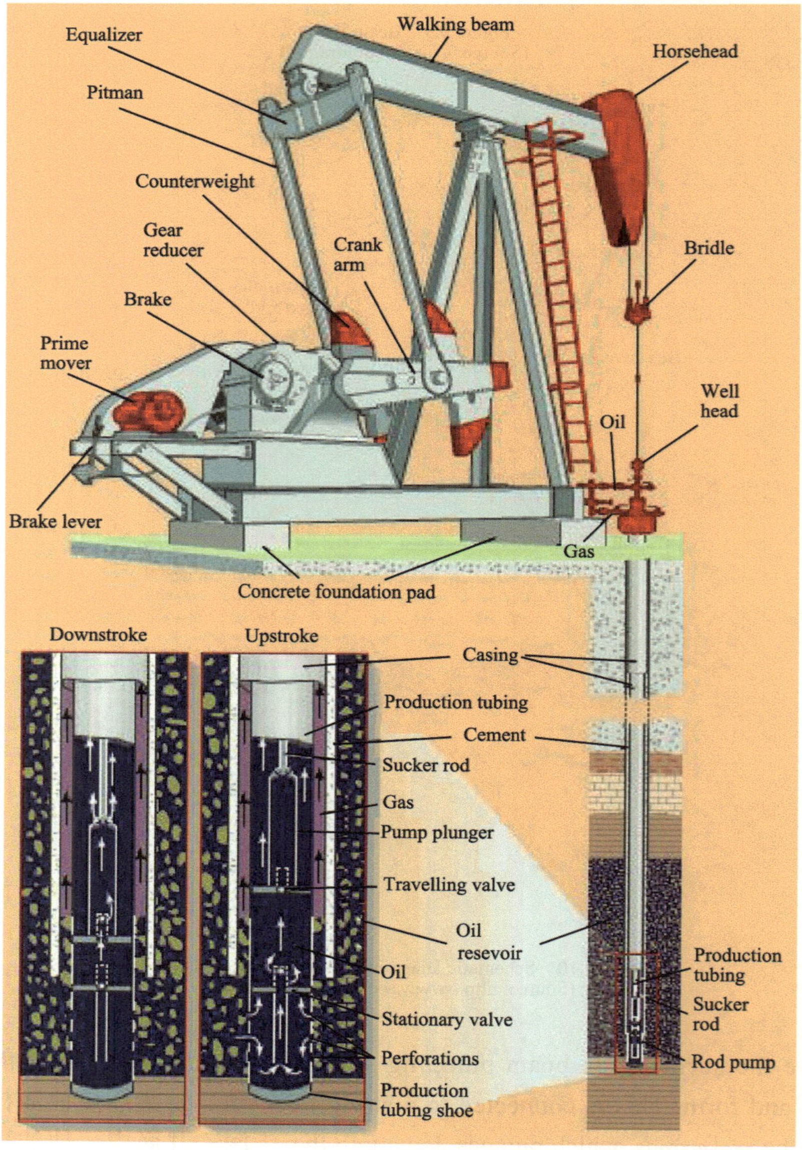

Figure 9.11 Artificial Lift Using a Beam-Pumping Unit
(Source: http://www.britannica.com/EBchecked/media/113919/The-artificial-lift-of-petroleum-with-a-beam-pumping-unit)

sucker or pull rods, moved up and down by a rocking arm driven by a diesel or electric motor at surface. These *Nodding Donkeys*, in all shapes and sizes, are very common in the older oilfield in the world. Rod Pumping Systems are high performance production systems that are ideal for many oil and gas applications. These systems will optimize production, reduce production costs and maximize well profitability(Figure 9.12).

游梁上的驴头连接抽油杆，在抽油杆带动下进行抽油泵活塞运动。驴头大小、形状各不相同，却在全世界的油田很常见。有杆泵采油表现稳定，是油田、气田的理想人工举升方式。这种举升方式能优化生产、减少采油成本、利益最大化（图9.12）。

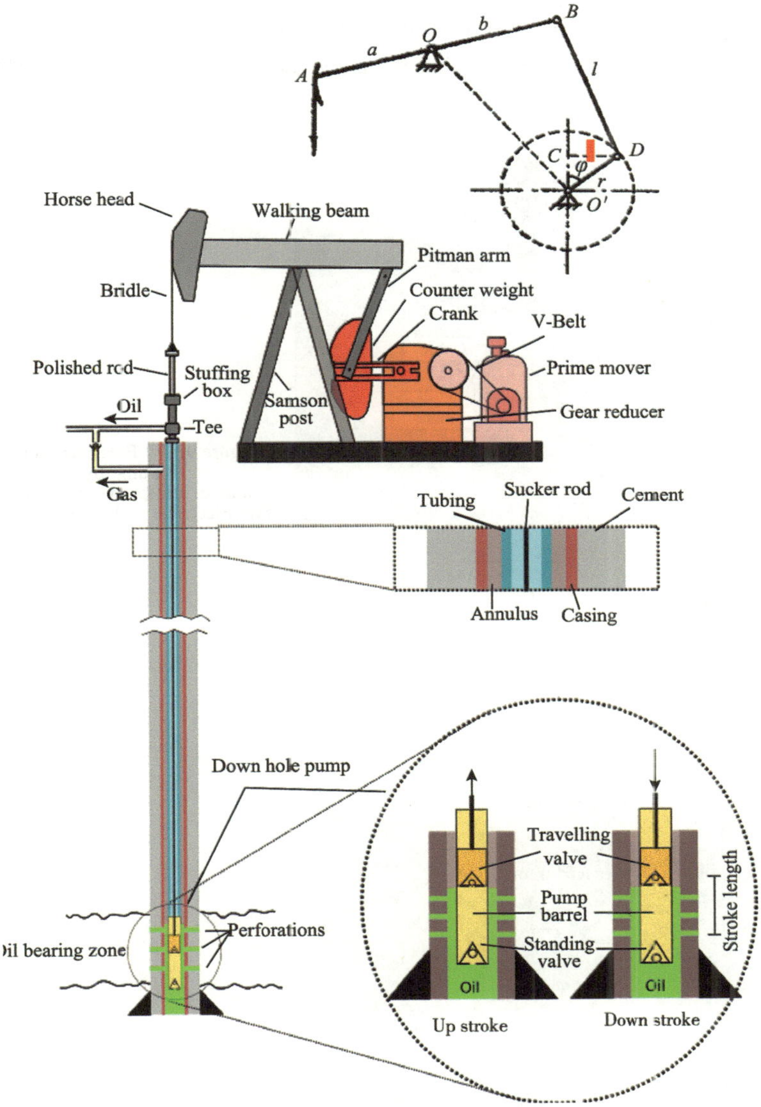

Figure 9.12 A Schematic of a Typical Oil Well Being Produced by a Pumpjack
(Source: http://en.wikipedia.org/wiki/Oil_wel)

9.2.3 Pumping Systems

9.2.3.1 Electrical Subsurface Pumps

Electrical Subsurface Pumps (EPS) commonly used in high-volume wells, is to lower an electrically-driven multi-stage centrifugal pump into the well(Figure 9.13). This converts electrical to hydraulic energy within the wellbore. These pumps have been used extensively in vertical shallow and cool wells, but require further development for the deviated deep and hot wells. The heat in the hot well has affected the electrical pumps, and lowering the cable through angled holes in the deviated wells has damaged the insulation(Figure 9.14).

9.2.3 泵系统

9.2.3.1 电潜泵

电潜泵一般用于高产油井中，指电动多级离心泵降入井底，从而在井筒中将电能转换为水力能（图9.13）。电潜泵一般用于直井和低温井，斜井和高温井需要电潜泵的进一步发展。高温井的温度会影响电机工作，斜井的井斜角会损害电缆的绝缘性（图9.14）。

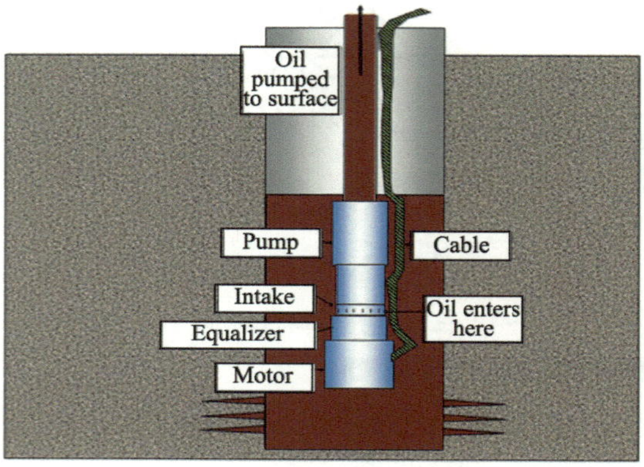

Figure 9.13 Electrical Submersible Pump System
(Source: https://en.wikipedia.org/wiki/File:Electrical_Submersible_Pump.png)

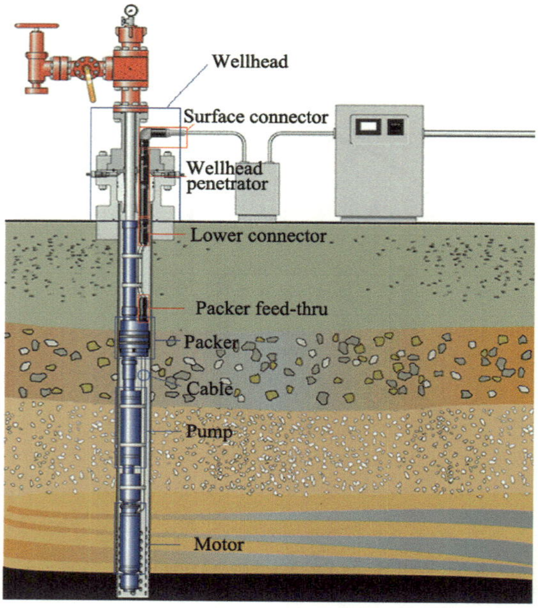

Figure 9.14 Electrical Submersible Pump
(Source: http://www.pftsys.com/products.php)

9.2.3.2 Hydraulic Pumps & Jet Pumps

A variant on this approach is to run the down-hole pump hydraulically. Hydraulic pumping equipment applies a downhole hydraulic pump, rather than sucker rods, which lift oil to the surface. Here, the production is forced against the pistons, causing pressure and the pistons to lift the fluids to the surface. Similar to the physics applied in waterwheels powering old-fashion gristmills, the natural energy within the well is put to work to raise the production to the surface(Figure 9.15).

Hydraulic pumps are generally composed of two pistons, one above the other, which are connected by a rod that moves up and down within the pump. Both the surface hydraulic pumps and subsurface hydraulic pumps are powered by power oil, or clean oil that has been previously lifted from the well. The surface pump sends the power oil through the tubing string to the subsurface hydraulic pump installed at the bottom of

9.2.3.2 水力泵和射流泵

电潜泵的另一个变种是水力泵，水力泵利用地面高压泵将动力液泵入井下泵，井下泵由一组成对的往复式柱塞组成，其中一个柱塞被动力液驱动，从而带动另一个柱塞将井内液体举升到地面（图9.15）。

水力泵一般由两个柱塞组成，上面的柱塞连着油杆在油泵内上下移动。地面水力泵和井下水力泵都是通过动力油或之前采出清洁原油驱动。地面泵通过油管向安装在油管柱底部的井下水力泵泵入动力油，从而驱动井下活塞来抽汲油藏流体到第

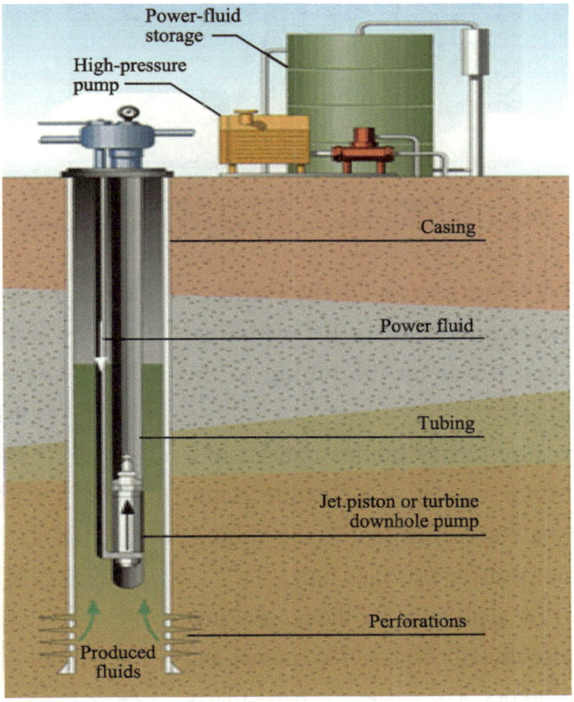

Figure 9.15 Hydraulic Pump
(Source: http://www.rigzone.com/)

the tubing string, the reservoir fluids are then sent up a second parallel tubing string to the surface.

二平行油管柱再到地面。

Hydraulic Pumping——uses power oil to stroke a downhole piston pump. The formation fluids and the returning power oil are pumped up the production tubing.

水力泵——利用动力油驱动井下柱塞，采出的地层流体和返回的动力油再被泵入生产油管。

Jet Pumping——forces the power oil through an orifice, providing lift to the produced fluids(Figure 9.16).

射流泵——迫使动力油通过喷嘴从而提供对原油的举升力（图9.16）。

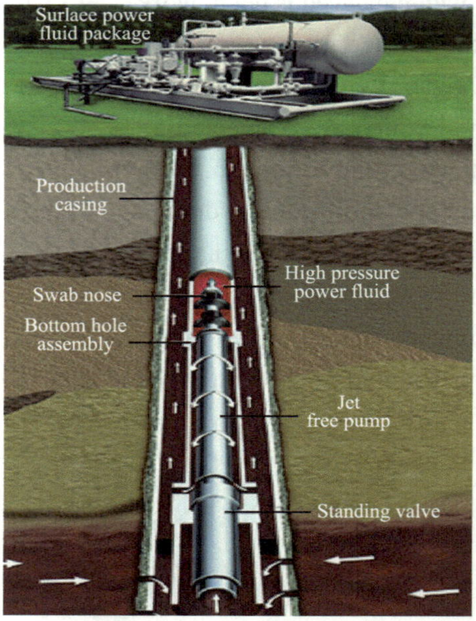

Figure 9.16　Subsurface Jet Pumps
(Source: http://www.rotatingright.com/artificial-systems.htm)

9.2.3.3　Plunger Lift & Progressing Cavity Pumps

Plunger-Lift——uses power oil to lift a plunger and its load of produced fluids up the tubing to the surface［Figure 9.17(a)］.

Downhole progressing-cavity pumps consist of a rotating worm-shaped rotor insider of a flexible stator. The rotor is driven by a rod string inside the tubing, which is rotated by a motor-driven surface unit.

This is a relatively new application for the progressing-cavity system, which is also used for both surface pumps and downhole drilling mud motors. The system is particularly good for pumping abrasive fluids［Figure 9.17(b)］.

9.2.3.3　柱塞泵和螺杆泵

柱塞泵利用动力油提升柱塞，从而举升油流入油管并到地面［图9.17(a)］。

井下螺杆泵由一个旋转的蜗形转子组成。转子由油管中定子驱动，这是一个由电机驱动的地表旋转单元。

有一种相对新的螺杆系统，也可以用于地面泵和井下泥浆马达。该系统对腐蚀性流体特别适用［图9.17(b)］。

(a) Plunger Lift System (b) Progressing Cavity Pumping System

Figure 9.17 Plunger Lift System & Progressing Cavity Pumping System
(Source: a.http://www.nov.com/Production/Artificial_Lift/Plunger_Lift_Systems.aspx;
b. http://www.nov.com/Production/Artificial_Lift/Progressing_Cavity_Pump_Systems.aspx)

9.3 Stimulation

Stimulation can be said to be all about permeability. Stimulation improves the permeability and makes a fluid to flow easily through rock. Many stimulation methods are used in oil industry. Two of the most common stimulation techniques are *acidizing and hydraulic fracturing.*

In many cases acidizing or fracturing is a routine part of the completion program. Either type of stimulation may also be applied soon after a well has been completed and has tested at lower production rates than expected. Stimulation may also be part of a remedial or workover program designed to improve productivity following a decline in production. Stimulation will often follow a formation pressure buildup test that was run to determine if the cause of

9.3 增产措施

增产措施说到底是渗透率措施。增产措施提高渗透率，使流体更容易在岩石中流动。石油行业中有多种增产措施，其中最常使用的是酸化和水力压裂。

在很多情况下，酸化和压裂是完井中的例行程序。当一口井完井并测试出产量比预期低时，这时候就需要进行增产。增产措施也可以是一种弥补或维护手段，用以提高油田后期的产量。增产措施一般伴随地层压力恢复测试，以确定低产是因为井筒附近渗透率减

low productivity was permeability reduction near the wellbore, low permeability throughout the reservoir, or low reservoir pressure. Acid stimulation can improve the first condition, whereas fracturing is necessary to significantly improve the second condition. Of course, the third condition can only be helped by pressure maintenance through injection of water or gas.

小，还是整个油藏渗透率，低下或油藏压力低下。酸化能提高井筒附近渗透率即改善第一种情况，而压力对第二种情况有用。第三种情况则需要注水或注气保持油藏压力。

Both acidizing and fracturing procedures involve the pumping of fluids down the tubing or drillpipe and into the formation. In fracturing, the objective is to apply enough pressure to actually split the formation apart, creating flow channels to the wellbore where either none or few previously existed. In most acidizing procedures, the objective is to squeeze the acid into the existing pore spaces of the rock matrix, where it can react to enlarge the flow channels and improve or restore the permeability of the region very near the wellbore [a radius of 8 to 24 in (20.3 to 61 cm)] without fracturing the producing formation. Acid-fracturing treatments create fractures. At the same time continued fluid injection increases the fracture' s length and width. Acid (normally 15% HCl) is then injected into the fracture to react with the formation and create a flow channel (by etching of the fracture surface) that extends deep into the formation. This allows more reservoir fluid to drain into the wellbore along the new fractures once the well is put back on production.

无论压裂还是酸化都需要将流体泵入油管或井筒中，从而进入储层。压裂的目的是利用足够大的压力压开地层，压出裂缝和通道使油流入井筒。绝大多数酸化的程序是将酸挤入岩石基质孔隙，从而扩大流通通道，提高或恢复井筒附近渗透率［半径为8～24in（20.3～61cm）］。酸化压裂则再产生裂缝。与此同时，注入的酸液增加裂缝的长度和宽度。酸液通常为15%的盐酸，注入裂缝中与地层反应，向地层深处创造新的流通通道（通过蚀刻的断面）。进而扩大油流面积，恢复生产。

9.3.1 Acidizing

Acid is used to remove damage near the wellbore in all types of wells (Figure 9.18). There are two categories of acidizing treatment in use: matrix acidizing and fracture acidizing. Matrix acidizing can be used in carbonate reservoirs and sandstone. Fracture acidizing, also known as acid fracturing is the most widely used acidizing technique for stimulating limestone or dolomite formations(Figure 9.19~Figure 9.21).

9.3.1 酸化

酸化用于去除井筒附近的损害（图9.18）。有两种类型的酸化：基质酸化和压裂酸化。基质酸化用于碳酸盐岩和砂岩油藏。压裂酸化用于灰岩和白云岩的最广泛的增产措施（图9.19～图9.21）。

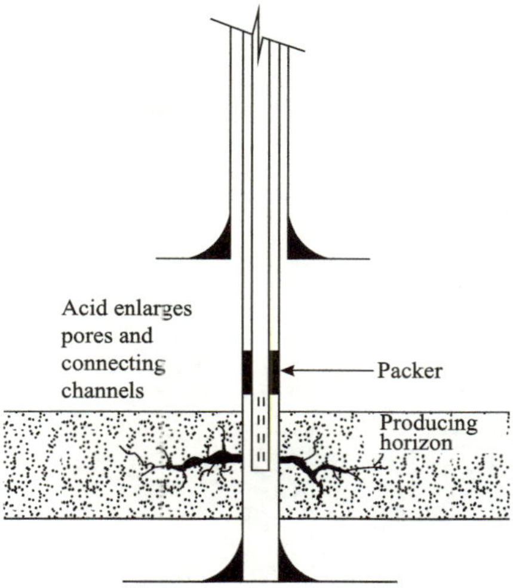

Acidizing operations involve the introduction of acids into the formation crevices to increase the flow capacity of the pay zone.

Figure 9.18 Acidizating
(Source: http://oilfielddirectory.com/oilfield/acidizing.htm)

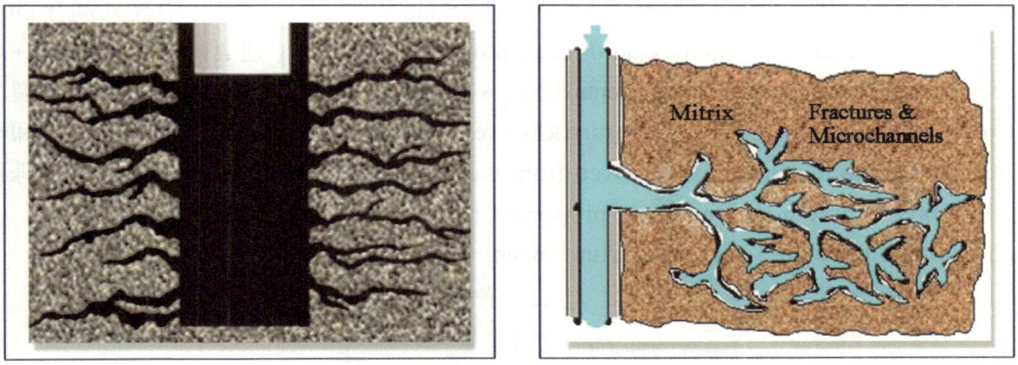

Figure 9.19 Acidization
(Source: http://www.yennai.com/acidization.htm)

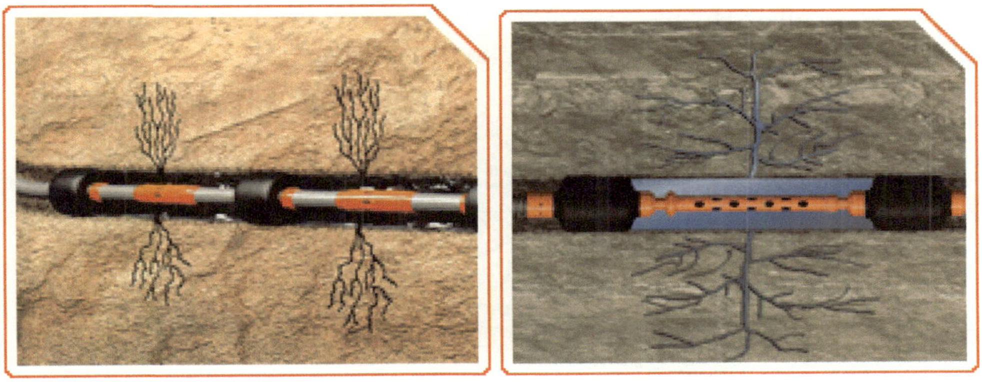

Figure 9.20 Multistage Fracture Acidization Treatment
(Source: http://www.tamintl.com)

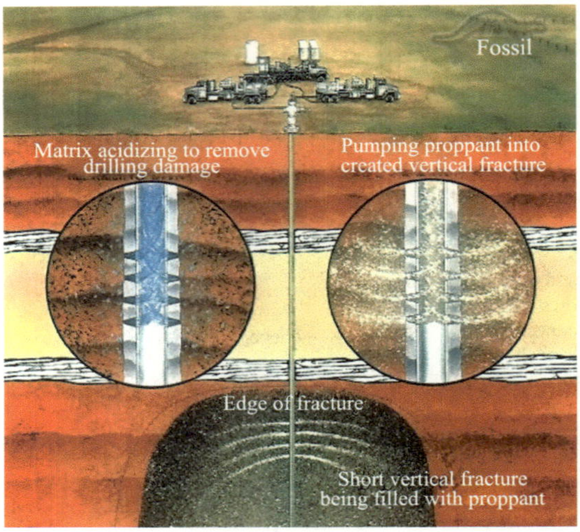

Figure9.21 Well Acidizating
(Source: http://www.rootforce.org/wp-content/)

9.3.2 Hydraulic Fracturing

Wells in low to moderate-permeability reservoirs are candidates for hydraulic fracture as a means of stimulating their performance. The high-pressure injection of 'fracking fluid' (primarily water, containing sand or other proppants suspended with the aid of thickening agents) into a wellbore to create cracks in the deep-rock formations(Figure 9.22). When the hydraulic pressure is removed from the well, small grains of hydraulic fracturing proppants (either sand oraluminium oxide) hold the fractures open.

9.3.2 水力压裂

当井钻到中低渗透率油藏时，适合用水力压裂作为增产手段。向井中注入高压的压裂液（通常是水+支撑剂+增稠剂），在井底憋压产生裂缝（图9.22）。当井内压力卸载后，裂缝中的支撑剂（砂砾或氧化铝）保持裂缝不闭合。

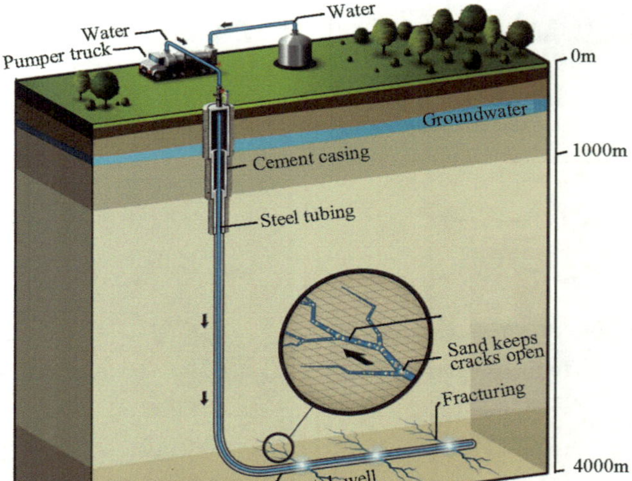

Figure 9.22 Hydraulic Fracturing
(Source:http://en.skifergas.dk/technical-guide/what-is-hydraulic-fracturing.aspx)

Hydraulic fracturing is a technique used primarily by the oil and gas industries as a tool for enhanced recovery, more specifically for reservoir stimulation, well connection, etc. The objective of hydraulic fracturing for well stimulation is to increase well productivity by creating a highly conductive path (compared to reservoir permeability) some distance away from the wellbore into the formation. Usually the conductivity is maintained by propping with sand to hold the fracture faces apart(Figure 9.23~Figure 9.25).

水力压裂是石油工业提高采收率、油藏增产、井间连通等的首选措施。水力压裂在井筒附近压裂出高导流能力的裂缝（与储层渗透率相比），增加油流通道从而提高井的生产能力。通常向裂缝中挤入砂砾等支撑剂，以保持裂缝不闭合从而增加导流能力（图9.23 ~ 图9.25）。

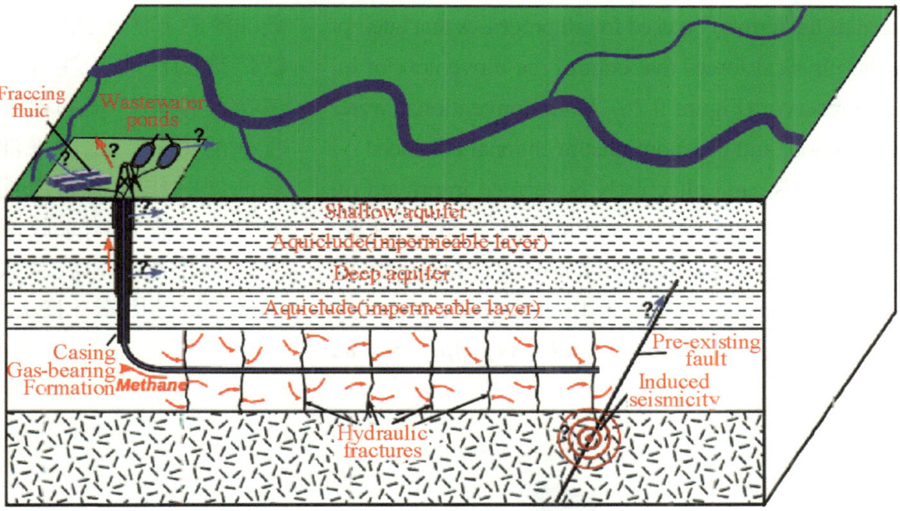

Figure 9.23 Waterflooding Operations
(Source:https://en.wikipedia.org/wiki/Hydraulic_fracturing)

Figure 9.24 Preparation for Hydraulic Fracturing
(Source:http://en.skifergas.dk/technical-guide/what-is-hydraulic-fracturing.aspx)

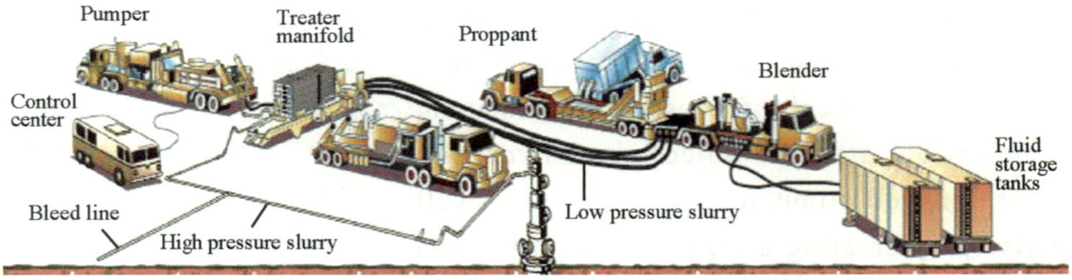

Figure 9.25 Fracturing Well Site
(Source: http://www.ohioenergyresource.org/topics/the-production-process)

9.4 Enhanced Oil Recovery

The termed enhanced oil recovery methods have been introduced as the term to be used for all processes that attempt to alter the physical forces that control the movement of oil within the reservoir(Figure 9.26). Both conventional water and gas injection, and the more unconventional enhanced oil recovery methods can collectively be termed improved oil recovery methods. Secondary and tertiary recoveries are together referred to a enhanced oil recovery (EOR)method. Among the numerous factors in fluencing the recovery efficiency, *displacement efficiency and sweep efficiency* are two key elements(Figure 9.27).

9.4 提高采收率

提高采收率指所有改变油藏油流的方法（图9.26）。既包括传统的注水、注气方式，也包括其他非传统的提高油藏采收率的方式。二次采油和三次采油同时作为提高原油采收率的方法。在影响采收率的因素中，驱替效率和波及效率是两个最关键的因素（图9.27）。

Figure 9.26 EOR
(Source:http://total.com/en/energies-expertise/oil-gas/exploration-production/strategic-sectors/eho/innovation/increasing-recovery-factors)

EOR
- Water Flooding
- Thermo Flooding
  - Fire Flooding
  - Steam Flooding
  - Cyclic Flooding
- Miscible Flooding
  - CO_2 Flooding
  - Nitrogen Flooding
  - Flue Gas Flooding
  - Liquefied Petroleum Gas Flooding
- Mobility Flooding
  - Polymers Flooding
  - Surfactant Flooding
  - Alkaline/Caustic Flooding
- Microbial Flooding

Figure 9.27 EOR Methods

9.4.1 Water Flooding

Water flooding is term used to describe the increase in oil recovery by injecting water into an oil producing reservoir. This is the most common secondary-recovery method and extends deeply into many tertiary recovery procedures throughout the producing lift of the reservoir. One problem with water flood is that it is difficult to push water through the formation as a vertical wall, that is, the water will spread out in the formation rather than move through it evenly. Gravity pulls the leading edge of the water down and causes it to move downward as it progresses through the reservoir. It can travel under the oil and leave a large amount of oil behind(Figure 9.28).

9.4.1 注水驱

注水驱指注水到油藏中增加采收率的方法。水驱是最普遍的二次采油方法，也沿用到三次采油的过程中。水驱的一个问题在于，注入的水并不均匀进入地层。重力使水驱前沿向下流动，而不是横扫油藏。因此，注入的水沉于油层下，导致大量的油驱替不了（图9.28）。

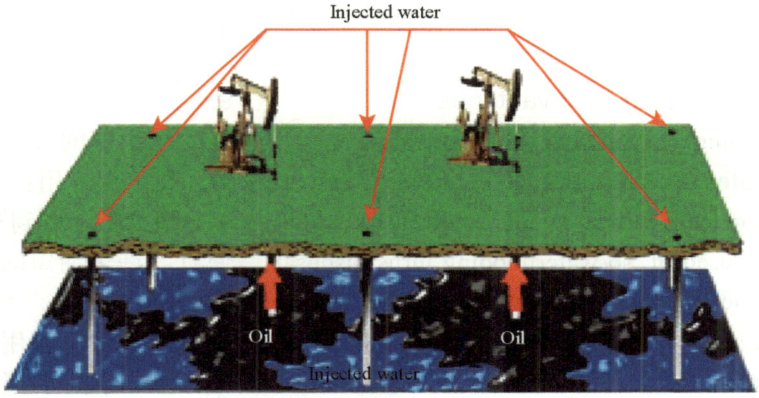

Figure 9.28 Waterflooding Operations
(Source:http://www.maverickenergy.com/operations/crawford County.aspx)

Waterflood recovery is dependent on a number of other technical variable. Some of the more common include: oil saturation at the start of water flooding, residual oil saturation, connate water saturation, and free gas saturation at the start of water injection. Also affecting recovery rates are the pore volume, oil and water viscosity, and effective permeability to water and oil. Other determining factors involve reservoir stratification, waterflood pattern, pressure distribution between injection and production wells, injection rate, and oil formation volume factor(Figure 9.29).

水驱采收率还受其他一些因素影响。比较常见的因素有：水驱前的含油饱和度、残余油饱和度和原生水饱和度、自由气饱和度等。其他的影响因素还有孔隙体积、油水黏度、油水有效渗透率。其他决定因素包括，油藏层理、水驱模式、注入井和生产井间的压力分布和油的地层体积系数（图9.29）。

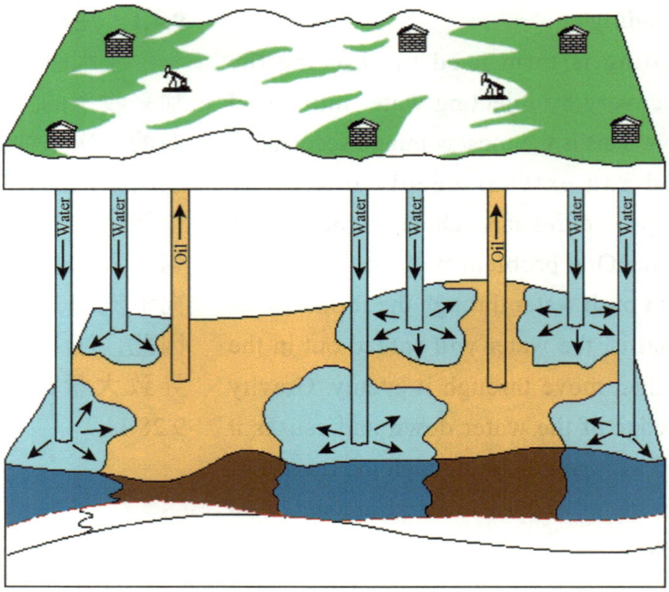

Figure 9.29 Water Flooding
(Source: http://www.kgs.ku.edu/Publications/Oil/primer13.html)

9.4.1.1 Waterflooding's Advantages

Waterflooding's many explain its popularity:

◆ Saltwater is readily available and inexpensive. Most fields produce significant volumes of saltwater that need to be disposed of, so waterflooding turns disposal cost into a benefit. The necessary "make-up" water (water needed in addition to produced water) can usually be obtained from saltwater zones downhole close to the pay zone.

◆ Since produced saltwater occurs naturally in the producing zone, it is chemically compatible with the reservoir fluids. If waters from other sources are used, incompatibility may cause precipitation of solids that plug the formation.

◆ Waterflooding equipment is relatively simple and inexpensive. Although saltwater is corrosive, stainless steel valve trims, plastic coating for pipe, etc. are able to handle it at reasonable cost.

◆ Waterflooding is a safe operation because saltwater is non-toxic and non-flammable and pressures are relatively low.

9.4.1.1 注水驱优势

注水驱的许多有利因素导致了其广泛应用：

◆ 地层水到处都有而且便宜。许多油田的产出水量很大，需要处理，注水工艺使得这些水得到了利用。在这些回注污水中加入的新鲜地层水也很容易从邻近地层中得到。

◆ 产出水由于也是从产层中采出的，因此与产层的配伍性好。如果采用其他的水源，可能会因配伍性不好，导致固体颗粒沉降堵塞地层孔道。

◆ 水驱设备相对简单、便宜，尽管盐水带腐蚀性，但不锈钢阀件、管壁塑料涂层等费用在相对可接受的范围内。

◆ 注水是一种较安全的工艺。因为地层水无毒、不燃，同时注水工艺也不复杂。

◆ The biggest advantage of waterflooding is simply that it works so well. Very few floods have turned out to be uneconomic(Figure 9.30).

◆ 注水工艺最大的优势是经济可行（图9.30）。

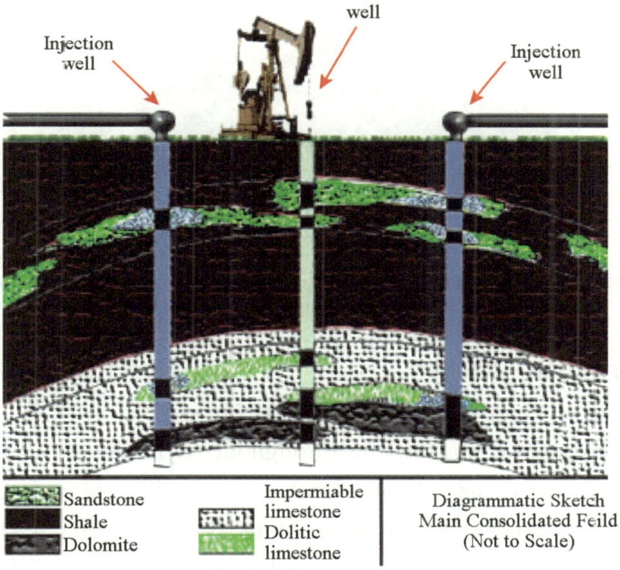

Figure 9.30 Injection Well & Oil Well
(Source:http://www.maverickenergy.com/operations/crawfordCounty.aspx)

9.4.1.2 Fluids Displacement

Saltwater and oil do not mix, so waterflooding is an immiscible displacement process(Figure 9.31, Figure 9.32).

9.4.1.2 流体驱替过程

地层水与油不相混合，因此水驱是一个非混相驱过程（图9.31、图9.32）。

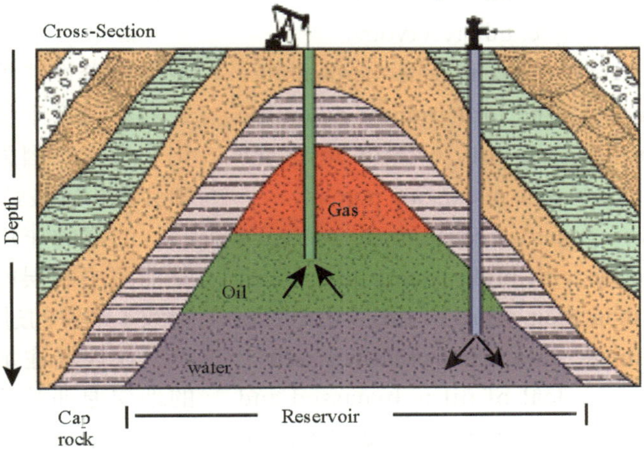

Figure 9.31 Cross-Section of Water Flooding
(Source: http://www.gg.uwyo.edu)

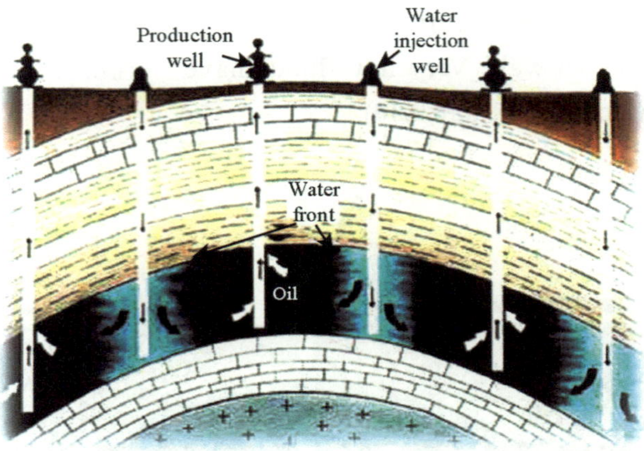

Figure 9.32 Water Used for Oilfield Injection Latest Updated Figures (Draft Information)
(Source: http://www.waterdropcycle.com/companies.html)

9.4.1.3 Reservoir Heterogeneity

There is a tendency to visualize sandstone reservoirs as having constant characteristics throughout. This is emphatically not the case. The natural forces that formed the reservoir were in a constant state of flux, with the result that heterogen-eities abound both vertically and horizontally.

Perhaps the most common heterogeneity encountered are barriers to vertical permeability. A momentary (in geologic time) change in depositional conditions often leaves almost invisible horizontal layers of clay within otherwise homogeneous sandstones. Instead of having one thick reservoir, the clay restrict vertical permeability, forming multiple, thin reservoirs.

Carbonate reservoirs have even greater permeability variation that sandstones. The formation of dolomitic porosity, for example, is highly erratic as is the deposition of calcium carbonate in existing porosity.

The various heterogeneities cause most reservoirs to waterflood unevenly with high-permeability thief zones taking more than their share of the water. This leads to early breakthrough of water into producing wells .

In practice, a great deal of oil is bypassed and early water breakthrough usually occurs, but the flood continues to operate profitably. The breakthrough

9.4.1.3 油藏非均质性

通常设想油藏是均质的，而事实并非如此。形成油藏的机理是千变万化的，因而形成的油藏无论在纵向上还是在横向上都是非均质的。

最常见的非均质性是垂向上的渗透隔层。由于沉积环境随时间上的变化，均质砂岩中会沉积有几乎难以察觉到的黏土层。这些黏土层将均质的厚砂层分割成一个个较薄的砂层。

碳酸岩油藏通常比砂岩油藏的渗透性非均质更强。就白云岩渗透率来说，其渗透率变化非常大，因为在已生成孔隙中会沉积碳酸钙。

油藏的非均质性导致了水驱油藏中贼层的形成。这样就导致了油井的过早水淹。

在现场注水实践中，大量的地层原油被注入水绕过，油井会因水窜过早见水。水窜层形成水通道。为保持足够压力

channel becomes a low pressure conduit for a high volume of water, so it' s often necessary to increase the water injection rate to keep pressure on the unswept formation. Larger pumping equipment is installed to handle the water. The water-oil ratio (*WOR*) of floods often reach 95% or higher before they become uneconomic(Figure 9.33).

来驱替未水淹层，注水速度要加以提高。这种可能需要安装更大的水泵。油井的含水不断加大，直到经济极限，即95%的含水率（图9.33）。

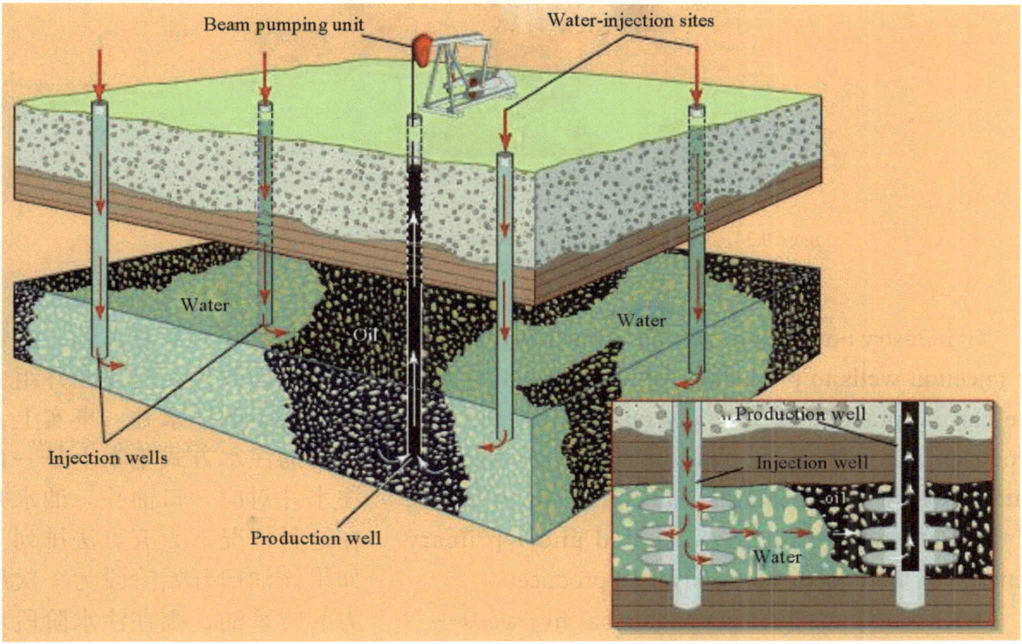

Figure 9.33 The Recovery of Petroleum through Waterflooding
(Source: https://www.britannica.com/technology/waterflooding)

9.4.1.4 Well Patterns

In the 1950s and 60s when waterflooding was relatively new, a number of peripheral floods were tried. The plan was for the injected water to gently encroach from the perimeters of the field, much like a water drive, and result in maximum recovery. A few injection wells were therefore located around the edge of the reservoir(Figure 9.34).

It soon became apparent, however, that with injection rates limited by so few injection wells, the projects would take too long. From a time-value-of-money perspective, this was not optimizing present worth.

9.4.1.4 井网

20世纪50～60年代，油井注水刚刚实行，对一些井组进行了边缘注水实验。在这些井组中，注水井从油田边缘均匀注水，以达到最大的采收率（图9.34）。

不过，生产结果表明，这种注水方式受效太慢，不能够达到最佳的经济效益。

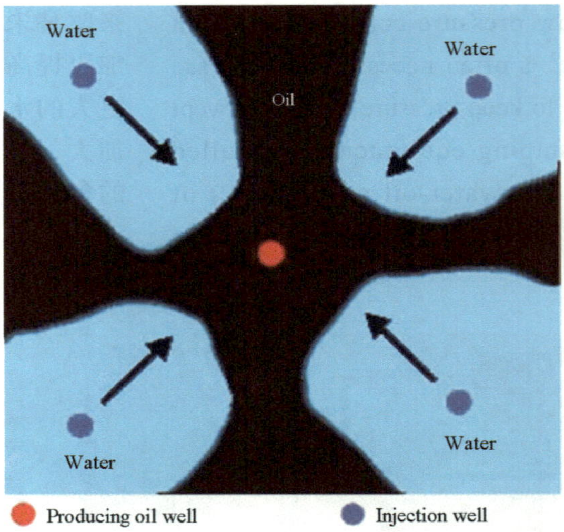

Figure 9.34 Water Pumped into Adjacent Wells Forces Oil to the Oil Well
(Source: http://nsioil.com/extracting-oil/)

In industry now recognizes that a much higher ratio of injection wells to producing wells is needed. In fact, the current practice is generally one injector for each producer. The wells are normally arranged in a five-spot pattern which has alternating injectors and producers. If the field was originally depleted under primary conditions, the injectors are converted producers.

Each individual five-spot pattern has five wells——one injector and four producers (or vice-versa). The injected water front moves preferentially in the direction of the pressure drawdowns created by the producers rather than outward in a strictly radial pattern. This results in a large portion of the reservoir——perhaps 50%——being bypassing.

Combining the oil that is bypassed in the vertical dimension with the oil that is bypassed in the horizontal dimension, suggests less than half of the reservoir is contacted by the flood water. These pockets of bypassed oil have been pressured by the flood and can often be produced profitably by infill drilling(Figure 9.35).

于是，大家意识到，要达到更好的效果，应该在井组中安排更多的注水井。事实上，当前的注水方式基本上是一口注水井对应一口油井。油水井在这种情况下为五点法排列。如果该油田开始阶段为一次压力衰减采油，则在注水阶段可以转部分油井为注水井。

对于五点法井网，每个井组有五口井，包括一口注水井和四口生产井。注入水主要向生产井压力降方向推进，而不是呈圆周放射状推进。这样一来，驱替水的波及系数差不多只有50%。

对比平面和垂直方向，只有一小半的原油得到注入水的驱替。注入水绕过的区域已经被注入水加压，可以通过打加密井的方法来采剩余油（图9.35）。

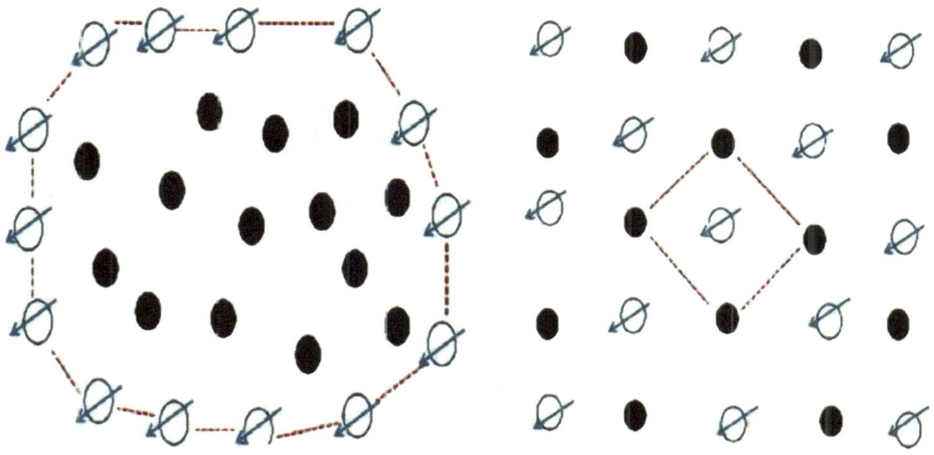

Figure 9.35 Peripheral Flood Pattern & Five-spot Flood Pattern
(Source: The Petroleum Industry, Charles F. Conaway)

9.4.2 Chemical Recovery Processes

By adding certain chemical substance to the injected fluid, its driving capacity can be greatly increase. The additives commonly used are surfactants, alkali and polymers. Surfactants and alkali are used to lower the interface tension of different underground phases, freeing the bounded oil(Figure 9.36). While polymers are used to optimize the flow the displacing phase so as to modify the mobility of the aqueous phase during displacement, drive evenly the front, weaken the high permeable stratum finger front, raise the efficiency of the displacement fluid, increase the pressure gradient and so on(Figure 9.37，Figure 9.38).

9.4.2 化学驱

在注入流体中添加某种化学物质，从而极大提高驱动能力，叫做化学驱。添加剂一般为表面化学剂、碱和聚合物。表面化学剂和碱用于降低不同地下流体相的表面张力，使束缚油自由流动（图9.36）。而聚合物能优化驱替相的流动能力，以便在驱替过程中改变气相流度比，使驱动更平稳，减弱高渗透地层的指进现象，提高驱替流体的驱替效率，增加压力梯度等（图9.37、图9.38）。

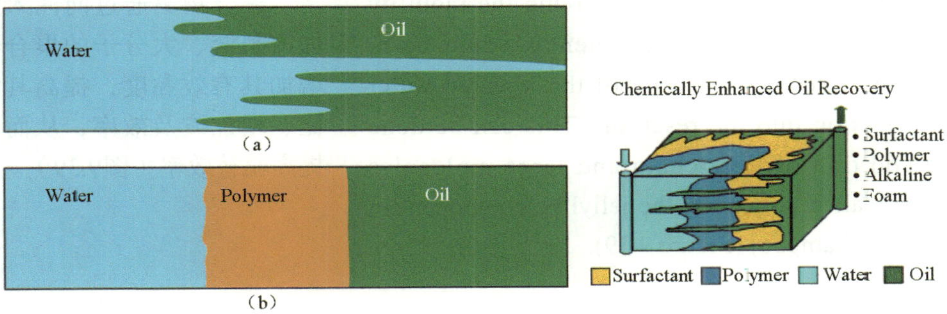

Figure 9.36 Polymer Flooding
(Source:http://extremetech-energy.com/)

Chemical Recovery Process
- Polymer Flooding
- Surfactant Flooding
- Caustic Flooding
- Combination Flooding

Figure 9.37　Chemical Flooding

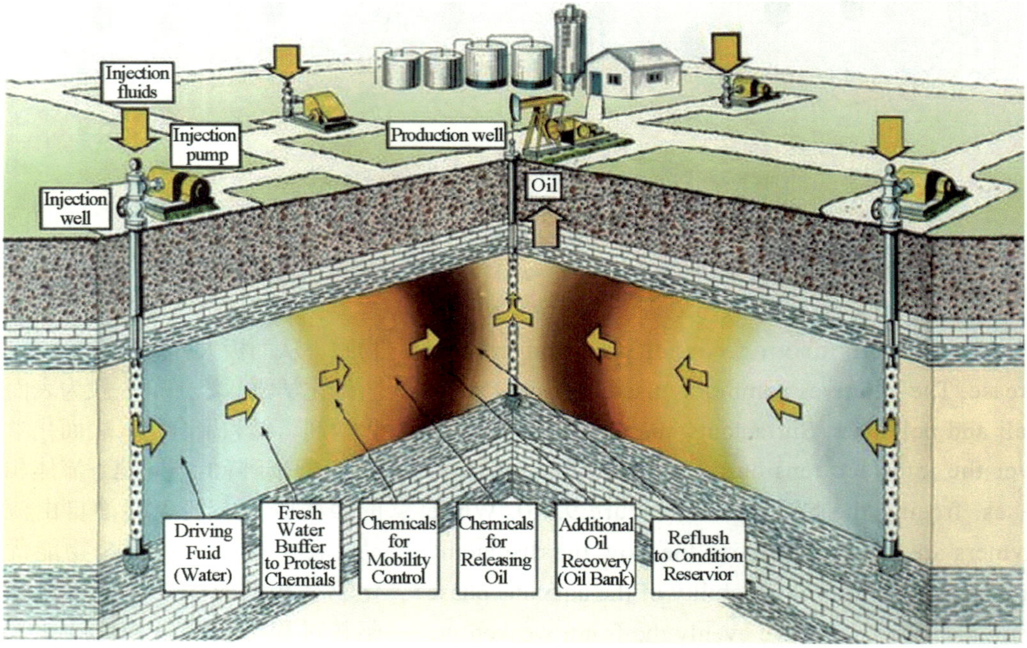

Figure 9.38　Chemical Flooding Operations
(Source: http://www.surtek.com/technologies.html)

9.4.2.1 Polymer Flooding

An obvious approach to improving the mobility ratio and enhancing sweep efficiency would be to increase the effective viscosity of the injected water before injection into the reservoir. This can be done by the addition of long chain and large molecules substance, called polymers (the jellylike component in your tube of shampoo) (Figure 9.39).

9.4.2.1 聚合物驱

聚合物驱通过向注入水中加入长链、大分子的聚合物，增加其有效黏度，提高其流度比、增加波及效率，从而明显提高驱替效率（图9.39）。

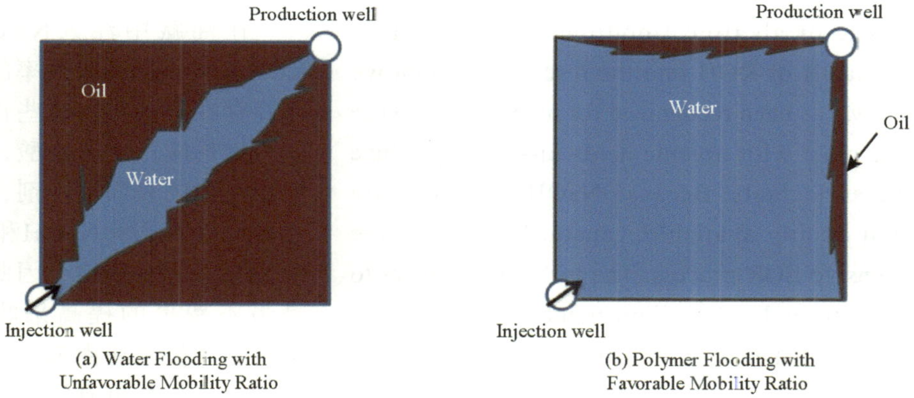

Figure 9.39　Polymer Flooding
(Source:http://large.stanford.edu/courses/2015/ph240/zerkalov1/)

9.4.2.2　Surfactant Flooding

Most of the oil remaining is left behind in the pores of the rocks, which capillary retention force is essentially responsible for resistance to further recovery displacement. The value of resistance is affected by the values of pore radius and interfacial tension, therefore reduction of the interfacial tension using surfactants becomes the most practical way of recovering the oil left after water-flooding.

Surface actives agents, or surfactants, are compounds that can act to reduce the interfacial tension at the interface between the oil and water in the reservoir. Detergents are examples of an everyday use of surfactants to allow water to displace oil (and the accompanying dirt) from clothing, dishes, etc. Reduction in interfacial tension improves the displacement efficiency of the flood an reduces the residual oil saturation. Surfactants are usually introduced to a waterflood as components in a water-oil-surfactant solution. This solution is also called a micro-emusion, or a micellar solution, because of the existence of micelles, aggregates of surfactant molecules surrounding microscopic oil droplets in water, or microscopic water droplets in oil.

9.4.2.2　表面活性剂驱

绝大多数岩石孔隙中的残余原油其毛细管力是进一步驱替的阻力。其阻力值受孔隙半径和界面张力影响，因此用表面活性剂减小界面张力是水驱后提高采收率最确切可行的办法。

表面活性成分，如表面活性剂是能在油藏中减少油水两相界面张力的化合物。清洁剂是日常生活中每天使用的表面活性剂，它促使水从衣服、盘碟等处将油（和附带污垢）驱除。减小表面张力提高驱替效率，从而降低残余油饱和度。表面活性剂也常被引入水驱中，作为油水表面活性溶液成分。这种溶解叫微破乳或胶束溶液，因为其中存在胶束，因而在水中微油滴附近聚合表面活性剂分子，或者在油中微水滴附近聚合表面活性剂分子。

9.4.2.3 Caustic Flooding

Caustic or alkaline flooding, which involves injecting NaOH or KOH into the reservoir to improve oil recovery, has been recognized for many years. These chemicals react with organic acids in-situ to produce soap-like surfactants. Because NaOH and KOH are cheap and readily available, caustic flooding is the least expensive EOR process. The technique has yet to establish itself and only a limit number of field trials have been undertaken with limited success.

9.4.2.3 碱驱

在油藏中注入NaOH或KOH溶液，提高采收率的碱驱已经存在很多年。这些化学物质就地反应产生有机酸，形成像肥皂般的表面活性剂，从而驱替原油。因为NaOH和KOH既便宜又容易获得，因此碱驱是最不费事的提高采收率方法。这项技术还尚未完全获得成功，只是数量有限的试验区块获得一定成功。

9.4.2.4 Combination Flooding

◆ Surfactant Polymer Flooding

This is a two-step process. The first step is injection of a surfactant slug. A surfactant is a wetting agent that breaks the surface tension between substances. The second step is the injection of a polymer mobility buffer. The surfactant slug may also be referred to as micellar solution, micro-emulsion, soluble oil, or swollen micelle. The purpose of the polymer is to provide mobility control for a more piston-like displacement.

◆ ASP Flooding

In the 1980s, Alkali/Surfactant/Polymer (ASP) ternary combination flooding which appeared as a new method for the tertiary recovery efficiency was effective in enhancing recovery efficiency in the later period of water flooding.

9.4.2.4 混合驱

◆ 表面活性剂聚合物驱

此过程分两步进行。第一步注入表面活性剂。表面活性剂作为湿相打开各物质间的表面张力。第二步注入聚合物。表面活性剂为胶束溶液、微破乳、可乳化油、膨胀胶束。聚合物的目的是提高活塞式驱替的流度控制。

◆ ASP驱

在20世纪80年代，碱/表面活性剂/聚合物（ASP）三元复合驱有效提高水驱后的三次采油效率。

9.4.3 Thermal Recovery Processes

Thermal processes attack the problem of an unfavorable mobility ratio by heating the reservoir and its fluids, either by adding heat via steam or hot water, or by generating heat within the reservoir by burning some of the oil in the formation(Figure 9.40，Figure 9.41). The most important effect of adding heat is the sharp reduction in the oil viscosity and the resulting decrease in the mobility ratio.

9.4.3 热力采油

或是通过蒸汽，或是通过热水，或是燃烧部分地层原油，加热油藏和流体的三次采油方式解决了地下流体不好流动的问题（图9.40、图9.41）。加热的最重要效果是减小原油黏度和降低流度比。

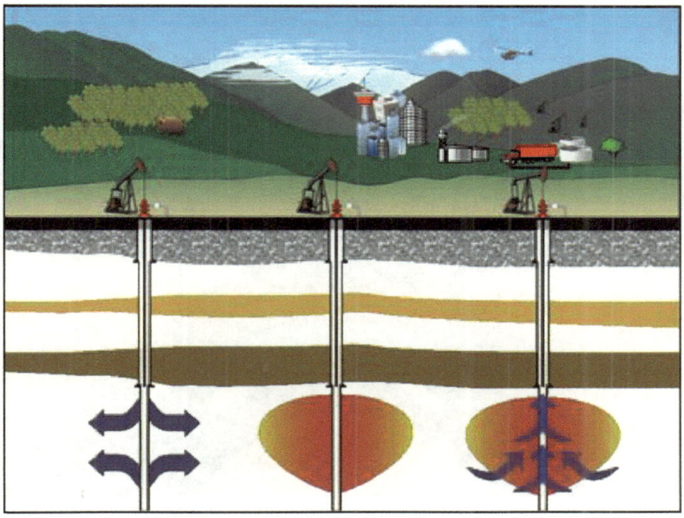

Figure 9.40 Thermal Recovery Processes
(Source:http://www.rogtecmagazine.com/)

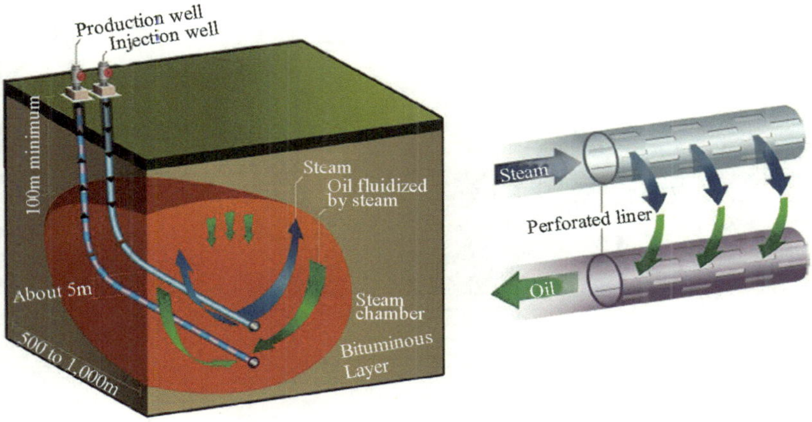

Figure 9.41 Thermal Recovery Processes-SAGD
(Source:http://total.com/en/energies-expertise/)

9.4.4 Miscible Recovery Processes

Miscible flooding works on the principle that some fluids are miscible with crude oil (methane, ethane, Carbon dioxide etc.) and can be used to displace oil with no capillary resistance. Most miscible flooding projects use CO_2 or nitrogen as solvents to increase oil recovery, but other injectants are sometimes used(Figure 9.42~Figure 9.47).

9.4.4 混相驱

混相驱的原理是部分流体能与原油（甲烷、乙烷、二氧化碳等）混相，从而无毛管阻力地驱替原油。大多数混相驱用二氧化碳或氮气作为溶剂，提高原油采收率，也有用其他注入剂的（图9.42～图9.47）。

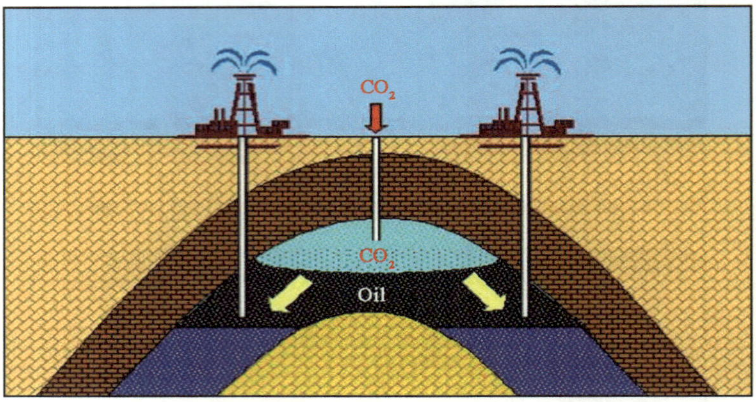

Figure 9.42　CO_2 Immiscible Displacement Processes(Plasynsky, 2008; Tzimas et al, 2009)
(Source:http://igutek.scripts.mit.edu/terrascope/?page=oilandgas)

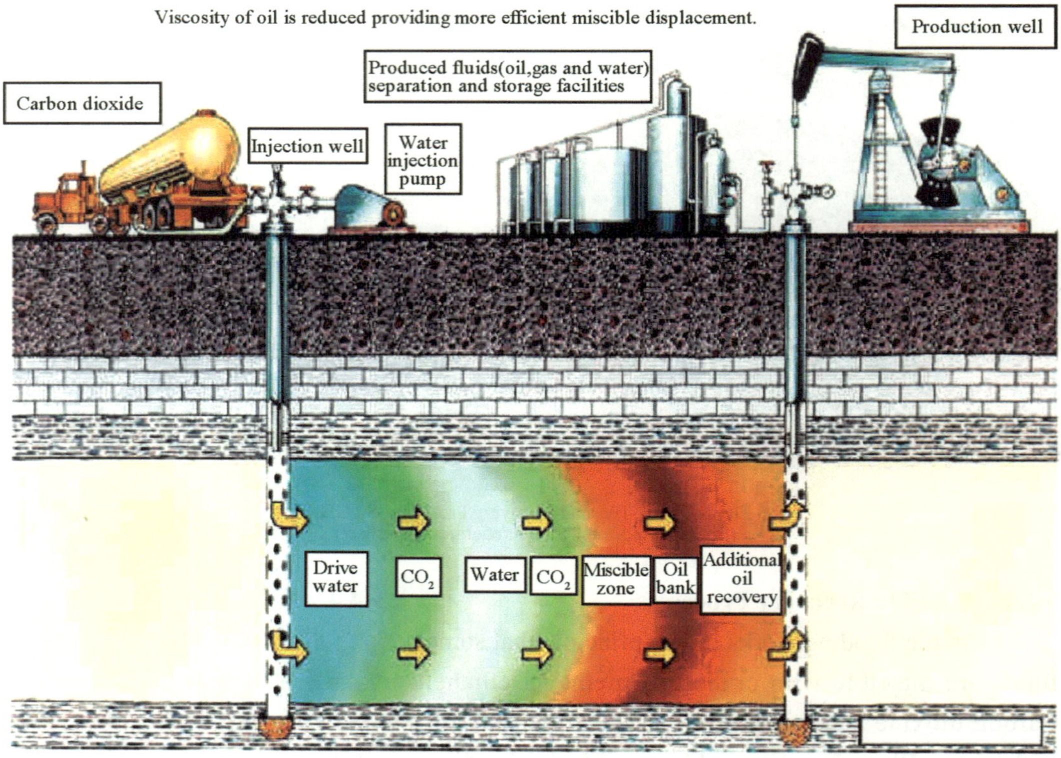

Figure 9.43　Miscible Recovery Processes
(Source:http://www.gasflooding.com/)

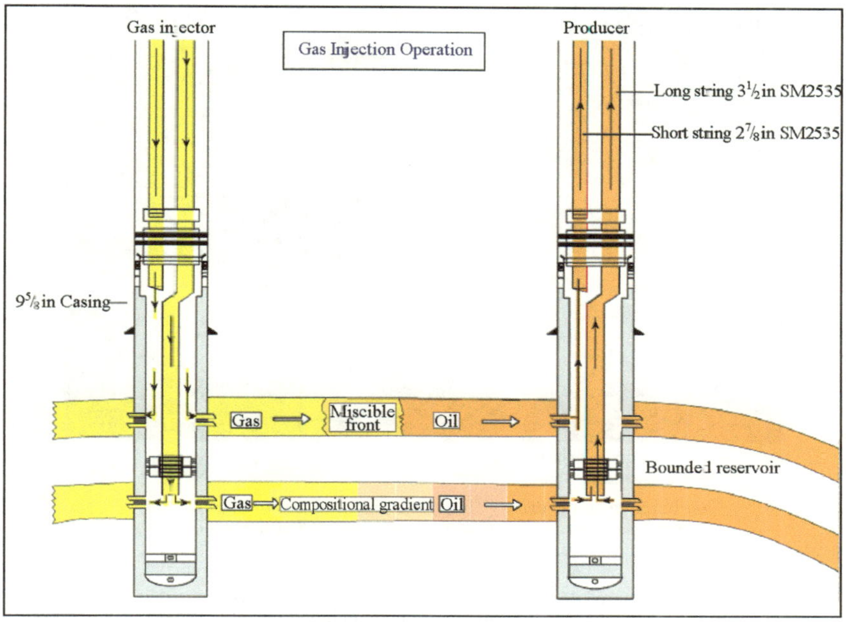

Figure 9.44 Oil Recovery by the Natural Gas Miscible Flooding and the Gas-Cycling Techniques
(Source: http://adocauh.cts-co.net/product/03/)

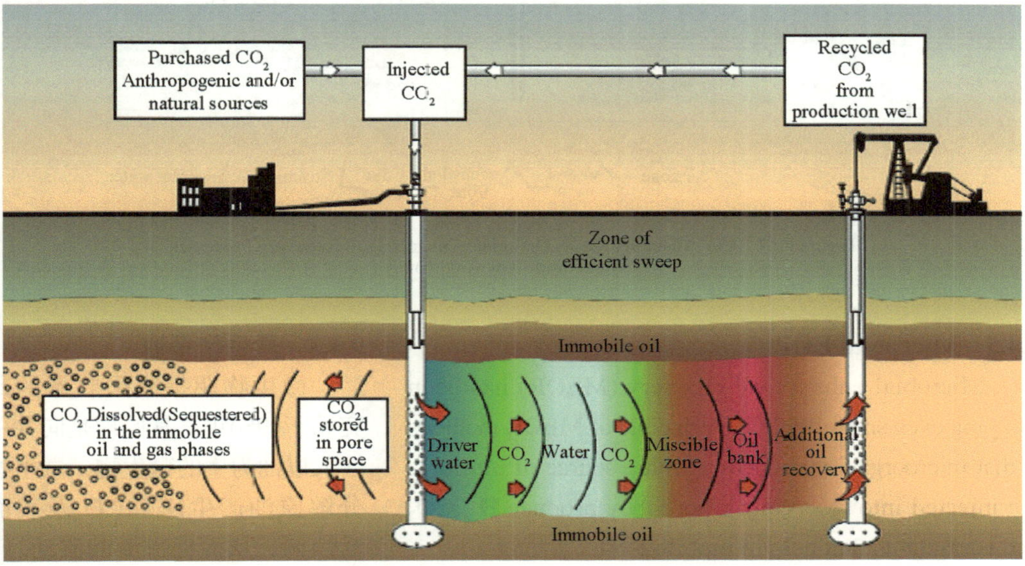

Figure 9.45 One-Dimensional Schematic Showing the Miscible CO_2-EOR Process
(Source:http://www.globalccsinstitute.com)

Figure9.46 Combined Miscible Recovery Processes
(Source: http://blog.wikipetro.org/tag/combined-miscible-drive/)

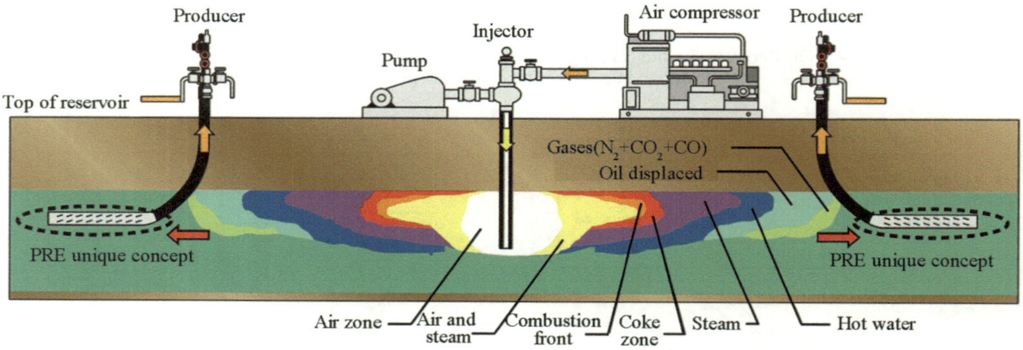

Figure 9.47 CO_2 Miscible, Steam Dominate Enhanced Oil Recovery Processes
(Source:http://www.ogj.com/)

9.4.5 Microbial EOR

Microbial enhance oil recovery (MEOR) has been studying even since 1950. The principle of MEOR methods is that microorganisms together with a source of nutrients are injected into reservoirs where they produce H_2, CO_2 and surfactants that help to mobilize oil.

9.4.5 微生物采油

微生物驱提高采收率的研究始于1950年。其原理是向井中注入以油为生的微生物，微生物分解原油产生H_2和CO_2及表面活性剂，从而促进原油流动。

SEPARATION &TRANSPORTATION & STORAGE 10

油气分离、储存与运输

Well fluids are often a complex mixture of liquid hydrocarbons, gas, and some impurities. It is necessary to remove the gas and some impurities from the liquid hydrocarbons before they stored, transported, and sold. Liquid hydrocarbons and objectionable impurities must also be removed from natural gas before the gas goes to a sale line. Impurities that might be found in some well streams are hydrogen sulfide, carbon dioxide, free water, water vapor, mercaptans, nitrogen, helium, and solids. Nearly all of the impurities cause various types of operating problems.

井内流体为碳氢化合物的液态和气态混合物，并包含很多杂质。在油气储存、运输、出售前，有必要去除其中的杂质和气相。同样，液相碳氢化合物和有害杂质也需在天然气出售前去除。在井流物中发现的杂质可能是硫化氢、二氧化碳、自由水、水蒸气、硫醇、氮气、氦气和固体杂质等。几乎所有的杂质都会导致操作问题。

The separation of natural gas, liquid hydrocarbons, and impurities is accomplished by various field-processing methods, depending upon the composition of the well stream and the desired end product. These methods include time, chemicals, gravity, heat, mechanical or electrical processes, and combinations of these.

根据井流物成分和产品最终要求，用不同的油田处理办法分离天然气、液态碳氢化合物和杂质，这些处理方法包括时间法、化学法、重力法、加热法、机械法和电法及其复合方法。

10.1 Oil Separation

10.1 原油分离

When crude oil is produced, various amounts of gas, water, and other impurities are mixed with the oil. Some of this mixture comes as free oil, some as free water, and some as a homogeneous mixture known as an emulsion. The gas, water, and other impurities must be removed before selling the oil(Figure 10.1). This separation process is called oil *surface treatment*.

生产原油时，不同数量的气体、水和其他杂质混在油里一起产出。这些混合物部分为自由油、部分为自由水，还有部分油水混合为均质状态的乳化油。这些气体和水及杂质需要在出售前去除。这个分离过

From the wellhead the oil with gas passes through an appropriately size pipeline to the train of separators. Onshore, with the wells spread all over the surface of the field and drilled vertically, these gathering lines may be some miles long to reach the separation plant(Figure 10.2). Offshore, where the wells are normally grouped on a platform and drilled directionally to spread out at reservoir depth, the pipework to the separation plant is relatively short(Figure 10.3, Figure 10.4).

The flow to the separator train in the early stage of the field' s life will normally be oil and gas. Reservoir water may also be produced increasing proportions over time. This water is saline and maybe salt-saturated

程叫原油地面处理（图10.1）。

从井口中出来的油混合着气体通过相应的管道到分离器（图10.2）。在陆上，油井遍布油田，输油管线离处理厂有很长距离。在海上，油井一般集中在海上平台上，定向钻井并扩展到油藏深度，输油管线相对短些（图10.3、图10.4）。

在油田生产早期，流入处理器的流体一般为油和气。油藏水随生产时间增加而增加。油藏水是含盐水，可能在油藏

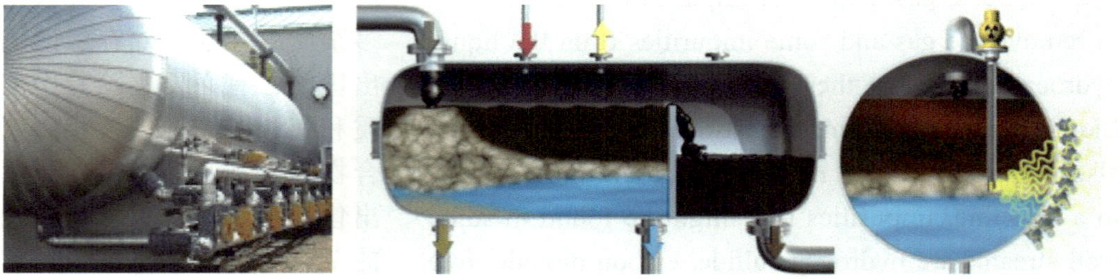

Figure 10.1 Oil/Water Separation Process
(Source: http://www.endress.com/eh/home.nsf/#products/profile-vision)

Figure 10.2 Onshore Processing Facilities
(Source: http://www.pacops.com/about/facilities/)

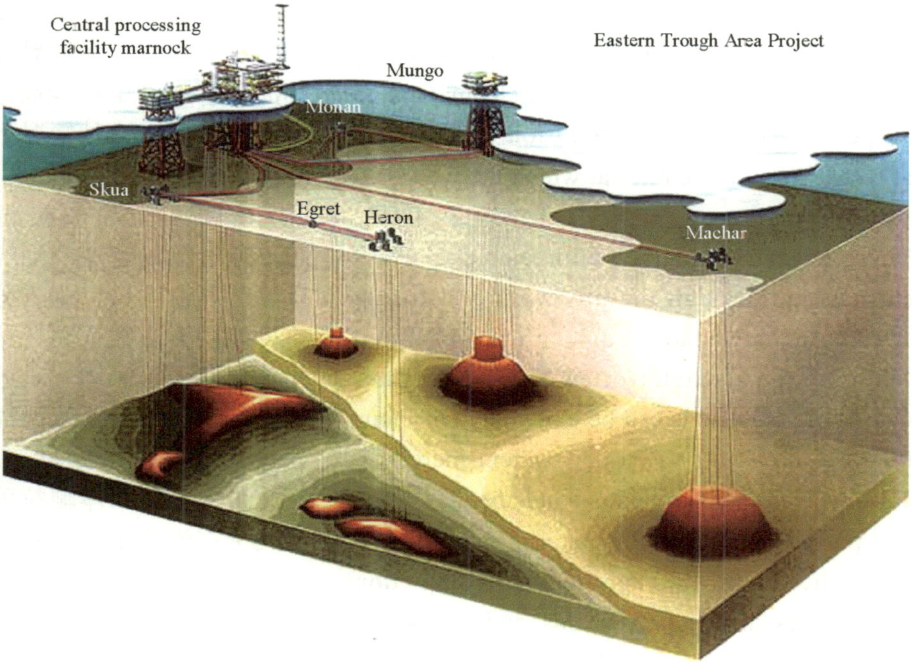

Figure 10.3 Layout Schematic of the Etap Fields
(Source:http://www.offshore-technology.com/projects/etaps/etaps5.html)

Figure 10.4 Offshore Central Processing Facility Where Natural Gas from the Ichthys Field Will Undergo Preliminary Processing
(Source: http://www.hydrocarbons-technology.com)

at reservoir temperature, so that it is supersaturated at surface conditions. In parts of the world the water produced with the crude oil from some wells may range up to 90 percent or higher, a condition that impose practical restrictions on the rates at which the wells can be produced.

温度下为饱和盐，在地面条件下为过饱和盐。在世界上，某些油田产出液中可能高达90%的水或更高，这种情况下需要对实际产量进行限制。

Where water is produced with the oil and gas turbulent passage through the pipework and fitting of the wellbore, wellhead, and separator manifold before entering the separator often creates oil/water emulsions. These emulsions can be extremely tenacious and difficult to break. As the water content increases, however, some water will remain unemulsified.

产出水与油气在湍流状态下通过输油管道和油嘴、井口、分离器进口，进入分离器前可能会形成油水乳化液。这些乳化液极为顽固、难缠，很难破乳。随着含水量增加，部分地层水保持未破乳状态。

Solid particles of reservoir material may also be entrained in the fluids, although the well completion will have been designed to prevent this, and if it occurs remedial work will be attempted using resins, etc. to stabilize the formation near the wellbore.

油藏中的固体颗粒可能会被夹带在流体中，虽然在完井时已尽量避免此类情况发生。如果发生此类情况，其补救办法是用环氧树脂加固井筒附近地层。

Taking advantage of their different densities, gas, oil and water are separated in separator. Separator may be vertical, horizontal, or inclined(Figure 10.5~Figure 10.8). Internally a number of techniques are used to encourage separation of the fluids. As the production stream traveling at high speed, centrifugal forces can be generated by appropriate entry and passage through the separator cylinder. Baffle plates, coalescing plates,

利用油、气、水之间的密度差可以将其在分离器中分离。分离器分为立式、水平式或倾斜式（图10.5 ~ 图10.8）。在分离器中采用多种措施促进分离。以高速进入分离器的产液流其离心力在入口处产生并流入分离缸。挡流板、凝结

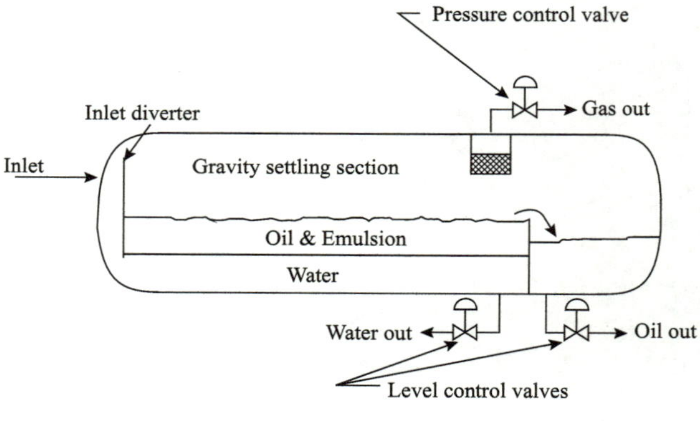

Figure 10.5 Horizontal 3-Phase Separator
(Source: http://www.oilngasseparator.info)

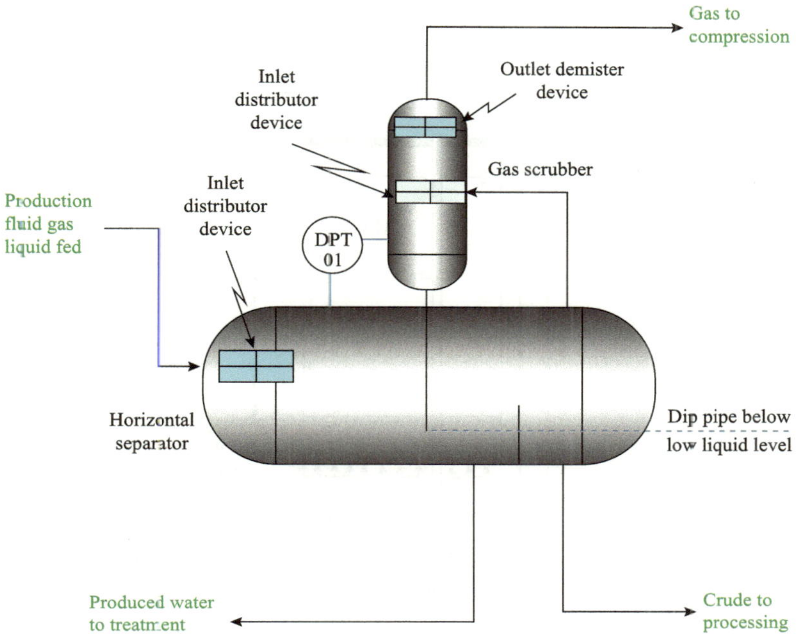

Figure 10.6 Horizontal 3-Phase-Separator-Vessels
(Source: http://www.ngltech.com/tech_swips.php)

Figure 10.7 Vessel, Separator, Vertical, 3-Phase, 24in × 7ft-6in
(Source: http://www.networkintl.com/)

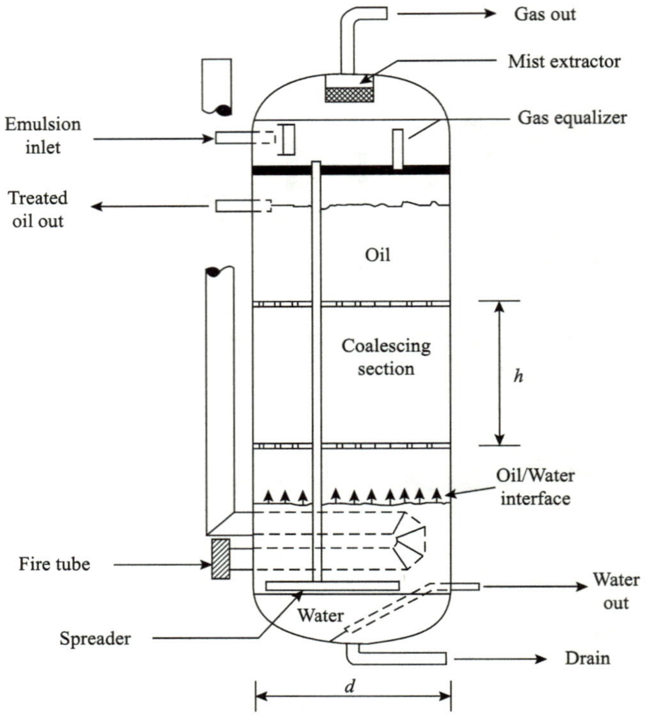

Figure 10.8 Vertical 3-Phase Separator
(Source: http://www.oilngasseparator.info)

mist pads, etc. are used to encourage oil to coalesce and fall while the gas rises. Including these items increases the weight of the separator. At the entrance to separator there is a vertical slot with an adjustable vane to allow velocity adjustment of the fluid film before it enters the cyclone tangentially. The oil film spirals down the walls of the cyclone, which is tapered to allow for energy loss during the oil's passage. Gas rises from the center of the cyclone, and additional gas that is released after the oil film cascades out of the cyclone is released through gas vent.

When the well stream reaches the separator at a high pressure it is common to have more than one stage of separation. In general the higher the pressure in the first stage separation, the higher will be the proportions of methane in the gas and the lower the proportion of the higher hydrocarbons. At subsequent, lower stage

板、除雾器等也促进油滴聚集下沉、气相上升分离。这些部件增加了分离器的质量。在分离器入口有个带可调节叶片的垂直槽缝，帮助调节液流进入旋叶切线的速度。油膜经旋叶旋转后沉降，在油流路上因能量损失而形成锥形。气从旋叶中心上升，多余的气由油膜经旋叶，经气口释放而来。

当高压井流物进入分离器一般需要进行多级分离。一般较高的压力级在一级分离，这级将得到较高比例的甲烷气和低比例的高分子碳氢化合物。而在低压力级，得到较低比例的甲烷气和高比例的高分子碳

pressure, the proportion of methane in the gas drops while that of higher hydrocarbons rises. Calculating the equilibrium between gas and liquid phases in any actual case is complicated, because of the large number of components involved and variations in the equilibrium constants with pressure, temperature, and concentration.

◆ Emulsion Breaking

Oil-field emulsions are usually of the water-in-oil type; however, a few of the emulsion are oil-in-water type and called reverse emulsions.

Where emulsions have been formed it is necessary to break them after free water and gas have been removed. There is an enormous range of complex chemicals on the market to perform this task. The emulsion is often heated to speed the process.

Economically it is impossible to remove all the water. If the crude oil produced following emulsion breaking and water separation is unacceptably saline for the refinery, fresh water may be added during the clean-up process to "wash-out" the salt. Whether this salt removal occurs at the production end or at the refinery depends on the area of the world involved; often crude oil desalt from part of refinery equipment, since they can remove salt introduced during tanker transport.

氢化合物。计算实际情况下的气液平衡相非常复杂，因为随着压力、温度和浓度变化，在平衡中大量的组分发生变化。

◆ 破乳

油田乳化物一般为水包油型，也有部分乳化物为油包水型，叫反转型乳化物。

当自由水和自由气去除后，就需要破除乳化物，这需要复杂的化学药品来完成这项任务。通常采用加热加速破乳过程。

从经济上考虑，不能去除所有的水。假如生产原油破乳并分离水后，其含盐量也不可接受，纯净水可能会在清洗过程中用于洗掉盐分。而去盐过程发生在生产后期，还是在炼化厂，则由当时具体情况决定。原油去盐很多时候发生在炼化厂，因为他们能一次性去除包括在油罐车里形成的盐分。

10. 2 Gas Separation

How the gas is treated depends on its economic value. Until the 1970s and because of lack of outlet much Middle East gas was flared untreated. With the rise in the market for natural gas liquids much is now treated to remove most liquefiable fractions. This is carried out by processing the rich (or "wet") gas streams in either a compression/absorption or a refrigeration/absorption type plant, to produce "lean" or "dry" gas to meet pipeline specifications for hydrocarbon dew-point, etc., and the raw NGL.

The raw gas from wellheads may pass through compressors to maintain pressure and may be heated to

10.2 天然气分离

天然气的处理方式由其经济价值决定。直到20世纪70年代，因为缺少外输管道，中东的天然气都是直接燃放的。随着液化天然气市场的成熟，现在天然气一般除去容易液化的部分。即采用压缩或吸收方式在压缩厂或吸收厂中去除天然气中的富气或湿气成分，生产干气或贫气达到输气管道要求低于露点的限制。

从井口产出的未进化气

prevent freezing and condensation. In the plants, the gas is chilled to separate methane from gas liquids. Then H_2S is removed by a chemical reaction in the presence of catalysts. The gas emerging from plants meets pipeline specifications for dry gas but may still contain small quantities of gas liquids, which are removed at straddle plants(Figure 10.9).

需进入压缩机保持压力并被加热以防止结霜和凝析。在处理厂，天然气被降温从气化液中分离出甲烷。而硫化氢则通过催化剂的化学反应去除。从处理厂出来要进入管道输送的干气中可能还包含少量气态碳氢化合物，这需要进一步去除（图10.9）。

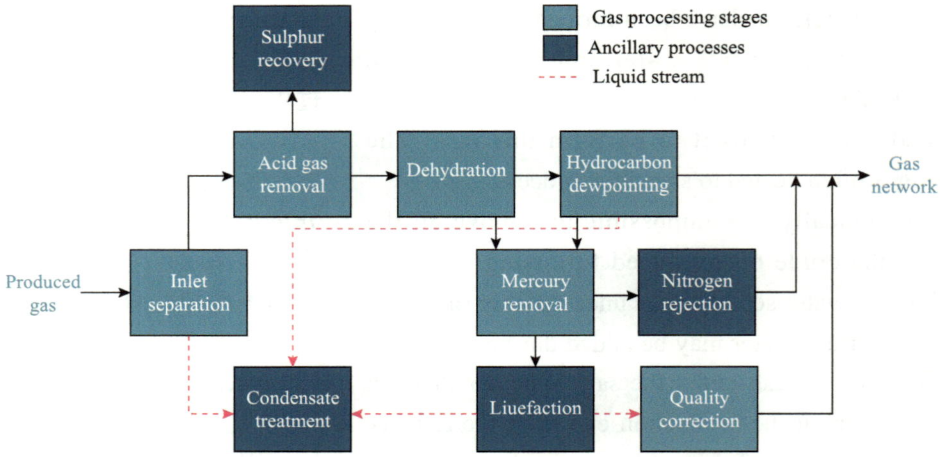

Figure 10.9 Gas Processing
(Source: http://www.gl-nobledenton.com/en/consulting/589.php)

The natural gas liquids are sold separately and used for enhanced oil recovery, as raw materials for oil refineries or petrochemical plants, or as fuels. Most of H2S is converted into elemental sulphur for use fertilizer manufacture and other industries.

Vary quantities of natural gas, known as associated gas or solution gas, are found and produced along with crude oil. This gas may be directed to processing plants or flared if quantities are too small to justify recovery.

天然气液化部分被单独出售，可以用于提高原油产收率，也可作为炼油厂或石化厂的原料或燃料。绝大多数硫化氢被转化为硫单质用于肥料生产或其他工业中。

不同数量的伴生气或溶解气也随原油一起产出，这些气体可能直接进入处理厂或因其量少不值得回收而燃烧掉。

10.2.1 Stages of Gas Processing

The number of steps and the type of techniques used in the process of creating pipeline-quality natural gas most often depends upon the source and makeup of the wellhead

10.2.1 气体处理过程

天然气处理过程一般主要由井流物成分、来源及输气要求决定。下面介绍一些天然气

production stream. In some cases, several of the steps shown in Figure 10.9 may be integrated into one unit or operation, performed in a different order or at alternative locations (lease/plant), or not required at all.

处理程序。

10.2.2 Gas–Oil Separators

In many instances pressure relief at the wellhead will cause a natural separation of gas from oil (using a conventional closed tank, where gravity separates the gas hydrocarbons from the heavier oil). In some cases, however, a multi-stage gas-oil separation process is needed to separate the gas stream from the crude oil. These gas-oil separators are commonly closed cylindrical shells, horizontally mounted with inlets at one end, an outlet at the top for removal of gas, and an outlet at the bottom for removal of oil(Figure 10.10). Separation is accomplished by alternately heating and cooling (by compression) the flow stream through multiple steps. Some water and condensate, if present, will also be extracted as the process proceeds.

10.2.2 气–油分离器

在很多时候压力在井口处下降导致天然气从油中分离（使用传统的封闭桶，重力使气态烃从油中分离），一般需要多级气-油分离过程将天然气从原油中分离出来。气-油分离器一般为封闭缸桶，水平放置，一端为入口，顶部出口用来排放分离后的天然气，底部出口用来排放分离后的原油（图10.10）。分离过程中加热和冷却（通过加压）交互进行着，流体通过多级分离设备。部分水和凝析液在此过程中析出。

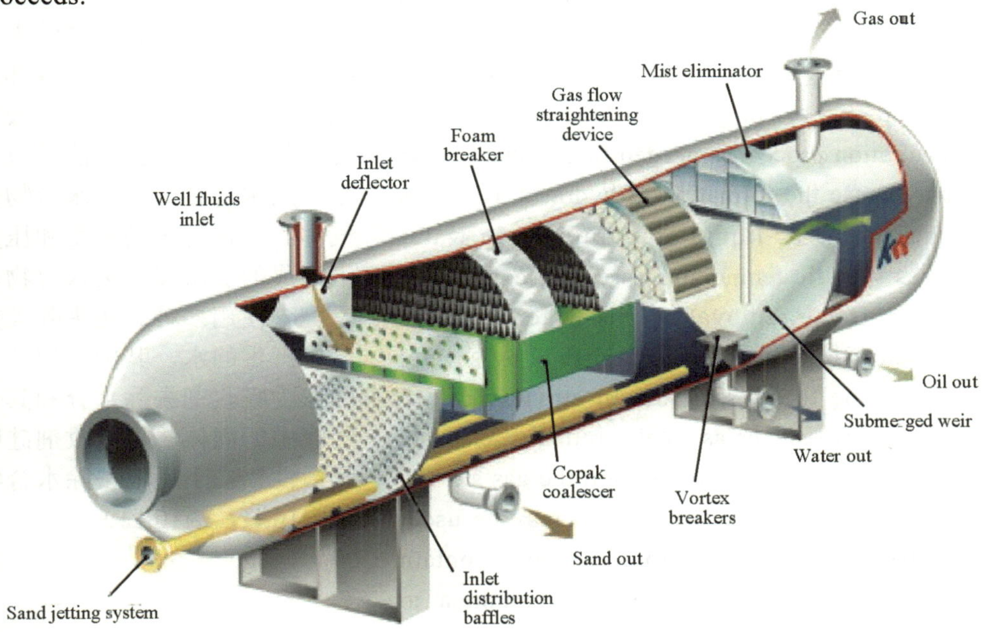

Figure 10.10 Typical Production Separator (Courtesy of Cameron Process Systems)
(Source:http://www.filtsep.com/view/7598/oil-and -gas-water-treatment-in-oil-and-gas-production-does-it-matter)

10.2.3 Condensate Separator

Condensates are most often removed from the gas stream at the wellhead through the use of mechanical separators. In most instances, the gas flow into the separator comes directly from the wellhead, since the gas-oil separation process is not needed. The gas stream enters the processing plant at high pressure (600 pounds per square inch gauge (psig) or greater) through an inlet slug catcher where free water is removed from the gas, after which it is directed to a condensate separator. Extracted condensate is routed to on-site storage tanks.

10.2.3 凝析分离器

通过机械分离器，凝析液可以从气体井流物中分离出来。在绝大多数情况下，气流从井口直接进入分离器，因此并不专门需要气-油分离过程。当气流以高压（等于或大于600磅每平方英寸）进入分离器入口，则需要直接进入凝析分离器。获取的凝析液进入井上储油罐中。

10.2.4 Dehydration of Natural Gas

Most natural gas contains substantial amounts of water vapor at the time it is produced from a well or separated from an associated crude-oil stream. Water vapor must be removed from the gas stream because it will condense into liquid and may cause hydrate formation as the gas is cooled from the high reservoir temperature to the cooler surface temperature. Liquid water almost always accelerates corrosion, and the solid hydrates may pack solidly in gas-gathering systems, resulting in partial or complete blocking of flow lines. So a dehydration process is needed to eliminate water which may cause the formation of hydrates. Hydrates form when a gas or liquid containing free water experiences specific temperature/pressure conditions. Dehydration is the removal of this water from the produced natural gas and is accomplished by several methods. Among these is the use of ethylene glycol (glycol injection) systems as an absorption mechanism to remove water and other solids from the gas stream. Alternatively, adsorption dehydration may be used, utilizing dry-bed dehydrators towers, which contain desiccants such as silica gel and activated alumina, to perform the extraction(Figure 10.11， Figure 10.12).

10.2.4 天然气去水

绝大多数从井口产出的天然气中，或原油中分离的天然气中包含水蒸气。水蒸气在输气前必须被去除，因为当温度由油藏中的高温度降到地面上的低温度时，水蒸气会凝析成液体并形成水合物。液化的水也会加速腐蚀，固体的水合物会堵塞输气管道，导致部分或全部管线瘫痪。因此，去水合物的过程需要去除能导致产生地层水合物的水。水合物是气体和水在特定的温度和压力下形成的。所以，去水合物就是配合使用多种方法去除天然气中产生的水。其中，用乙二醇作为吸收剂去除水分和其他气流中的固体。用吸收剂硅胶和活性铝在干燥塔去除水合物（图10.11、图10.12）。

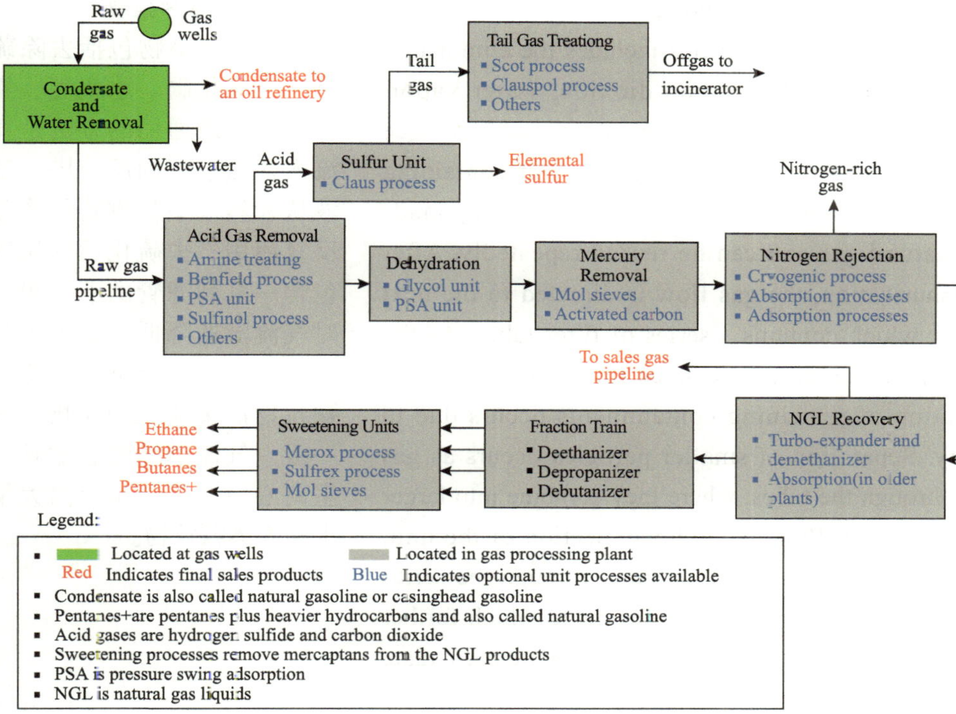

Figure 10.11 Natural Gas Processing
(Source: http://en.wikipedia.org/wiki/Natural-gas_processing)

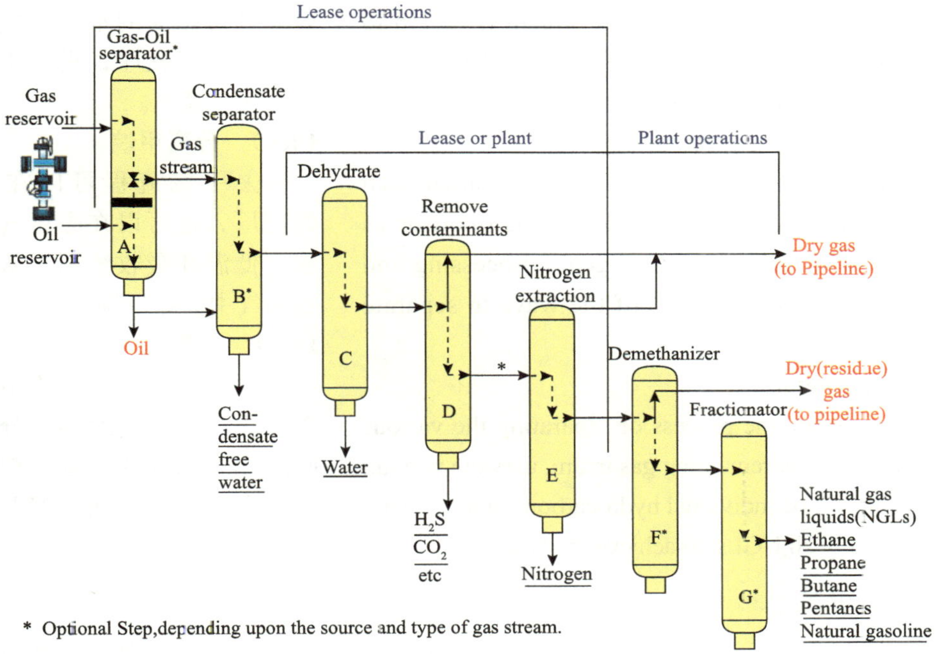

Figure 10.12 Generalized Natural Gas Processing Schematic
(Source: http://www.unigas.org/stages.html)

10.2.5 Contaminant Removal

Removal of contaminates includes the elimination of hydrogen sulfide, carbon dioxide, water vapor, helium, and oxygen. The most commonly used technique is to first direct the flow through a tower containing an amine solution. Amines absorb sulfur compounds from natural gas and can be reused repeatedly. After desulphurization, the gas flow is directed to the next section, which contains a series of filter tubes. As the velocity of the stream reduces in the unit, primary separation of remaining contaminants occurs due to gravity. Separation of smaller particles occurs as gas flows through the tubes, where they combine into larger particles which flow to the lower section of the unit. Further, as the gas stream continues through the series of tubes, a centrifugal force is generated which further removes any remaining water and small solid particulate matter.

10.2.5 去除污染物

去除污染物包括去除硫化氢、二氧化碳、水蒸气、氦气或氧气。最通用的方法是直接流入含氨溶液的反应塔。氨溶液从天然气中吸收硫化物并能反复使用。去硫化后气流进入包含一系列滤管的下一级。这时气流流速降低，靠重力分离剩余污染物，当进入滤管时小颗粒物汇集成大颗粒物流入下一级。气流继续流进滤管，产生离心力，进一步分离大量的水和固体颗粒物。

10.2.6 Nitrogen Extraction

Once the hydrogen sulfide and carbon dioxide are processed to acceptable levels, the stream is routed to a Nitrogen Rejection Unit (NRU), where it is further dehydrated using molecular sieve beds.

10.2.6 去除氮气

一旦硫化氢和二氧化碳被处理到可接受的水平，气流将进入去氮气程序，进一步用分子筛去除氮气。

10.2.7 Methane Separation

The process of demethanizing the gas stream can occur as a separate operation in the gas plant or as part of the NRU operation. Cryogenic processing and absorption methods are some of the ways to separate methare from NGLs.

10.2.7 分离甲烷

去甲烷流程可以在分离处理厂，也可以在去氮气时完成。低温和吸收法用于从液化石油气中去除甲烷。

10.2.8 Fractionation

Fractionation, the pracess of separating the various NGLs present in the remaining gas tream, uses the varying boiling points of the individual hydrocarbons in the stream, by now virtually all NGLs, to achieve the task.

10.2.8 分馏法

分馏法是利用气流中不同碳氢化合物不同的沸点分离不同液化石油气的方法，现在几乎所有的液化石油气都要完成这个流程。

10. 3 Water Separation

Disposal of produced water can prove a problem. Both onshore and offshore it may be possible to inject

10.3 产出水处理

处理产出水是个难题。海上或陆上可能将油性的产出水

the oily production water into a disposal well. However, in many cases it is treated to reach oil content acceptable to authorities and discharged at surface.

注入报废井中。但是很多情况下，当产出水含油量在当地可接受范围内，一般直接排放到地面。

The characteristics and quantities of this water vary considerably, both from field to field, and in any one field from time to time, often in near-random fashion. It may contain suspended solids such as clays, corrosion by-products and heavy hydrocarbon complexes, such as tar or wax.

产出水的性质和数量变化相当大，不同的油田，或同一油田的不同时期都不一样。产出水中可能含黏土、腐蚀性副产品和重质碳氢化合物，如焦油或蜡。

The use of demulsifiers, biocides, corrosion inhibitors, and other chemicals in the production process and secondary and tertiary recovery methods all can affect the characteristics of the water. Some produced water contains significant quantities of oil droplets smaller than 25 microns, while some components of crude oil may be significantly soluble in the water. These soluble components are usually also the more volatile and better biodegradable components.

在生产过程和二次、三次采油过程中，用反乳剂、抗生物剂和腐蚀性阻抗剂及其他化学品会影响产出水的性质。一些产出水包含大量直径小于25μm 的油滴，而一些产出水包含大量溶解油。这种溶解大量原油的乳状液通常更容易挥发和更好的生物降解。

The simplest method of gravity separation is a settling tank, sump, or caisson. Offshore, where long residence time would be unacceptable, plate interceptors are often used to reduce the effective depth of the fluid column and to aid separation. These systems of densely packed plates, like gas/oil separators, come in a variety of styles. Corrugated, parallel, and tilted plate separators are all in use(Figure 10.13). They appear to be effective on droplets of diameter above 60 ~ 150 microns.

最简单的油水分离方法是置一沉降罐、沉降坑、沉降箱进行重力分离。海上油田长时间的沉降不可接受，拦截板常用于减少液柱有效深度从而辅助分离。如油水分离器里有各种各样的密集拦截板。波状、平行状、倾斜状的拦截板都在使用（图10.13）。拦截板对直径为60 ~ 150μm的油滴有效。

A method of accelerating the process is gas or air flotation, where separation is stimulated by the small bubbles adhering to the oil droplets and speeding their ascent. Other methods used onshore include coalescers, filters, electrostatic precipitators, membrane processes, and biological treatment, but none are common offshore.

另一种加速油水分离的方法是气体悬浮法，利用小气泡黏着油滴加速油滴上升从而刺激油水分离。陆上油田还有其他分离方法，比如聚集法、过滤法、静电分离法、膜分离法和生物分离法，但这些方法在海上油田却不常用。

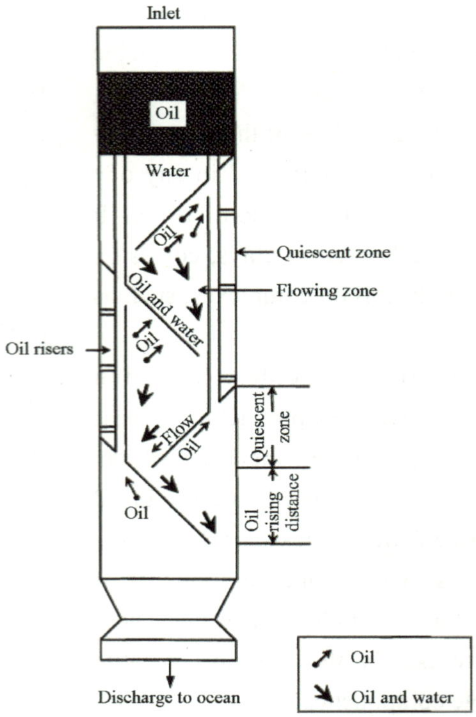

Figure 10.13 Vertical 3-phase Separator on Offshore
(Source: http://www.netl.doe.gov/technologies/pwmis/techdesc/physep/)

10.4 Transportation & Storage

Petroleum pipeline and network are the fundamental facilities in petroleum storage and transportation.

10.4.1 Petroleum Transportation

Many oil fields have been discovered in places that are far from the point at which the oil will be refined and used. It is the liquid state of oil that offers the great advantage over coal as the primary energy source in today's industrialized world. No matter how abundant coal is, it is bulky and heavy and therefore difficult to ship. Oil can flow through pipelines to the market at a relatively low cost. Of course, the oil must be pumped because the pipelines go up and down the hills.

From the wellhead to the consumer, petroleum pipeline systems provide us with energy fuels and products we use every day(Figure 10.14). There are

10.4 油气储运

石油管道和管网是油气储存和运输的基础设施。

10.4.1 石油运输

很多油田在远离原油炼化和需求的远方被发现。在当今工业世界中，液态油作为主要能源比煤更具优势。不管煤的储量有多丰富，因其笨重而难运输。而油可以通过管道以花费相对低廉的方式流入市场。当然，油必须泵入管道以完成上山和下山。

从井口到消费者，石油管道为我们提供能源燃料和日用产品（图10.14）。主要有三

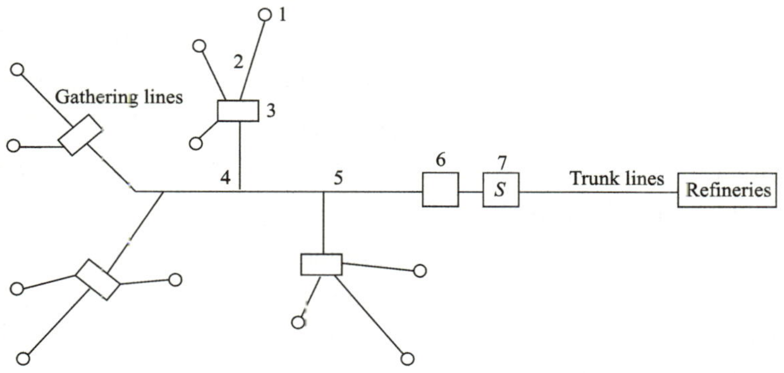

Figure 10.14 Gathering Line to Trunk Line to Refinery
1—Wellhead; 2,4,5—Gathering lines; 3—Separation; 6,7—Separation &Treatments

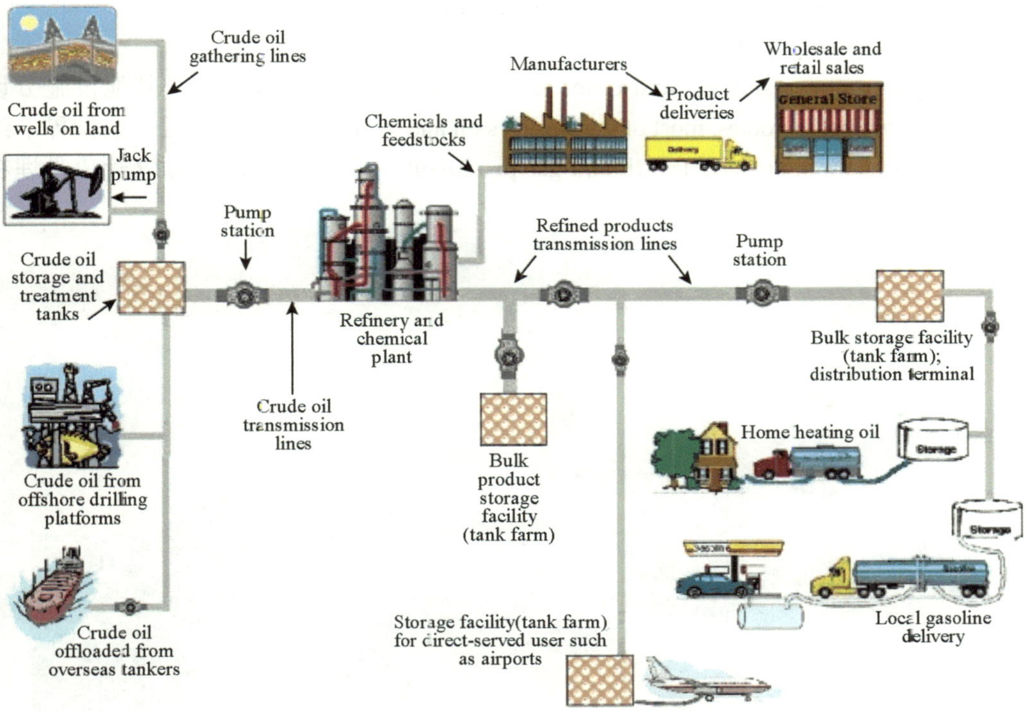

Figure 10.15 Petroleum Pipeline Systems
(Source: http://primis.phmsa.dot.gov/comm/PetroleumPipelineSystems.htm)

essentially three major types of pipelines along the petroleum transportation route: gathering systems, crude oil pipeline systems (trunk/transmission), and refined products pipeline systems (distribution). Gathering pipeline systems gather crude oil from production wells (Figure 10.15). Crude oil pipeline systems transport crude oil from the gathering systems to refineries. Crude oil systems can be tens to hundreds of miles in length and cross state and continental borders. Refined products pipeline systems transport refined products such as gasoline, kerosene and many industrial feedstock petrochemicals from refineries to the end user or to storage and distribution terminals. Refined products pipelines can extend tens to thousands of miles and cross state and continental borders. The pipe used in oil pipeline systems can range in size from 2 inches to 42 inches in diameter(Figure 10.16).

大类型的石油运输管道：收集线、运输线和炼化产品线。收集线从生产井井口收集产出原油（图10.15）。运输线将原油从联合站输运到炼化厂。运输线可延伸几十或几百公里长，也可能跨国或跨洲。炼化产品线将炼化产品如汽油、煤油或其他工业原料从炼化厂运输到终端用户或储集分配站。炼化产品线可能有几十或几千公里，也可能跨国或跨洲。这些输油管线直径从2in到42in不等（图10.16）。

Though pumping stations and other facilities are scattered along pipeline routes, most of the world' s oil and gas pipeline system is not visible. Oil and gas produced

尽管泵站和其他设施分设在输油管线附近，但世界上绝大多数输油、输气管道是埋在

Figure 10.16 Building a Petroleum Pipeline
(Source: http://www.pipechina.com.cn)

from offshore wells are brought to shore by pipeline, often through water several hundred feet deep. Oil and gas pipeline systems are remarkable for their efficiency and low transportation cost(Figure 10.17).

地下而不可见的。海上油井中生产的油气通过水下几百英尺的海底管道输送到陆地上。油气输送管道以其高效、低成本闻名（图10.17）。

Figure 10.17 Onshore Processing Facilities & Pipeline
(Source: http://www.pipechina.com.cn &http://pipelinepsc.com/)

A recent, more thorough investigation concludes that crude trunk lines consume about 0.4% of the energy content of the crude transported per 1,000 km. Products pipelines use about 0.5% of the energy content of the products moved per 1,000 km. 1.0 % of the energy content is used for oil movement by rail; 2.5% for natural gas pipeline; 3.2% for oil trucks.

The link between pipeline size and economy is apparent. A 36-in diameter line can carry up to 17 times more oil or gas than a 12-in diameter pipeline, but consumption and operation costs do not increase at nearly the same ratio.

A device called pig cleans pipelines. This mechanism has metal blades that scrape the inside of the pipe to keep it clears of the tar like substance that forms in it. The pressure of the oil itself forces the pig to move through the pipe. The pig can only go from one pumping stating to the next, where it is taken out and cleaned.

The pipelines must be constantly checked. Each

最近，一项调查研究得出，石油主干线管道每输送1000km石油消耗其0.4%的能量。生产管道每输送1000km石油消耗其0.5%的能量。而用火车输送相应距离的原油要消耗其1.0%的能量；用火车输送相应距离的天然气要消耗其2.5%的能量；用油罐车输送相应距离的原油要消耗其3.2%的能量。

管道的尺寸与其成本效益的关系是明显的。一段36in直径的管道输送能力是12in直径管道的17倍，但是其消耗和运营成本却没有相应增长。

使用一种名为“管道猪”的装置清理管道。其原理是用金属叶片刮管道内壁从而去除管内焦油类物质。管内油压促

section is patrolled by a pipeline walker who looks for evidence of a leak or other damage in the line. Nowadays, many "walkers" do their inspecting from low-flying airplanes, since traces of seepage can be seen from the air.

The transportation of oil has also led to the spectacular development of giant ships, the largest ships that have ever sailed the oceans of the world (Figure 10.18). They weigh as much as 500000 tons. Monsters weighing 200000 tons, which would have seemed impossible not many years ago, have become common.

使管道猪在管道中移动。管道猪只能从泵站点的始端到下一末端，从末端取出并清理。

输油气管道必须不断检查，每一段管线都需要检查是否有泄漏或损害。现在，许多地方用低空飞行器从空中检查是否有管道泄漏痕迹。

今天，原油运输已经发展到用巨型油轮将其运输到世界各地（图10.18）。巨型油轮重达500000t，大油轮重达200000t，这在很多年前是不可想象的，现在却成为普遍现象。

Figure 10.18 Crude Oil Supertanker
(Source: http://www.digital-camera.gr/)

Other special types of carries have been developed for the transportation of oil or oil products. On water, special barges are used for transporting oil on sheltered or inland waterways likes lakes, rivers, or canals. They are built on the same principle as the supertanker——the wall of the ship is also the wall of the oil tanks (Figure 10.19).

原油和其产品还可通过其他类型的装置输送。如在湖泊、河流、运河水上用驳船走水路将原油运到庇护所或海岛。驳船像油轮一样用船体建造超大储油罐（图10.19）。

Figure 10.19 2000 tones Oil Barge
(Source: http://www.bonnyfair.com.hk/deliveries.htm)

On land, the tank cars that were developed early in the history of the oil industry to transport petroleum by rail(Figure 10.20). Today, one of the familiar sights on our highways is the tank truck, which is a special kind of automobile used to carry petroleum or petroleum products.

在陆地上，油罐火车在石油工业早期就开始运油（图10.20）。今天，在高速公路上我们还可以看到油罐车运油及石油产品。

Figure 10.20 Oil Tank Cars
(Source: http://www.newsobserver.com)

The tank truck is ordinarily used for short-haul transportation-distances of a few hundred miles at the most(Figure 10.21). It usually carries oil products from the refinery to the distributor or from the distributor to the customer.

油罐车通常用于最多几百公里的短途运油（图10.21）。一般把油产品从炼化厂运输到分散站，再由分散站运往消费者。

Figure 10.21　Oil Tank Cars
(Source: http://en.wikipedia.org/wiki/Tank_truck)

10.4.2　Petroleum Storage

Production, refining, and distribution of petroleum products require many different types and sizes of storage tanks(Figure 10.22). Small bolted or welded tanks might be ideal for production fields while larger, welded storage tanks are used in distribution terminals and refineries throughout the world. Product operating conditions, storage capacities, and specific design issues can affect the tank selection process.

10.4.2　石油储存

生产、炼化和分销石油产品需要多种不同类型、尺寸的储油罐（图10.22）。小的焊接油罐可能适合油田矿场，而大的焊接油罐用于世界各地的分销站和炼化厂。产品的操作情况、储存设备和特殊设计会影响储油罐的选择。

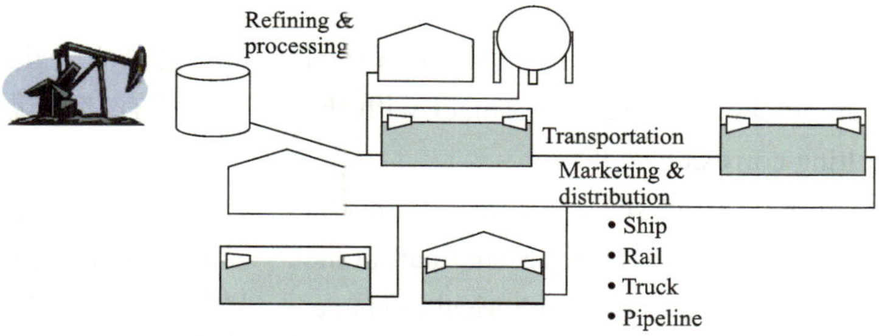

Figure 10.22 Oil Tank
(Source: http://petrowiki.org/Oil_storage)

10.4.2.1 Storage tanks

Properly designed petroleum storages must prevent leaks and potential contamination of soils, surface water or groundwater. Even a very small leak is enough to contaminate millions of liters of water.

The installation of underground storage tanks is not recommended.

Most farm petroleum storage tanks are above-ground, single-walled, horizontal storage tanks. Such tanks should have secondary containment systems to retain any leaks.

Double-walled, vacuum monitored tanks have a vacuum between the two walls. If the gauge indicates that the vacuum has been lost between the two steel walls, then there is a leak in one of the walls.

Double-walled tanks are frequently found to be a good option for smaller farm operations.

All tanks must have a corrosion protection coating and/or be primed and painted to prevent rusting.

All fuel tanks should be located outside of building

10.4.2.1 储油罐

合适的储油罐设计必须防止泄漏、土壤污染、地下水污染和地面水污染。即使很小的泄漏也会污染百万升的水。

因此，不推荐安装地下储油罐。

大多数现场储油罐是直接立于井场上，单壁，水平储存罐。一些储油罐应该有二级密封系统以防止任何的漏泄。

双层、真空储油罐在两层壁中有一可监测的真空层。假如测压表显示两层之间的真空层被破坏，说明任何一层可能存在漏泄。

双层储油罐常认为更适合小井场。

所有储油罐必须有防腐、防锈保护层和优先层。

used for other purposes.

为安全或其他原因，所有油罐必须立于建筑外。

All tanks should be raised to at least 15cm on steel feet with concrete anchors to prevent contact with the ground and resulting corrosion.

所有储油罐必须用钢支架支撑至少离水泥地15cm以上，防止接触地面造成腐蚀。

Storage tanks come in all sizes and shapes. Special applications might require tanks to be rectangular, in the form of horizontal cylinders, or even spherical in shape. Horizontal cylinders and spheres are generally used for full pressure storage of hydrocarbon or chemical products. The atmospheric or low-pressure storage tanks widely are used from the production fields to the refinery. The most common shape used is the vertical, cylindrical storage tank. Gross capacities can range from 100 bbl to over 1.5 MMbbl in a single storage tank. Corresponding tank sizes range from approximately 10 ft in diameter to over 412 ft in diameter for some of the largest floating-roof tanks ever constructed.

储油罐的形状、大小各异。在一些特殊情况下，要求方形的储油罐，一般储油罐为圆筒形甚至为球形。水平圆筒和球形罐一般用于高压储存碳氢化合物和石化产品。低压储罐一般用在矿场和炼化厂，且一般为垂直圆筒形。一个储油罐的储油能力从100bbl到1.5MMbbl不等。相应的储油罐直径为10～412ft，还有更大的浮顶罐。

The general types of atmospheric storage tanks (AST) in use may be open top tanks (OTT), fixed-roof tanks (FRT), external floating-roof tanks (EFRT), or internal floating-roof tanks (IFRT). Depending on the product, a closed floating-roof tank (CFRT) may even be selected.

大气压下，储油罐一般分为开顶式、封顶式、外浮顶式和内浮顶式。根据油产品需要，还有封闭内浮顶油罐。

The above ground storage tank has evolved with time. Figure 10.23 illustrates this trend, which has emphasized improved safety and improved product loss control. Production facilities generally rely on either open-top tanks or fixed-roof tanks operating at or slightly above atmospheric pressure.

以上为地面储油罐。图10.23展示现今的储油罐，它们强调提升安全和控制产品损失。矿场上一般用开顶式或固定顶式油罐，其操作压力略高于大气压。

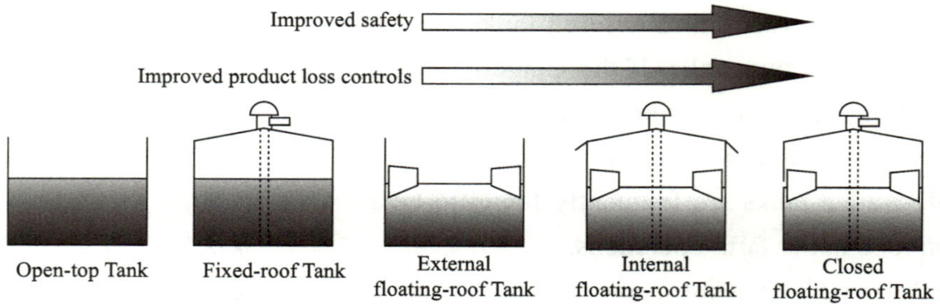

Figure 10.23 Current Storage Options
(Source: http://petrowiki.org/Oil_storage)

◆ Open Top Tanks (OTT)

The OTT was one of the first tanks used to store petroleum products. While it provides liquid containment, direct exposure of the liquid surface to the atmosphere assures high evaporative losses, product odors, and increased potential for fires. The OTT has only limited use, primarily for collection of contaminated run-off or wash water and wastewater processes.

◆ 顶部开放性储油罐(OTT)

顶部开放性储油罐是最初用于储存石油产品的油罐。它提供了液体的储存容器，但是液体表面直接暴露在空气中，导致高油气的蒸发损失，增加了潜在发生火灾的风险。因此，顶部开放性储油罐的使用有限，主要用在收集被污染液流或废水的过程中。

◆ Fixed Roof Tanks (FRT)

The FRT provides improved containment of product vapors and reduces the potential for fires. The FRT still exposes the liquid surface to the tank vapor space, producing significant product evaporative losses. This increases the possibility of forming a combustible gas mixture in the vapor space for certain more volatile petroleum products. For this reason fixed roof tanks in refineries are generally used for products with vapor pressures less than 1.5 psi.

◆ 固定顶储油罐(FRT)

固定顶储油罐更好地控制油气产品的蒸汽，减少火灾风险。固定顶储油罐仍然将液体表面暴露在储油罐蒸汽空间里，造成巨大的产品蒸发损失。对于某些易挥发油品而言，这增加形成可燃性气体混合物的可能性。因此，在炼油厂中，固定顶储油罐通常储存蒸汽压小于1.5psi的产品。

◆ Floating Roof Tanks

Although not normally used in production operations, floating roof tanks are often used in pump stations or terminals where the crude oil has been stabilized to a vapor pressure of less than 11.1 psi.

◆ 浮顶罐

在生产操作过程中不常用，但浮顶罐常用于泵站或终点站，在那里，原油稳定的蒸汽压小于11.1psi。

10.4.2.2 Oil Depot

An oil depot (sometimes called a tank farm, installation or oil terminal) is an industrial facility for the storage of oil and/or petrochemical products and from which these products are usually transported to end users or further storage facilities. An oil depot typically has tankage, either above ground or underground, and gantries (framework) for the discharge of products into road tankers or other vehicles (such as barges) or pipelines.

Oil depots are usually situated close to oil refineries or in locations where marine tankers containing products can discharge their cargo(Figure 10.24). Some depots

10.4.2.2 油库

油库（有时叫罐区或油码头）指储存原油或石化产品的地方，以便于将其运输到下级用户或下一个储存地。油库一般有地面或是地下的储油罐和起重机架，以方便向油罐车、油轮或管道中倾倒原油。

油库一般离炼油厂或海上油轮港比较近，这样方便装卸原油及其产品（图10.24）。一些油库连接油管从而得到原

Figure 10.24 An Oil Depot in Kowloon, Hong Kong around the Mid-1980s. (The Depot was Redeveloped into a Residential Area Laguna City in the Late 80s and Early 90s)
(Source: https://en.wikipedia.org/wiki/Oil_terminal)

are attached to pipelines from which they draw their supplies and depots can also be fed by rail, by barge and by road tanker (sometimes known as "bridging").

Most oil depots have road tankers operating from their grounds and these vehicles transport products to petrol stations or other users.

An oil depot is a comparatively unsophisticated facility in that (in most cases) there is no processing or other transformation on site. The products which reach the depot (from a refinery) are in their final form suitable for delivery to customers. In some cases additives may be injected into products in tanks, but there is usually no manufacturing plant on site. Modern depots comprise the same types of tankage, pipelines and gantries as those in the past and although there is a greater degree of automation on site, there have been few significant changes in depot operational activities over time.

油，一些油管和铁路、驳船、油罐车连接。

绝大多数油库由油罐车从地面运输油产品到加油站或其他用户。

在多数情况下，油库中并不加工石油，油产品到达油库（从炼油厂）并最终转运到用户手中。一些油产品需要加入添加剂后再运输，但这不是就地生产工厂。现代油库综合了油罐、油管、起重机架等过去的多种部门，从而实现了更大程度的自动化。

PETROLEUM REFINING AND ITS PRODUCTS 石油炼化及其产品 11

Crude oil is a raw material, like iron ore, which must be processed before it can be used (Figure11.1). In fact, there are several different processes through which oil is processed, but together they are called refining.The first step is called distillation.

原油是一种原始材料像铁矿石，需要经过加工才能利用（图11.1）。原油的加工程序有多种，这叫炼化。而炼化的第一步叫蒸馏。

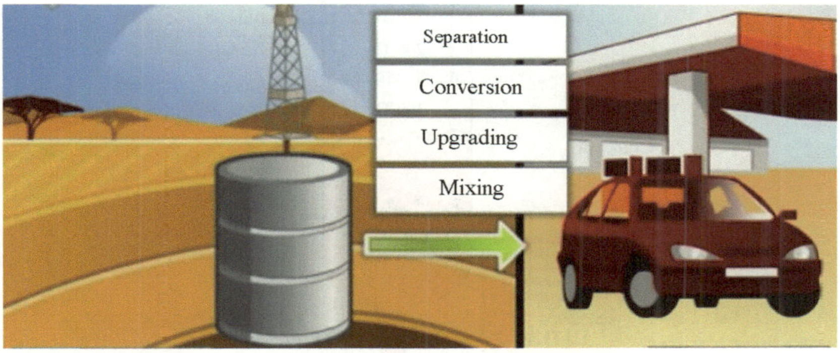

Figure 11.1 Oil from the Well to Customer
(Source: The Principal of Refining, TOTAL)

11.1 Distillation

The theory of distillation is very easy. Most liquids boil when they reach a certain temperature. When crude oil boils, it sends out vapor and condenses. All the different hydrocarbons in crude oil are at different temperatures, and their vapors all make separate liquids. So, in order to separate the hydrocarbons, the scientists distill the crude at different temperatures, their vapors must also condense at different temperatures. When

11.1 蒸馏

蒸馏的原理很简单，大多数液体的沸点不同，当原油沸腾，它汽化并凝析。原油是多种碳氢化合物的混合物，因此不同的沸点温度能分离不同的液体。科学家以某一温度蒸馏原油，它们的蒸汽在某一温度下凝析，从而分离不同组分得

the hydrocarbons boil, their vapors condense into its fractions and forms different liquids.

The oil is distilled in a very large steel tower, the technical name of which is the *fractionating tower*(Figure 11.2). The tower is thirty to fifty meters tall and its diameter is one to three meters. It is divided into chambers, each of which contains a layer of trays. There are holes in the trays. The vapors go through these holes and condense on the trays. The chambers are at different heights, and the temperature at each height is different. The temperature inside the tower is highest at the bottom, and it is lowest at the top. So each layer of trays is cooler than the layer below it. There are not enough chambers of different temperatures in one tower, so two or three towers are needed.

到不同液体。

原油蒸馏在一个叫分馏塔的很大的铁塔中进行（图11.2）。分馏塔一般有30～50m高，直径为1～3m。分馏塔里分很多层，每一层有塔盘。塔盘上有很多洞眼，蒸汽经过洞眼而凝析在塔盘上。塔内每一层高度不同，温度也不同。底部温度最高，顶部温度最低。每一塔盘都比它下面那层温度低。如果一个分馏塔中分馏层不够，则需要2～3个分馏塔。

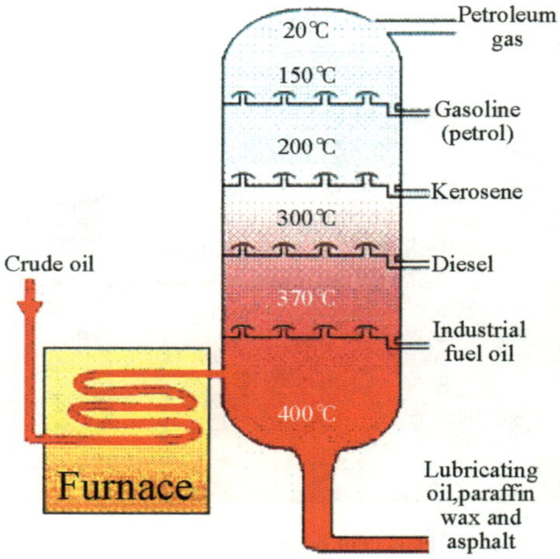

Figure 11.2 Fractionating Tower
(Source:www.iciba.com)

Beside the first tower there is very large furnace, the pre-heater, where the temperature of the oil is raised to 340 degrees Centigrade. The boiling oil and vapors then flow from the furnace into the fractionating tower. They flow into a chamber near the bottom. Some of the oil with the heaviest hydrocarbons makes no vapor. This oil-the residue -drops to the bottom and flows out of the tower again. Later bitumen and other heavy

在第一个分馏塔旁有一个巨大的火炉，叫加热炉。加热炉能将油温升高到340℃。沸腾的油和其蒸汽从加热炉流入分馏塔。它们从分馏塔近底部进入，部分最重质的碳氢化合物加热后没有蒸汽，这是油渣。油渣沉积在分馏塔底部，会再

materials will be made from the residues. All the other hydrocarbons in the oil make vapors that float up the tower to different chambers(Figure11.3).

次进入分馏塔蒸馏。沥青和其他重油材料由油渣而来。原油中其他碳氢化合物沸腾上升达到不同塔盘（图11.3）。

Figure 11.3 Typical Industrial Fractionating Tower
(Source:www.iciba.com)

When the vapors rise into the cooler part of the fractionating tower, they lose heat. The vapors rise until they are just below their own boiling temperature. Then they condense on a tray and turn into a liquid again. The liquid on each tray flows into a separate small tower there the liquid breaks down into smaller fractions. The separation of the hydrocarbons is then more accurate. The distilled then flow through separate pipes into separate storage tanks(Figure11.4).

当油蒸汽上升冷凝在分馏塔较低温度的塔盘上，它们失去热量，除非它们再次获得沸点温度，否则它们不会继续上升。因此，它们在塔盘上再次冷凝成液体。每一层塔盘上冷凝的液体进入更小的塔，以便继续分离出更小组分。后续的碳氢化合物分离更准确。蒸馏后的液体通过隔离的管道进入不同的储存罐中（图11.4）。

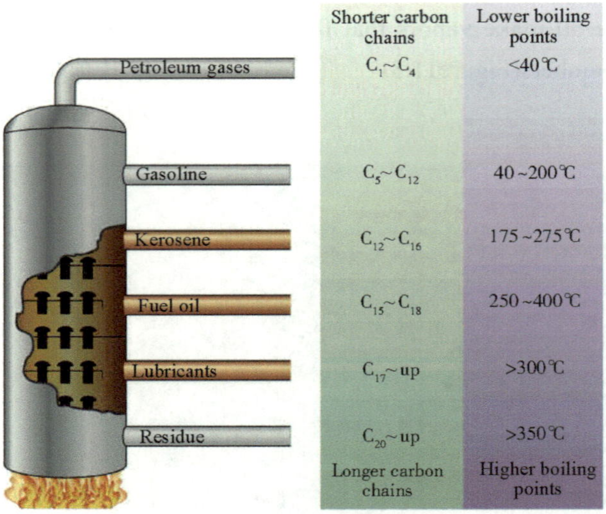

Figure11.4 Crude Oil Distillation Tower
(Source: http://www.acwadam.org/images/hydropedia/big/porosity_permeability_big.jpg)

When cuts with boiling-points over 350℃ are required it is necessary to do the distillation under the vacuum to reduce the temperature at which the material boils, as some high-boiling compounds start to decompose when heated to their boiling-point at atmospheric pressure.

当温度高于350℃时，有必要进行真空蒸馏，以降低高沸点组分在大气压中的沸点。

11.2 Primary Refining and Secondary Refining

The primary refining process is the distillation of the feed stock into its basic fractions, and then the re-distillation of these, in separate towers, into highly concentrated intermediates. In a simple refinery, the split would probably consist of these six basic cuts: petroleum gas (refinery gas and LPG), gasoline, naphtha, kerosene, gas oil, and residue.

The secondary processes are designed to remove any impurities from the distilled fractions, and to convert some of the distilled hydrocarbons into different molecular forms. Conversion processes can produce hydrocarbons that do not exist in reservoir crude(Figure 11.5).

11.2 初次加工和二次加工

初次加工是将原油蒸馏成基本成分，然后再进入另一中塔蒸馏，直到得到高浓度的中间产品。一般简单蒸馏可以将原油分出六大基本成分：石油气（炼厂气和液化石油气）、汽油、粗汽油、煤油、柴油和油渣。

二次加工的目的：去除蒸馏产品中的杂质；改变蒸馏产品中的分子结构。转化过程是为了生产原油中没有的碳氢化合物（图11.5）。

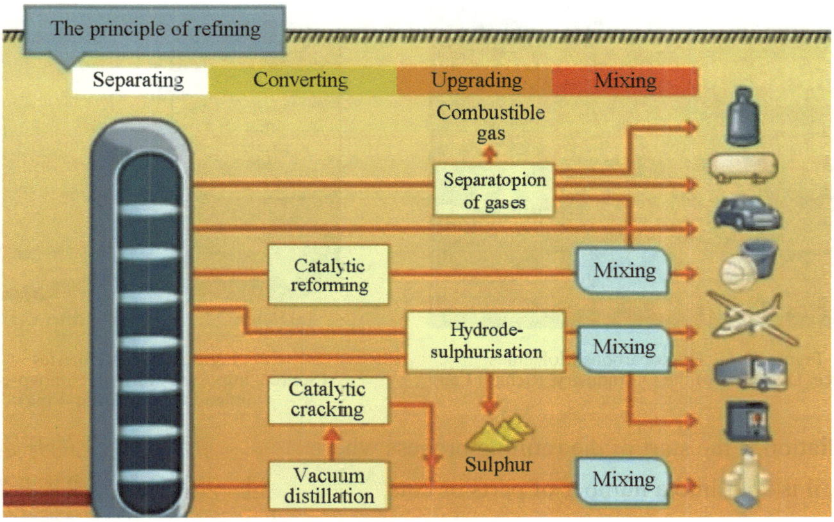

Figure 11.5 Refining
(Source: The Principal of Refining, TOTAL)

All crudes contain organic sulphur compounds (e.g., H_2S, mercaptans) which will be carried out over from the column into the resulting gases, distillates and residue. The higher the density of a crude, the greater its sulphur content. The secondary refining process for the treatment of toxic, corrosive and evil-smelling sulphur-compound impurities is known as sweetening. Caustic washing and hydrodesulphurization are finishing processes designed to remove H_2S and mercaptan impurities.

所有的原油都包含有机成分硫化物（例如硫化氢、硫醇），硫化物会最终形成气体、液体和固体残渣。原油密度越高，含硫量越高。二次加工能去除这种有毒、腐蚀性的难闻硫化复合物，这叫甜化过程。碱洗和加氢脱硫是现行的去除硫化氢和硫醇的方法。

In conversion processes, the structures of natural hydrocarbon molecules are changed. In one such processes, the known as cracking, large hydrocarbon molecules are cracked or broken to form two or more smaller molecules(Figure 11.6). This can be done by the action of heat and pressure alone (thermal cracking) or by heat in the presence of a suitable catalyst (catalytic cracking). The main purpose of cracking is to increase the yield of lighter, more valuable fractions from medium and residual cuts(Figure 11.7). When a particular feedstock has been cracked, the various products resulting from the process can be separated out in a fractionating tower.

在转化过程中，天然碳氢化合物分子结构被改变，这叫裂化和亘整。裂化指碳氢化合物的大分子被裂化和打断，形成2个或多个小分子（图11.6）。这能单独通过加热、加压改变（加热裂化）或通过催化剂改变（催化裂化）。裂化的主要目的是为了从中质、重质成分中增加轻组分的产量（图11.7）。当特定的成分裂化后，不同的产品从分馏塔中分离出来。

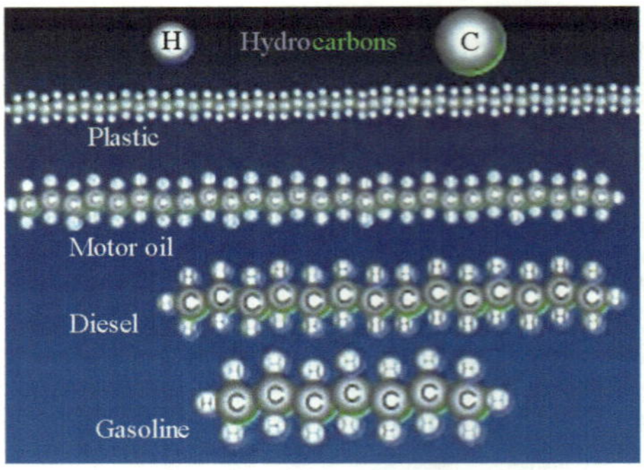

Figure 11.6 Hydrocarbons Molecules
(Source: Overview of the Oil Industry, Richard Lau)

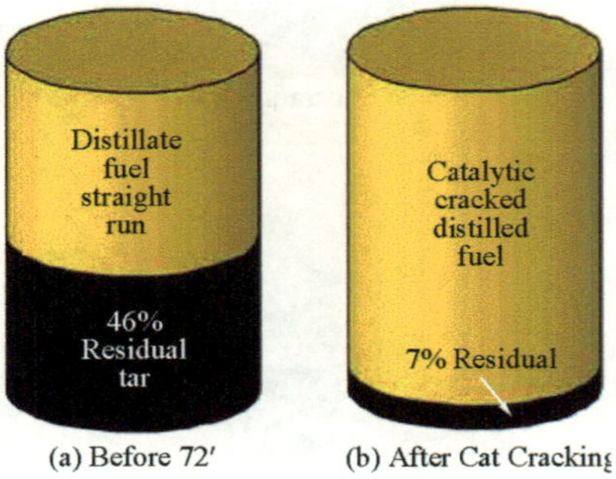

Figure 11.7 Refineries
(Source: http://www.drfuel. com/wp-content/uploads/2009/12/faq-4.jpg)

Distillation is the start of the refining process where the crude oil is split into a number of parts or cuts.

蒸馏是炼化的开始，炼化将进一步把原油分成若干个小成分。

11.3 Refinery Products

11.3 石油炼化产品

The fuels refined from crude oil can be divided into two general types: ① fuels that are exploded, when vaporized with air, to provide primary moving power, and ② those fuels that are either burned directly for heat and light or are converted into secondary energy sources such as electricity(Table11.1).

原油炼化产品可以分为两大类：①用于第一能源的燃料；②直接燃烧获得热和光或用于转化为电能的第二能源（表11.1）。

Table11.1 Representative Examples of Petroleum Constituents

Name of Fraction	Representative HCS	Approximate BP/℃
Natural gas	CH_4	-161
Liquefied gas (LP gas)	C_3H_8, C_4H_{10}	-44 ~ 1
Petroleum ether	C_5H_{12}, C_6H_{14}	30 ~ 60
Aviation gasoline	C_5 ~ C_9	32 ~ 150
Auto gasoline	C_5 ~ C_{12}	32 ~ 210
Naphtha	C_7 ~ C_{12}	100 ~ 200
Kerosene	C_{10} ~ C_{16}	177 ~ 290
Fuel oil	C_{12} ~ C_{18}	205 ~ 316
Lubricating oils	C_{15} ~ C_{24}	250 ~ 400

(Source: http://www.wiredchemist.com/environmental/instructional/a-case-study-of-the-exxon-valdez-oil-spill-of-1989)

Conventional oil refinery products:

常规炼油产品有：

◆ Motor Spirit

◆ 车用汽油

The boiling range of a motor spirit is generally from 30℃ up to 200℃. However, a straight-run fraction of this range, cut directly from crude oil, would give violently

车用汽油的沸点范围通常为30 ~ 200℃。然而，这个沸点范围内从原油中直接得到的直链馏

heavy knock if used in a modern car. Containing a high percentage of normal (straight chain) paraffins (as for example normal heptane, already mentioned) of low anti-knock value, it would have a research octane number well below, say, the level of about 100, now common for premium grades in many countries. It is, therefore, necessary in a refinery to install process units to manufacture concentrates of those hydrocarbons that have a high octane number. Such hydrocarbons are aromatics, olefins and isoparaffins.

分却不适用于现今汽车中，反而使现今汽车不抗爆震。含有高比例的直链烷烃（如前所提的正庚烷）有较低的抗爆值，而研究表明辛烷值大约为100时有较好的抗爆震性能，在很多国家是优级油品。因此，炼油厂有必要安装处理设备集中生产高辛烷值的碳氢化合物。如芳烃、烯烃和烷烃这类碳氢化合物。

◆ Kerosene

Kerosene, also known as "paraffin" in the UK, is manufactured as a straight-run fraction from crude oil generally in the boiling range 150~250℃. Although the demand for kerosene as an illuminant is diminishing, it finds widespread use in central-heating installations and domestic heating stoves, and is also a convenient fuel for incubations, blow-lamps and camping stoves of the "primus" type.

◆ 煤油

煤油在英国称为链烷烃，沸点范围通常为150～250℃，是原油的直链组分。尽管现在以煤油照明的需要日益减少，但煤油在中央供热和家庭取暖方面广泛应用，也是孵化器、喷灯、野营燃料炉的便携燃料。因此，评价煤油首先关注燃烧品质。

◆ Diesel Oil

Generally speaking, diesel oil includes petroleum distillates, in the boiling range 200～350℃. In manufacturing diesel fuels particularly for automotive use, the main behavior and sulphur content. Although good ignition characteristics are necessary in a diesel fuel to ensure easy starting and smooth running, they are by no means so critical as is the case with a motor gasoline. For a diesel fuel the important factor is the shortness of the period between the injection of the fuel into the cylinder and the start of combustion. A good diesel fuel should begin to burn at several places in the combustion chamber within a very short time of injection; if the delay is too great the whole charge could fire simultaneously, causing violent combustion and heavy knock.

◆ 柴油

一般而言，柴油包含沸点范围为200～350℃的石油馏分。根据汽车性能和硫含量要求生产特定的汽车需要柴油。尽管易燃性是柴油确保易启动和驾驶平稳的必要条件，但汽油的易启动和驾驶平稳性比柴油更重要。从柴油注入气缸到开始燃烧的间歇时间是柴油油品重要的判断因素。好的柴油应该在注入后极短时间内在气缸不同位置处开始燃烧，假如燃烧大大延后整个燃烧可能同时开火，导致剧烈的燃爆造成汽车剧烈震荡。

◆ Lubricating Oils

Typical classifications of lubricating oils include automotive lubricants, diesel-engine lubricants, steam-

◆ 润滑油

典型的润滑油包括汽车润滑油，内燃机润滑油，蒸汽涡

turbine oils, gear and transmission lubricants, metal-working lubricants (i.e. cutting oils), rolling oils (for metal-working) and compressor oils.

轮油，齿轮传输润滑油，机械润滑油（即冷却油），机油和压缩油。

◆ Waxes

Wax, although regarded as an undesirable material when manufacturing a lubricating oil, is nevertheless of intrinsic value as a separate product. The wax that is removed from the lighter lubricating oil stocks has relatively large crystals and is known as "paraffin wax"; that from the heavier fractions has much smaller crystals and is usually referred to as "micro-crystalline wax". Paraffin waxes are normally marketed on the basis of a melting-point test. Control of this characteristic is clearly of importance in connection with the traditional use of such wax for the manufacture of candles and tapers. It is also very relevant in relation to the use of wax for the impregnation of paper (e.g., for good cartons). If the wax melted too readily, a stack of such containers would lead to stick together on a hot day. Paraffin wax is also used for electrical insulation, in polishes, for the "defeathering" of poultry and as waterproofing agent. This latter use also provides a major outlet for microcrystalline wax.

◆ 石蜡

尽管石蜡被认为是生产润滑油过程中不受欢迎的产品，但以内在价值作为一个独立产品。石蜡从较轻的润滑油中分离得到，此类石蜡有相对大的结晶称为烷烃蜡，从较重组分分离得到的小结晶石蜡称为微晶蜡。烷烃蜡通常以熔点测试确定其市场价格。掌握石蜡熔点关系到传统石蜡行业——用石蜡生产蜡烛和蜡纸。假如石蜡太易融化，一堆如此的石蜡可能会在某个炎炎夏日粘到一起。烷烃蜡也被用于电气绝缘，抛光剂，家禽去毛和防水剂。后者也可以使用微晶蜡。

◆ Bitumen

The residue obtained after distillation of crude oil to remove materials boiling up to the end of the gas-oil range (about 350℃), can be further distilled, under vacuum, to give waxy distillates for the preparation of lubricating oils, or for feeding to units such as a catalytic cracker. The "vacuum residue" left behind from his second stage of distillation is known as "bitumen". Bitumen is a dark viscous material and in fact is normally regarded as a solid at ambient temperature. Because of its adhesive, plastic nature and waterproofing qualities, it is widely used for road-making purposes, generally in the form of a binder for stone aggregates.

◆ 沥青

原油蒸馏后获得的残余油去除杂质再沸腾到油气沸点顶段（大概350℃）在真空条件下能进一步蒸馏，产出蜡质馏分为生产润滑油作准备，或作为下一步催化裂解的进料。而这二次蒸馏后的真空残余就是沥青。沥青是黑色黏状物质，在室温下实际上常被看成是固体。因为沥青具有黏性、可塑性及防水性，广泛用于道路铺设，常作为基石间的联合剂。

References
参考文献

[1] Conaway, Charles F.. The Petroleum Industry: A Nontechnical Guide [M] . PennWell Publishing Company,1999.

[2] 高群峰. 石油工程概论 [M] . 北京：石油工业出版社，2009.

[3] 江淑娟. 石油科技英语 [M] . 北京：石油工业出版社，2007.

[4] 陈铁龙，孙艾茵. 石油工程专业英语 [M] . 北京：石油工业出版社，2006.

[5] 崔树清，杨志龙. 石油工程英语 [M] . 北京：石油工业出版社，2007.

[6] 任晓娟，徐波. 石油工业概论（第二版）[M] . 北京：中国石化出版社，2012.

[7] 张继红，艾池. 石油工程专业英语 [M] . 北京：石油工业出版社，2009.

[8] 康勇. 石油科技英语基础教程（第二版）[M] . 北京：石油工业出版社，2012.

[9] 江淑娟，吴松林. 油田开发英语 [M] . 北京：石油工业出版社，2002.

[10] 王红秀. 石油工程专业英语（第2版）[M] . 哈尔滨：哈尔滨工业大学出版社，2012.

[11] 李璗，陈军斌. Mechanics of Oil and Gas Flow in Porous Media（英文版）[M] . 北京：石油工业出版社，2012.

[12] 聂翠平，刘贵喜. 石油工程专业英语教程 [M] . 北京：石油工业出版社，2009.

[13] Frank Jahn, Mark Cook，Mark Graham. Hydrocarbon Exploration and Production (2nd Edition) [M] . Elsevier Scientific Publishing Company, 2008.